Land, Innovation, and Social Good

Land, Innovation, and Social Good

Editors

Kwabena Asiama
Rohan Bennett
Christiaan Lemmen
Winrich Voss

MDPI • Basel • Beijing • Wuhan • Barcelona • Belgrade • Manchester • Tokyo • Cluj • Tianjin

Editors

Kwabena Asiama
Geodetic Institute
Leibniz University Hannover
Hannover
Germany

Rohan Bennett
Swinburne Business School
Swinburne University of
Technology
Melbourne
Australia

Christiaan Lemmen
Faculty of Geo Information
Science and Earth Observation
University of Twente
Enschede
The Netherlands

Winrich Voss
Geodetic Institute
Leibniz University of Hannover
Hannover
Germany

Editorial Office
MDPI
St. Alban-Anlage 66
4052 Basel, Switzerland

This is a reprint of articles from the Special Issue published online in the open access journal *Land* (ISSN 2073-445X) (available at: www.mdpi.com/journal/land/special_issues/land_innovation).

For citation purposes, cite each article independently as indicated on the article page online and as indicated below:

LastName, A.A.; LastName, B.B.; LastName, C.C. Article Title. *Journal Name* **Year**, *Volume Number*, Page Range.

ISBN 978-3-0365-1911-1 (Hbk)
ISBN 978-3-0365-1910-4 (PDF)

Contents

About the Editors

Kwabena Asiama

Kwabena Obeng Asiama is currently a research scientist/lecturer at the Geodetic Institute of the Leibniz University of Hannover, Germany. He received his Ph.D. (2019) and M.Sc. (2015) from the Faculty of Geo-Information Science and Earth Observation (ITC) of the University of Twente with a focus on Land Administration. He completed B.Sc. in Land Economy (2012) at KNUST, Kumasi, Ghana. In 2018, he received the FIG-Survey Review Prize at the XXVI FIG Congress. He was also named one of the 40 under 40 motivated and accomplished young surveying professionals by the xyHt magazine. Kwabena's research interests span real estate valuation (with a focus on 3D cadastres, large-scale land acquisitions, and areas without land markets and unregistered lands), land governance, as well as innovative approaches to land administration and land management activities on customary lands of the Global South.

Rohan Bennett

Rohan Bennett is an Associate Professor in Information Systems with the Swinburne Business School, Melbourne, Australia; and a Geodetic Advisor with Netherlands Kadaster. He specializes in spatial information systems and land information management. He has previously held posts with the University of Twente (NL), Kadaster International (NL), and University of Melbourne (AU). He holds degrees in Science, Engineering, and a Ph.D. in Land Administration from the University of Melbourne. His pracademic activities span approximately 20 years and include project work across countries in Europe, Oceania, Africa, Asia, and the Middle-East –on topics including digital transformation, UAV application, blockchain, and AI.

Christiaan Lemmen

Christiaan Lemmen is a Full Professor at ITC, the Faculty of Geo-information Science and Earth Observation, University of Twente, The Netherlands (part-time). He is also a Geodetic Advisor with Kadaster International, the international branch of the Netherlands Cadastre, Land Registry and Mapping Agency. Christiaan is an Editor of the International Standard for the Land Administration Domain Model (LADM), published as ISO –19152. The second edition of LADM is under development since 2018. He is a Co-Chair of the Land Administration Domain Working Group of the Open GeoSpatial Consortium. Christiaan is the Director of the OICRF, which is the Office International du Cadastre et du R´egime Foncier (the international Office for Cadastre and Land Records), a documentation centre for land administration, and a permanent institution of the International Federation of Surveyors (FIG). He holds a Ph.D. in Land Administration from Delft University of Technology, The Netherlands.

Winrich Voss

Winrich Voss is an University Professor in Land and Real Estate Management at GIH, Geodetic Institute Hannover, Faculty of Civil Engineering and Geodetic Science, Leibniz University Hannover (LUH), Germany. He is chairman of the Leibniz Research Centre TRUST on Transdisciplinary Rural and Urban Spatial Transformation. His research interest combines three main topics: information management in real estate markets (market transparency), development of rural areas and spatial transformation in urban-rural linkage in national as international context. Winrich holds a degree in geodesy and a Ph.D. in Spatial Planning from TU Dortmund, Germany. He brings in 15 years of practical experience in urban-rural development and valuation as well and 15 years of academic experience.

Preface to "Land, Innovation, and Social Good"

The administration of land tenure, value, and use has changed significantly over recent decades. Change has been driven by efforts to better balance rural and urban land use, improve recognition of previously informal land rights in certain geographical regions, and to acknowledge the social and cultural values relating to land. These changes have been underpinned by the new wave of technological innovations that have expanded the ways and means of doing land administration. Running parallel to these seemingly positive developments are the land-related societal problems including land tenure insecurity, food shortages, slum formation, and environmental degradation. This edited volume brings together articles spanning innovative approaches to securing land tenure, land valuation, and the planning of land use, with the aim of deepening the understanding of the nexus between societal challenges and land, and further enriching the knowledge on how current technological advances can be used to meet the societal challenges in the area of land administration.

This edited volume comprises nine research articles, three review articles, and one project report, developed by multidisciplinary teams from Africa, Asia, Europe, the Oceania, and North America. This also reflects the spread of the geographical focus of the articles. Researchers, both early and seasoned, will benefit from the insights on the ongoing and current research, in the field.

A big thanks goes to the team at MDPI for their guidance through all the steps of the editing this special collection, the anonymous reviewers, and the authors, who in more ways than one, made this book possible.

Kwabena Asiama, Rohan Bennett, Christiaan Lemmen, Winrich Voss
Editors

Editorial

Land, Innovation, and Social Good

Kwabena Obeng Asiama [1,*], **Rohan Bennett** [2,3], **Christiaan Lemmen** [3,4] **and Winrich Voss** [1]

1   Geodetic Institute, Faculty of Civil Engineering and Geodetic Science, Leibniz University of Hannover, 30167 Hannover, Germany; voss@gih.uni-hannover.de
2   Swinburne Business School, Swinburne University of Technology, Hawthorn, VIC 3122, Australia; rohanbennett@swin.edu.au
3   Kadaster International, Cadastre, Land Registry and Mapping Agency of the Netherlands, 7311 KZ Apeldoorn, The Netherlands; chrit.lemmen@kadaster.nl
4   ITC Faculty, University of Twente, 7500 AE Enschede, The Netherlands; c.h.j.lemmen@utwente.nl
*   Correspondence: asiama@gih.uni-hannover.de or kwabena.asiama@gmail.com; Tel.: +49-(0)-511-762-2406

Citation: Asiama, K.O.; Bennett, R.; Lemmen, C.; Voss, W. Land, Innovation, and Social Good. *Land* **2021**, *10*, 503. https://doi.org/10.3390/land10050503

Received: 3 April 2021
Accepted: 30 April 2021
Published: 8 May 2021

Publisher's Note: MDPI stays neutral with regard to jurisdictional claims in published maps and institutional affiliations.

The administration of land tenure, value, and use is undergoing a new wave of technological innovation. The need for faster, more affordable, accessible, fit for purpose approaches to undertaking land administration functions has led to the push of applicable technological innovations into the arena. The maturation and scaled implementation of crowdsourced data capture techniques (i.e., ubiquitous mobile devices), imagery-based mapping approaches (i.e., HRSI, UAVs, LiDAR), and cloud storage options are all adding to the expanded land administration toolbox. Further from these developments, there is a mix of even more novel developments that are being advanced, including blockchain technology, smart contracts, computer-assisted land use planning decision support systems for smart cities, and automatic valuation models.

Running in parallel to these technological advances in the land administration domain are global-level societal challenges. These manifest as key livelihood problems within national borders and at the community level—and include issues of land rights inequality, slum formation, food insecurity, natural disaster risk, and exposure. These societal challenges are not new, but have been exacerbated in the past two decades: More people are increasingly exposed to the risks, most of the time, in more places. However, as with all social issues, societal challenges are rooted in their respective legal, social, and cultural peculiar contexts. Hence, attempts to resolve these challenges are also context specific. This notwithstanding, land issues are very appropriate to create social innovation, based on technological evolutions and combined with civic society activities. The papers presented in this Special Issue further contribute to social innovations via the improvement in land administration processes.

The relationship between the two issues, technological innovations and societal challenges, which are related to the administration of land in this Special Issue, form two theories about the sources of innovation. On the one hand, the aspect of technological innovation, which makes science and technology the central theme—the driver of innovation—and highlights its roles in the development of solutions to societal challenges. This has been described by many as a 'supply side' motivation for innovation or more commonly 'technological push'. On the other hand, the pull of the societal demands, where innovation is defined by the conditions or needs of the recipient community of users. Here, there is recognition of the demands of the community within a broader socio-political context, and more importantly characteristics of the end market, especially the end users, as well as the total makeup of the community, including the social, economic, legal, and political characteristics, are all considered important.

Therefore, the focus of this Special Issue lies at the nexus of the exploration of this interaction between technological innovation and social challenges, in the context of land administration. The Special Issue aims at providing an overview of the trending developments in the technologies aiding the functions of land administration, the societal demands that drive the developments or the applications of these technologies, and how these forces interact in terms of designed systems, impacts, and outcomes. The Special Issue, in terms of the trending technological developments in the sphere of land administration, on the one hand, presents articles relating to current developments and better implementation of currently used technologies—GNSS, and smartphone apps, for example—as well as the modelling of workflows to support these technologies—such as process modelling, cadastral evaluation services, and participatory land administration. On the other hand, the Special Issue presents a retinue of the societal demands that necessitate the adoption of the technologies and techniques indicated. These societal demands include gender equality, land tenure security, environmental protection, food security, and sustainable cities and settlements (Figure 1).

Figure 1. The aspects of land, innovation, and social good.

The collection of peer-reviewed articles in the Special Issue number thirteen, include nine research articles, three review articles, and one project report. These articles were developed by multidisciplinary teams, drawn from Africa, Asia, Oceania, North America, and Europe, with a consequential spread in the geographical focus of these studies across the same areas, summarised in Table 1 below.

Table 1. Overview of the papers presented in the Special Issue.

Source	Title	Country/Geographic Context	Technological Innovation and Techniques	Societal Demand
Aditya et al.	Participatory Land Administration in Indonesia: Quality and Usability Assessment	Indonesia	Participatory Land Registration, GNSS	Land Tenure Security
Ameyaw et. al.	Transparency of Land Administration and the Role of Blockchain Technology, a Four-Dimensional Framework Analysis from the Ghanaian Land Perspective	Ghana	Blockchain Technology	Land Tenure Security
Asiama et al.	Towards Responsible Consolidation of Customary Lands: A Research Synthesis	Ghana	Smartphone Apps	Food Security, Land Tenure Security
Auziņš, Armands	Capitalising on the European Research Outcome for Improved Spatial Planning Practices and Territorial Governance	Europe	-	Sustainable Cities and Settlements
Bennett et al.	Hybrid Approaches for Smart Contracts in Land Administration: Lessons from Three Blockchain Proofs-of-Concept	Australia (New South Wales)SwedenCanada (British Columbia)	Smart Contracts, Blockchain Technology	Land Tenure Security
Choei et al.	Improving Infrastructure Installation Planning Processes using Process Modelling	South Korea	Infrastructure Installation Planning, Process Modelling	Sustainable Cities and Settlements

Table 1. *Cont.*

Source	Title	Country/Geographic Context	Technological Innovation and Techniques	Societal Demand
Deng et al.	Land Registration, Adjustment Experience, and Agricultural Machinery Adoption: Empirical Analysis from Rural China	China	-	Food security, Land Tenure Security
Home et al.	History and Prospects for African Land Governance: Institutions, Technology, and 'Land Rights for All'	Africa	-	-
Ntihinyurwa et al.	Farmland Fragmentation, Farmland Consolidation and Food Security: Relationships, Research Lapses and Future Perspectives	-	-	Food Security
Paradza et al.	Could Mapping Initiatives Catalyze the Interpretation of Customary Land Rights in Ways that Secure Women's Land Rights?	Sub-Saharan Africa	-	Gender Equality
Salifu et al.	Innovating Along the Continuum of Land Rights Recognition: Meridia's 'Documentation Packages' for Ghana	Ghana	Evaluation of Meridia Land Documentation	Land Tenure Security
Sousa et al.	Evaluation of the Objectives and Concerns of Farmers to Apply Different Agricultural Managements in Olive Groves: The Case of Estepa Region (Southern, Spain)	Southern Spain	-	Environmental Protection, Food Security
Trystuła et al.	Evaluation of the Completeness of Spatial Data Infrastructure in the Context of Cadastral Data Sharing	Eastern Poland	Evaluation of local cadastral network services	Land Tenure Security
Twum et al.	Gender, Land and Food Access in Ghana's Suburban Cities: A Case of the Adenta Municipality	Ghana	-	Food Security, Land Tenure Security, Gender Equality

This following presents the works that make up the Special Issue. The above table shows the spread in the focus of the papers in this Special Issue, with a number of the papers focussing on either technological innovation, societal demands or both. The papers in the issue are presented below in the alphabetical order of the authors' last names. This is deemed a straight forward and equitable way of arranging the equally weighted articles: All contribute meaningfully and directly to the overarching theme of the Special Issue.

Aditya et al. [1] present a project report on a quality and usability analysis of participatory land registration (PaLaR) in the rural areas of Indonesia, focusing on data quality, cost, and time. PaLaR is described by the authors as a community-focused systematic land titling project that is aimed at collecting spatial and legal data. The work, based on a pilot study comparing PaLaR and the regular systematic land registration, in terms of their spatial accuracy, conducted in two rural communities, finds that though PaLaR resulted in a lower spatial accuracy, it was a better fit locally, providing the needed spatial and legal information.

Ameyaw and de Vries [2] add on the growing arena of blockchain technology, and its application to land administration. Drawing on the Ghanaian perspective, the authors review the potentials of using blockchain technology to enhance the transparency of land administration functions. The authors, after an examination of the land tenure registration, valuation, use planning processes, find that it is possible to have a permissionless public blockchain across all the processes of land administration. This integration of the land administration processes, responsible departments, and stakeholders, thus could foster openness, availability, access to information, and promote transparency and participation in the land administration processes. However, the authors are quick to point out the possible threats and pitfalls—such as limited storage of data and scalability, and the electricity consumption needed for the operations.

With the limited use of land consolidation, and a decreasing level of food security in the Sub-Saharan African (SSA) region, Asiama et al. [3] synthesize four studies that describe the development of a responsible land consolidation strategy, with one fit for the region. In this paper, the authors show the comparison of three countries with a responsible land consolidation strategy to one without. This is done to determine the factors that inhibit the development of a responsible land consolidation strategy. Further studies in the synthesis describe the development of the participatory land administration, as well as a land valuation approach and a land reallocation approach to address the factors inhibiting the use of land consolidation on SSA's customary lands, respectively. Though the synthesis concludes that the developed land consolidation strategy will reduce land fragmentation towards increasing food productivity, local customs in the area can be an impediment.

Auziņš Auziņš [4] distinguishes between planning systems and practices, with the aim of determining the pre-conditions and the challenges that need to be considered to improve spatial and development practice in the future. The author explores the trends and directions in the evolution of spatial and territorial governance, with a focus on the connections between the EU policies and the diverse national planning perspectives. The key results from a literature review and a comparative study were that, firstly, there is the need for an agenda-setting for comprehensive evidence gathering (CEG) if exploring spatial planning practices and territorial governance in selected European countries, and secondly, that a set of objectives for a values-led planning (VLP) approach needs to be introduced for improvement of land use management.

Bennett et al. [5] further explore the implications on the use of blockchain technologies on the functions of land administration, with a specific focus on smart contracts. With three proof of concept cases from Sweden, Australia (New South Wales), and Canada (British Columbia), the authors examine the hybrid approaches being used to introduce smart contracts into the current land registration technology infrastructure, in order to keep the land registry as the ultimate decider of the validity of transactions. The results from the comparison shows that the hybrid approach can enable the adherence to land administration business requirements, with further scaling implementation requiring a more holistic view of the sector, with attention being given to the remaining issues, that among others, include business model analysis, stakeholder acceptance, and trust building.

Choei et al. [6] present an analysis of the time and costs constrained in implementing a development impact fee (DIF) in South Korea. In this study, the authors use a case study in Jeju South Korea to compare the efficacy of a proposed approach, procedural modelling method using CityEngine, to the traditional method which uses AutoCAD. The study finds that that procedural modelling provides real-time 2- and 3-dimensional modelling and design evaluation. Moreover, it allows for a more efficient assessment of plan quality and calculation of DIF and further uncovers the need to diffuse procedural modelling to better support local planning practices.

Deng et al. [7] link land tenure security to advance factor inputs in agricultural modernisation in developing countries, with an aim of clarifying their relationships from the perspective of property rights theory and endowment effects. The study finds that though land registration does not have an effect on the rate of farm machinery adoption, and a negative experience in adjusting to new technologies exists, the interaction between land registration and adjustment experience has a positive impact on the adoption of agricultural machinery.

Providing a bird-eye view of the past, and present situation of African land governance, as well as proffering suggestions for the future direction, the communication article provided by Home [8] explores, through a multi-disciplinary approach, the recent advances in geo-spatial technology on the African continent. Zooming down to four Anglophone countries in the region, the article draws upon a range of sources to discuss how among other things, the colonial legacies of these countries have influenced the culture of their national land administration systems.

Ntihinyurwa and de Vries [9] in this review article interrogate the relationship between farmland fragmentation and farmland consolidation, and the variations of food security, to develop a model that will indicate the conditions under which farmland fragmentation needs to be maintained, resolved or controlled for the purposes of food security. The study finds that whilst the best management of farmland fragmentation for food security purposes can be achieved by minimizing the problems associated with physical and tenure aspects of farmland fragmentation along with the optimization of its potential benefits, various agriculture intensification programs, agroecogical approaches, and land saving technologies can be the most suitable strategies to maximize the income from agriculture on fragmented plots under the circumstances of beneficial fragmentation.

The initiatives towards securing the land tenure rights of womenin Sub-Saharan Africa's customary lands are often undermined by among others, inadequate community awareness and the challenges of recording the land rights. Paradza et al. [10] analyze case studies of selected mapping initiatives in SSA to determine the extent to which mapping both as a cadastral exercise and emerging practice in the initiation of participatory land governance initiatives, catalyze the transmission of customary land rights in ways that have a positive impact on women's access to land in customary land tenure areas. The study finds that some of the bottlenecks of these initiatives include the expensive soft- and hardware, illiteracy of women, legal status of the maps in the community, and in the country, among others. This work makes a significant contribution to our understanding of what instruments in the land registration toolbox can strenthen women's land rights.

Salifu et al. [11] explore how the land Documnetation Packages from Meridia recognises the continuum of land rights in its innovative land registration processes within the institutinal setting of Ghana. The authors describe the processes of registration, as well as the actors involved and the nature of their encounters, both in terms of the conventional registration process and that of Meridia. The study finds that though Meridia's process reflects the continuum of land rights, it also poses questions for future research regarding the political economy of land tenure certification and regarding the actual uses and benefits of issued certificates.

Sousa et al. [12] investigate the drivers and concerns that condition farmer's choices of a given olive groove management model. In a case study in the Estepa region of Spain, the authors find that most of the concerns of the olive farmers were directed towards conservation objectives. It was further found that organic and educated olive farmers are more likely to share this view.

Trystuła et al. [13] assess the completeness of National Spatial Data Infrastructures (NSDIs) containing the core land administrationd dataset—the cadastral dataset- also a key part of the EU's INSPIRE project. The authors develop an assessment framework which enables the identification of websites that publish cadastral data through INSPIRE network services, and those with a high development potential. The authors recommend the results of the assessment to be used in the ongoing construction of NSDIs and to improve the quality of network services and their availability for end users.

Twum et al. [14] explore the underlying gender disparity in the access to food and land in the suburban cities of Ghana, and its evolution through the years as a result of settlement expansion and urban growth. Using a case study in the Adneta Municipality of Ghana, the authors find that though women engage with the power structures on a daily basis, the result is either a burden, a benefit or both, depending on their socio-cultural status and other factors in terms of access to food and land.

Overall, this Special Issue brings together a wide range of landed societal demands and attempts to meet them using contemporary technology innovations. It shows a deepening recognition of, and need to explore, the societal demands of food security, land tenure security, sustainable cities and communities, and environmental protection, in the context of land administration. Morover, technologies such as blockchain technology, smartphone apps, and GNSS are also shown to assist in dealing with such societal demands, albeit each bringing its own flow-on challenges, and potentially unintended consequences. The collection also brings to the fore, the need to have the views of the local people, users, and stakeholders taken into account during the design and ongoing upkeep of land administration systems.

Author Contributions: Conceptualization and writing—original draft preparation, K.O.A.; writing—review and editing, R.B., C.L. and W.V. All authors have read and agreed to the published version of the manuscript.

Funding: This research received no external funding.

Acknowledgments: The authors acknowledge the support of colleagues and peers from their affiliated institutions and those working more broadly in the land administration domain.

Conflicts of Interest: The authors declare no conflict of interest.

References

1. Aditya, T.; Maria-Unger, E.; Bennett, R.M.; Saers, P.; Lukman Syahid, H.; Erwan, D.; Wits, T.; Widjajanti, N.; Budi Santosa, P.; Atunggal, D.; et al. Participatory Land Administration in Indonesia: Quality and Usability Assessment. *Land* **2020**, *9*, 79. [CrossRef]
2. Ameyaw, P.D.; de Vries, W.T. Transparency of Land Administration and the Role of Blockchain Technology, a Four-Dimensional Framework Analysis from the Ghanaian Land Perspective. *Land* **2020**, *9*, 491. [CrossRef]
3. Asiama, K.O.; Bennett, R.; Zevenbergen, J. Towards Responsible Consolidation of Customary Lands: A Research Synthesis. *Land* **2019**, *8*, 161. [CrossRef]
4. Auziņš, A. Capitalising on the European Research Outcome for Improved Spatial Planning Practices and Territorial Governance. *Land* **2019**, *8*, 163. [CrossRef]
5. Bennett, R.M.; Miller, T.; Pickering, M.; Kara, A.-K. Hybrid Approaches for Smart Contracts in Land Administration: Lessons from Three Blockchain Proofs-of-Concept. *Land* **2021**, *10*, 220. [CrossRef]
6. Choei, N.-Y.; Kim, H.; Kim, S. Improving Infrastructure Installation Planning Processes using Procedural Modeling. *Land* **2020**, *9*, 48. [CrossRef]
7. Deng, X.; Yan, Z.; Xu, D.; Qi, Y. Land Registration, Adjustment Experience, and Agricultural Machinery Adoption: Empirical Analysis from Rural China. *Land* **2020**, *9*, 89. [CrossRef]
8. Home, R. History and Prospects for African Land Governance: Institutions, Technology and 'Land Rights for All'. *Land* **2021**, *10*, 292. [CrossRef]
9. Ntihinyurwa, P.D.; de Vries, W.T. Farmland Fragmentation, Farmland Consolidation and Food Security: Relationships, Research Lapses and Future Perspectives. *Land* **2021**, *10*, 129. [CrossRef]
10. Paradza, G.; Mokwena, L.; Musakwa, W. Could Mapping Initiatives Catalyze the Interpretation of Customary Land Rights in Ways that Secure Women's Land Rights? *Land* **2020**, *9*, 344. [CrossRef]
11. Salifu, F.; Abubakari, Z.; Richter, C. Innovating Along the Continuum of Land Rights Recognition: Meridia's "Documentation Packages" for Ghana. *Land* **2019**, *8*, 189. [CrossRef]
12. Rodríguez Sousa, A.A.; Parra-López, C.; Sayadi-Gmada, S.; Barandica, J.M.; Rescia, A.J. Evaluation of the Objectives and Concerns of Farmers to Apply Different Agricultural Managements in Olive Groves: The Case of Estepa Region (Southern, Spain). *Land* **2020**, *9*, 366. [CrossRef]
13. Trystuła, A.; Dudzińska, M.; Źróbek, R. Evaluation of the Completeness of Spatial Data Infrastructure in the Context of Cadastral Data Sharing. *Land* **2020**, *9*, 272. [CrossRef]
14. Twum, K.O.; Asiama, K.; Ayer, J.; Asante, C.Y. Gender, land and food access in ghana's suburban cities: A case of the adenta municipality. *Land* **2020**, *9*, 427. [CrossRef]

 *land*

MDPI

Article

Hybrid Approaches for Smart Contracts in Land Administration: Lessons from Three Blockchain Proofs-of-Concept

Rohan Bennett [1,2,*, Todd Miller [3], Mark Pickering [1] and Al-Karim Kara [4]

[1] Swinburne Business School, Swinburne University of Technology, Hawthorn 3122, Australia; mpickering@swin.edu.au
[2] Kadaster International, Netherlands' Cadastre, Land Registry and Mapping Agency, 7311 KZ Apeldoorn, The Netherlands
[3] ChromaWay, S-111 64 Stockholm, Sweden; todd.miller@chromaway.com
[4] Land Title and Survey Authority of British Columbia, Victoria, BC V8W 9J3, Canada; Al-Karim.Kara@ltsa.ca
* Correspondence: rohanbennett@swin.edu.au or rohan.bennett@kadaster.nl

Abstract: The emergence of "blockchain" technology as an alternative data management technique has spawned a myriad of conceptual and logical design work across multiple industries and sectors. It is also argued to enable operationalisation of the earlier "smart contract" concept. The domain of land administration has actively investigated these opportunities, albeit also largely at the conceptual level, and usually with a whole-of-sector or "big bang" industry transformation perspective. Less reporting of applied case applications is evident, particularly those undertaken in collaboration with practicing land sector actors. That said, pilots and test cases continue to act as a basis for understanding the relative merits, drawbacks, and implementation challenges of the smart contract concept in land administration. In this vein, this paper extends upon and further refines the existing discourse on smart contracts within the land sector, by giving an updated, if not more nuanced, view of example applications, opportunities, and barriers. In contrast to the earlier works, a hybrid solution that mixes smart contract use with existing technology infrastructure—enabling preservation of the role of a land registry agency as the ultimate arbiter of valid claims—is proposed. This is hypothesised to minimise disruptions, whilst maximising the benefits. Examination of proof-of-concept work on smart contract and blockchain applications in Sweden, Australia (State of New South Wales), and Canada (Province of British Columbia) is undertaken. Comparative analysis is undertaken using several frameworks including: (i) business requirements adherence, (ii) technology readiness and maturity assessment, and (iii) strategic grid analysis. Results show that the hybrid approach enables adherence to land dealing business requirements and that the proofs-of-concept are a necessary step in the development trajectory. Furthering the uptake will likely depend on again taking a whole-of-sector perspective, and attending to remaining issues around business models, stakeholder acceptance, partnerships and trust building, and legal issues linked to data decentralisation and security.

Keywords: blockchain; smart contracts; land administration; land registration; cadastre; technology readiness levels; land conveyance; mortgage discharge

 **check for updates**

Citation: Bennett, R.; Miller, T.; Pickering, M.; Kara, A.-K. Hybrid Approaches for Smart Contracts in Land Administration: Lessons from Three Blockchain Proofs-of-Concept. *Land* **2021**, *10*, 220. https://doi.org/ 10.3390/land10020220

Academic Editor: Volker Beckmann

Received: 29 January 2021
Accepted: 17 February 2021
Published: 22 February 2021

Publisher's Note: MDPI stays neutral with regard to jurisdictional claims in published maps and institutional affiliations.

1. Introduction

Subsequent to the initial and overly simplified hype around the potential application of blockchain technology to the domain of land administration, a more circumspect discourse on the relative merits and immediacy of its implementation is emerging [1–3]. This is not to suggest that critical assessments were not forthcoming [4], or that the overstatements have abated fully: works continue to espouse conceptual blockchain land registry designs and the imminent benefits [5–7]. That said, such conceptual works appear to be fewer in number, and the level of attention apportioned them is more aligned to the depth of technical inquiry underlying them.

Meanwhile, one of the key discernments from the more balanced discourse is that the notion of placing an entire title or deed registry, and all related transactions and processes, on the blockchain is fanciful, at least in the short term [1]: any full tokenisation of property in a given jurisdiction is still likely to be many years away. Anecdotal evidence from practice suggests full uptake is something that governments, the conventional custodians of these records, will only consider in the context of a more matured technology offering; one that can be shown to fully satisfy cyber-security considerations, and that has had the commensurate policy, legislative, and regulatory development attention of government.

This does not necessarily preclude the utilisation of blockchain technology in the land sector in the short term. A more nuanced appraisal suggests that more targeted and smaller-scale applications of the technology to specific land dealings might be more realistic, not to say more useful, in the land administration domain.

Examples of specific land dealings, or subset use cases of land administration, could include land conveyance, mortgage creation and discharge, or off-plan development approvals. These transactions could utilise blockchain technology by integration with existing land administration technology infrastructures. That is, rather than seeking to fully re-engineer and shift all land administration processes, data and document storage onto the blockchain, the focus in the short term could be on capitalising on the immediate benefits of the technology to make more efficient specific land administration processes.

Land dealings, such as land transfer or mortgage discharge, are those processes that enable the transferring of rights over a spatial unit of land, from one party to another [8]. In all cases, they generally require at least two transacting parties, that is a buyer and seller, and depending on the local legal and financial systems, numerous third-party actors to support the transaction. A contract, or deed, signed by both transacting parties, accompanied by a statutory legal instrument, is usually an essential component.

These specific land dealing blockchain use cases are enabled by the customisable programming logic stored in blockchains. This enables the operationalisation of "smart contracts", a conceptual idea predating blockchain technology by more than a decade [9,10]. Whilst this convergence of theoretical concept (smart contracts), emerging technology (blockchain), and potential application (land dealings) in the land sector is not entirely novel [11,12], these specific but important transaction subsets of land administration are all too often bundled up, if not lost, in the more radical sector-wide digital transformational visions, foreseen to be underpinned by blockchain, but that have, to date, lacked the combination of political will, public finance, and land agency impetus to implement.

For the smaller use cases of specific land dealings, this need not be the case: smart contracts supporting land conveyance, for example, can be implemented on the blockchain, and could be enacted as a somewhat independent technology layer, enabling interaction between transacting parties, the land registry, financial institutions, attorneys, and other parties, whilst not requiring the wholesale disruption of embedded existing technology arrangements.

The hypothesis is that this "hybrid" approach would simply provide a more efficient layer or interface for making and enforcing land-related agreements between actors; ones that could provide shorter-term solutions for the provision of a more secure, auditable, and distributable solution for supporting property record changes amongst buyers and sellers. Unlike earlier larger-scale sector-wide digital transformation visions, the relatively simple smart contract concept could be implemented without the need for significant land agency disruption, complete IT infrastructure rebuilds, or full database redesign.

To this end, this paper explores the potential for the application of smart contracts, implemented using blockchain technology, for the specific land dealings inherent to land administration. The overarching aim is to contribute to the more nuanced understandings of the potential role, benefits and drawbacks, of blockchain technology in the land sector— by beginning from a more incremental or targeted mindset with regards to implementation.

To achieve this aim, a comparative analysis is undertaken of findings from three proof-of-concept studies undertaken in Sweden, the Australian State of New South Wales, and the

Canadian Province of British Columbia. For each study, smart contracts were applied, and proofs-of-concept developed for specific land dealings within those jurisdictions. Several theories are used to guide the case analyses, thereby enabling case comparison and subsequent results triangulation. These frameworks included the core business requirements of land dealings, technology readiness and maturity level analysis, and strategic grid analysis.

The remainder of the paper is structured as follows. A contemporary update is provided on the principles of, and relationships between, land dealings, land administration, smart contracts, blockchain technology, technology readiness/maturity, and hybrid solutions. This leads to an outline of the comparative study methodology, and subsequent presentation of the results. Each case is first presented separately before the discussion section delivers the synthesis. This section, along with the conclusion, makes predictions on the likely future development and research trajectories of blockchain and smart contracts applied to the land administration domain.

2. Background

2.1. Land Dealings and Conveyance

Before examining the potential role of smart contracts for specific land dealings, it is necessary to provide a brief overview of evolution of land dealings theory and practice. This informs, justifies, and provides criteria upon which to assess the relative benefits and drawbacks of any subsequent smart contract-based land dealing proof-of-concept.

Procedures for enabling the transfer of immovable property developed over millennia. Notwithstanding vastly different social and environmental contexts, and the land transfer mechanisms they espoused [13], suggests that for any situation, a means of transaction and evidence piece can be identified. Using these constructs, he suggests four major developments can be observed—particularly if a Western standpoint is taken.

Initially, transfer processes were linked to localised social customs, involving symbolic gestures to validate transfers, and witnesses as the key form of evidence. Paper-based systems followed, with private conveyancing between two consenting parties being supported by the creation of a deed (or legal document), again witnessed by a third party with the appropriate social, religious or juridical status. These documents formed an evidence base to not only support land transfer, but could also enable use of land as collateral, to support claims in cases of dispute, and for supporting inheritance.

As nation states further developed after the medieval period, beyond the European Renaissance, and into the Enlightenment, more organised systems or repositories for the registration of deeds were developed. That is, the trusted third party (i.e., state or religious institution), not only witnessed deeds, but also made and stored copies of the deeds at a central location [14]. The input or responsibility of the third party, in terms of verification of the legal validity of the transaction, and the associated liability assigned the third party in the case of fraudulent activity instigated by the transacting parties, may have been rather limited.

In cases where third party involvement developed to include the writing of a government-backed title or certificate, terminology evolved to contrast "**deeds registration**" from "**title registration**" [15]—the "Torrens" system being one variation. The key differentiating feature of title registration is the certificate transfer. Unlike a signed and witnessed deed, that merely acted as one piece of evidence that a transaction had occurred, the certificate of title provided a legal point of "truth", to the point that, if later it was discovered that a certified transaction was actually a fraudulent one, with the fraud committed by a seller without the authority to transfer, an unwittingly defrauded land holder would receive compensation from the State, rather than receiving their land rights back, under certain conditions.

The period of European colonisation resulted in those deeds and title registration systems being transferred globally [15,16]. Consequently, the statutory, if not formal, land transfer processes, in most nation states find antecedence in either deeds registration, title registration, or a combination of the two (e.g., Trinidad and Tobago, where land parcels in the same jurisdiction might be recorded and administered in separate systems) [17]. Whereas deeds registration systems record legal fact, title registration systems record legal consequence.

The required higher levels of government involvement oversight in title systems sees them sometimes referred to as "**positive systems**", in contrast to "**negative**" deeds systems[1]. As pointed out by [18], "improved" deeds registration systems now appear very similar in practice and process to title registration systems. Indeed, modern theorists tend to argue that types of registration system can be considered on a continuum, rather than fitting clearly into one of the two historical and theoretical categories [19].

Regardless of the statutory system in use, transacting parties are required to compile a set of legal documents (or instruments) to get the transaction onto the register [18]. Specifications for these documents, and perhaps complexity in requirements, typically increased over time—dependent on the drivers and problem cases (e.g., fraud) experienced within a jurisdiction. In many cases, it will require the completion of a prescribed transfer form, a contract of sale, and a mortgage creation instrument.

Modern deeds and title registration systems were responses to identified weaknesses in earlier transfer methods. The strength of both lies in their simplicity with regards to four principles [15,18]. First, the **registration principle** (sometimes called "curtain" in Torrens literature) demands that, in order for a transaction to be considered "legal fact", it needs to be recorded in the authoritative "book" (or database in modern systems) [11]. This means that person-to-person, without government oversight, transactions are not recognised legally. The major reasoning[2] here is to stop a land unit (or parcel) being transferred multiple times by a single party: the authoritative book would reveal that the land has already been transferred.

Second, the **principle of publicity** demands that the book and transactions within, must be available and accessible for the public to view. The principle helps to remove information asymmetries and ensure transparency in conduct, for both transacting parties, and the government alike. In practice, books are not fully open, and may be considered semi-open: privacy and security controls are placed on the ways and means for accessing transacting data [18].

Third, the **consent principle** articulates that for any changes to be recorded in the book, relating to a person or parcel, the impacted parties must give consent. This principle builds on the previous two, with anti-fraud being a major motivation.

Finally, the **principle of speciality** declares that both parties and land units must be unambiguously defined. This is increasingly the focus of international standardisation efforts [20], and usually achieved through person and parcel identification (ID) systems, and the use of cadastral maps and field sketches. The IDs make transaction processes simpler, and seek to minimise identity fraud. It should be noted that literature on Torrens and other titling systems expand the four principles to also include: curtain, mirror, and insurance[3]. Each of these generally seeks to increase the power of the body doing the registration and fast track dispute resolution.

Despite the successes of both systems, they carry limitations—or at least perceived weaknesses. First, regarding **time**, in many systems, transfer processes generally take weeks, if not months, to complete [21]. This lag between transaction instigation and completion has been argued as inefficient by the property sector and related actors [22], and ultimately supports weak governance, if not corruption and informality in the sector.

Second, **cost** to transfer, particularly in developing contexts, is argued as prohibitive [21]. A transfer usually involves a range of professionals and several parallel processes, each

---

[1]   It should be noted that it is usually Anglophone literature that makes these distinctions, with preference for title registration perhaps being transferred to the subsequent terminology.

[2]   The registration principle also supports value capture via land taxation by government agencies.

[3]   Torrens and other titling systems add to the four above mentioned principles, including the insurance principle, whereby the authority responsible for the book (i.e., often government) will provide compensation to parties judged to have been defrauded of property, due to inadequate checks by the registry, at the time of registration.

attracting fees or duties. In theory, modern IT should reduce costs of storage, processing, and transparency provision [23]—yet, have these gains been passed on, or is a level of rent seeking persistent to some systems? Arguments and examples can be presented from both sides. What is clearer, is that if cost to register is too high, informal non-statutory transactions will occur, "outside of the books"—undermining the utility and value of the register, and the first principle above mentioned [24].

Third, like all administrative developments, **complexity** seeps into processes over time—in terms of parties, processes, and systems involved [8,21]. Regulatory reform is often a response [25], yet regulatory reform has often proved difficult to deliver in the land sectors in many jurisdictions.

Fourth, the **duplication** of effort is also evident in many systems [20,23]. This may include repeating data entry, superfluous checking of documents, and so on. Duplication could be considered a subset of the cost issue.

Finally, and perhaps most importantly, despite the best efforts of both deeds and registration systems, **fraud** is still possible and certainly occurs [26]. This can be actioned by buyers, sellers, other actors, or even the registry officials: the systems and controls are still penetrable with loopholes relating to instruments, documents and processes available for exploitation.

Like most sectors, over the last five decades (approximately), technology in the form of digital systems, databases, internet, and web services has greatly impacted upon registration systems—in terms of function and service delivery [27]. These have served to reduce existing limitations in both systems in terms of time, cost, complexity and duplication: many cases of cost reduction and process simplification (or access) can be observed. However, the new technological approaches also opened up new opportunities for fraud—and for this reason, amongst others[4], most land administration systems have tended to take a conservative stance and have been later technology adopters. Contemporary systems tend to still use a mix of digital and paper-based processes and documents. Whilst the developments are yet to fundamentally challenge or alter the underlying theoretical principles inherent to both deeds and registration systems, emerging concepts and tools relating to blockchain, including smart contracts, create interesting questions, if not opportunities [1].

2.2. Smart Contracts

It is also necessary to provide a contemporary overview on smart contracts in terms of theory and application, and to distinguish it from those of blockchain, recalling that the latter refers to one mechanism of how and where data in a smart contract is stored, validated, and viewed. This background supports understanding the smart contract proof-of-concepts developed in Sweden, Australia, and Canada, and subsequent assessment.

"Contracts" are agreements that can be enforced by law [28]. They are generally legally binding documents (although, can be merely verbal), agreed upon by at least two parties, prescribing the rights and duties of the parties involved [29]. Groupings of the elements can vary; however, contracts are generally considered to require offer; acceptance; intention to create legal relations; consideration (or value); legal capacity; and consent [28].

Translating from Hemmo, "Contracts enable organized collaborative activity and are used to carry out economic activity" [10,30,31]. This view allows the contracting mechanism to be positive and actionable, versus the common perspective that contracts are primarily designed to manage risk and exposure, thus resulting in the limiting of business activity.

Envisioning the impact and consequences of information technologies on contracts, business, and legal practice, the "smart contract" concept emerged in the 1990s. Whilst agreement on the scope of the "smart contract" concept has become more difficult on account of significantly increased attention across domains, Szabo's definition [9] of the concept, constituting a "set of promises, specified in digital form, including protocols within which the parties perform on these promises" remains prominent, and essentially

---

4 Rent seeking, enabled by manual processes, for example, is recognised as another reason for land sector inertia.

foresaw the conversion of traditional paper-based contracts, elements, and associated manual processes, into digital self-executing ones.

That is, the smart contract concept saw the "if, then, else" statements of conventionally legal contracts, fitting comfortably with the constructs used in computer programming: legal agreements could be translated and executed as computer code. This could include all the binding elements and specific clauses of a contract. This digitisation then opens the opportunity for the digitalisation of workflows, and completion of contractual actions by computers that previously required human action or involvement. It would make transactions between geographically and even socially disparate (i.e., non-trusted) parties far easier.

Taking this broad definition, smart contracts were already in play in the 1990s. Szabo, for example, referenced vending machines as operating on an implicit contract: a dollar is exchanged for a can of soda. Moving forward, parties now routinely sign digital agreements with online service providers such as Netflix, Apple and Google, approving debiting of accounts, in return for use of an asset. If the party fails to live up to the terms of contract (e.g., failure to pay subscription), the party's online account, and access to the asset, may automatically be suspended. It should be noted that these examples are not "trustless" transactions. They require the users to inherently trust that online service providers will deliver the services, as all require payments in advance. That is, consumers have to trust service providers like Netflix to provide the movie when they pay in advance. This means it is an asymmetrical trust relationship. Netflix say they have "Ben Hur", and the consumer "hopes" that it will be provided after payment. Netflix validates the consumer through a credit provider (e.g., Mastercard), but the consumer has no way of validating Netflix. Continuing the analogy, a true "trustless" transaction would be where the consumer can actually see on an independent blockchain that Netflix "owns" the rights to "Ben Hur", and an exchange is recorded on the blockchain, which gives me access to the movie.

At any rate, whilst the "smart contract" concept and even its application can be considered decades old, for some applications, technology limitations (amongst those of a more institutional nature) stymied scaled and decentralised implementation: creating verifiable public and decentralised agreements, on the order in which a series of digital transactions had occurred, was an unsolved technical challenge.

Enter Satoshi Nakamoto's 2008 bitcoin currency [32], underpinned by blockchain technology. It resolved the order of transaction issue [33], and thus paved the way for recorded and completed decentralised and verifiable online public agreements. Blockchain technologies enabled non-trusting parties to record and execute agreements, on a distributed peer-to-peer network, without the need for a trusted intermediary [34]. In this way, "smart contracts" can be distinguished from "blockchain": the former, initially conceived in 1997, predates the latter by 11 years [32]. Put simply, the combination of blockchain technology and smart contract concepts enabled a new form of transparent "trustless" transactions.

Mainstream blockchain development platforms emerged, including Ethereum and Hyperledger, and consequently, the smart contract concept experienced a revival in development attention [35], albeit still with more limited scaled application [10]. Essentially, smart contracts take the form of code, residing on the blockchain, and these codes can be used to automatically verify and enforce contracts, digitally, without central authorisation [35].

Putting the above into practice, Table 1 reveals the current state-of-play with regards to actual implementation of smart contracts, as compared to more conventional contracts. Several features are worthy of mention. Firstly, on specification, in the smart contract situation, the contract, including the terms of agreement, has been converted to computer code. This is the key characteristic that subsequently enables the downstream execution of many of the contract terms and tasks can be achieved through automated processes. These processes can include the transfer of property title, automated payment of duties or fees, or payment credits to cover escrow accounts.

Table 1. Comparison of traditional contracts against smart contracts.

Criteria	Conventional Contracts	Smart Contracts
Specification	Natural language and legal prose	Code
Identity and Consent	"wet" signatures	Digital Signatures
Dispute Resolution	Judges, adjudicators, arbitrators	Consensus via blockchain
Nullification	Parties via legal enforcement Process of breached terms	Parties via Agreed Upon Digital Nullification workflow and block consensus
Payment	Independent third-party Process	Automatic, based on executed terms (Built into Contract)
Escrow	Independent third-party Process	Automatic, based on executed terms (Built into Contract), or not even required

Second, in terms of identity and consent, digital signatures, using asymmetric cryptography dating back to the 1970s, are fundamental to smart contracts. Every transaction connected to a smart contract must be signed. The integrity of the system rises and falls on the level of confidence that the network has about each party to the contract. In essence, this is the very reasoning for the principle of speciality in conventional registration systems.

Every person required to execute their part of the contract must have a digital key. The challenges for architects of a smart contract network are to balance the relative simplicity of a central authority that issues credentials versus challenges of self-sovereign identity and key management. Key management is certainly a non-trivial issue, but multi-signature frameworks and custody models have emerged to address both security and consumer adoption concerns.

Third, in terms of dispute resolution, nullification, payments, and escrow, it can be seen that these tasks are largely automated, programmed as workflows, and thus remove the need for trusted third party decision making and action in the smart contract solution.

2.3. Technology Readiness Levels

Having defined smart contracts, differentiating them from both conventional contracts and blockchain, an introduction is now given to analysing and understanding the relative readiness and maturity of the technology, in the context of its potential utilisation in land administration, and specific land dealings.

The concept of "**Technology Readiness Levels**" (TRL) is an approach to classifying technology from basic principles and conceptualisation through to implementation in an operating environment. The TRL framework was developed by NASA for the development of technology in the Space Program in the 1970s [36]. The framework has been adapted and utilised for information technology systems development [37] and new product development [38]. The nine phases in the original NASA framework as adapted to commercial development are (adapted from [38]): TRL 1: Principal research into key properties of a technology; TRL 2: Conceptualisation of a new potential application for the technology; TRL 3: Develop analytical "proof of concept" of core functionality; TRL 4: Component and/or breadboard validation in a laboratory environment; The focus of technology development is on achieving project objectives; TRL 5: Validation of basic technological capabilities in a relevant environment; TRL 6: "High-fidelity alpha prototype demonstrated in a relevant environment"; TRL 7: Beta prototype demonstrated in an operational environment; TRL 8: System completed and qualified to relevant project requirements/industry standards through test and demonstration; and TRL 9: System proven to achieve all project requirements in operational environment. Preliminary appraisal of blockchain applications in land administration, as uncovered in [1], suggests that existing developments lie between TRL 2 and TRL 3: most reported developments are conceptual or limited proofs of concept.

While useful to explore the technology readiness of blockchain and smart contracts to perform land registry functions, TRL has some limitations. TRL ignores organisational considerations of the organisation developing or implementing the technology and the environment in which the organisation operates [36] and fails to consider integration of the new technology/system with existing systems [37]. The TRL framework finishes with the successful operation of systems in the target environment and does not progress to commercialisation and diffusion throughout an industry or market.

A parallel can be drawn between the limitations of TRL and previous studies on the maturity of blockchain: most studies on blockchain focus on the technology and ignore the organisational considerations of adopting new technology [39]. While an argument for the adoption of blockchain is the elimination of intermediaries [40], there is still the need for the governance of organisations within a blockchain network [41].

Therefore, to support TRL analyses and evaluations of blockchain and smart contracts in the land dealings, it seems pertinent to also include **technology adoption theories**. The technology, organisational, and environmental (TOE) framework examines antecedents to technology adoption [42]. This model has recently been used in a systematic review of the literature exploring blockchain adoption considerations [43]. The major technological considerations identified were complexity, perceived benefits, security, compatibility, maturity, relative advantage and smart contract coding. Most referenced organisational capabilities were organisational readiness, including value chain readiness, appropriate knowledge and financial resources; top management support; organisational size, business model readiness and innovativeness. Major environmental considerations included market dynamics encompassing competitive pressure and market standards, the regulatory environment, government and stakeholder support, business use cases and industry pressure.

In a similar approach to TOE, [39] identified that institutional and market factors need to be considered as well as the technological factors. From an institutional perspective, cultural resistance by industry incumbents needs to be overcome; knowledge and understanding need to be developed amongst businesses, customers and government around the potential use and implications of use of blockchain; and how the technology can be integrated into existing strategies and processes [39]. Market factors include the changing role of intermediaries and associated potential disruption; the need to embed smart contracts in software; and impact on business processes [39].

Another supportive theoretical framework for blockchain application evaluation is those linked to **capability maturity models**. Typical engineering capability maturity models [44,45] have been adapted to develop a maturity model for the engineering of distributed ledgers as shown in Table 2 [46]. These can also be useful for assessing maturity with respect to smart contract approaches to facilitating land dealings.

Table 2. A maturity model for blockchain and smart contract application in land dealings.

	Maturity Phase	Intention	Artifact	Scope
1.	Initial	Discovery of the potential benefits and how the replacement of intermediaries may impact process and governance structures	Development of Minimum Viable Product (MVP) blockchain prototypes	Selection of blockchain platforms is not systematic and the roles of the blockchain and existing database technology are indistinct
2.	Structured	Use a structured technology-selection process to identify appropriate platform	An appropriate platform selected and the design of a partner network and governance frameworks	Specific criteria have been used to select blockchain use cases and to distinguish the solution from existing database technology
3.	Automation	Moves toward process automation based on smart contracts	Smart Contracts—use the platform to go beyond distributed transaction management	The scope of smart contracts is limited to "single dependencies between data or business processes"
4.	Business Collaboration	Distributed autonomous organisation	Complex relationships and automated processes across	A network of visible partners is expressed by inter-linked smart contracts.
5.	Verification	Formally proven automation	Correctness of smart contracts and DAOs checked	Verification by known model checkers

Adapted from [46].

In summary, analyses of the progress of the use of blockchain and smart contract technologies in land dealings could benefit from identifying the TRL phase of the selected projects, exploring the degree to which TOE factors have been considered, and the blockchain maturity phase that the project represents.

2.4. Towards Hybrid Solutions

Putting all the above together, despite the perceived benefits of smart contracts applicable in many sectors, beyond crypto currencies such as Bitcoin, scaled implementation of the smart contract concept via blockchain within mature industries, such as those with heavy government oversight or regulatory control, still remains limited.

This applies equally to the land sector: despite much hype and conceptual design work, evidence of fully operationalised blockchain-driven solutions in the land sector remains scarce [1]. Land dealings are complex transactions encumbered with the body of sometimes very old legislative and regulatory controls and processes. Existing legal systems and accompanying administrative procedures need adaptation to incorporate the smart contract concept, at least if the transaction is to be considered "legal" under any form of deeds or title registration system). Moreover, there persists the notion that smart contracts might operate beyond human control and bind the parties to agreements or expose them to unintended liabilities through malicious behaviours.

That said, more circumspect conceptual thinking on "hybrid" approaches is emerging. In [47] the term "hybrid" refers to designs that tend towards semi-private and more permissioned write access, aligning them with conventional land administration processes. Only identified and authenticated actors would be permitted to write. Going further, in this work, the term is adapted to include the combined use of conventional database technologies, integrated with blockchain technology. Additionally, the idea is to veer away from earlier whole-of-sector digital transformations designs, and focus in on specific dealings, activities, and actors. The aim is for minimal disruption to existing institutions and infrastructures, or even to demonstrate integration with those systems. These hybrid approaches seek to deliver the benefits of smart contracts to land dealings, whilst minimising risk and resistance—however, they also require rigorous scrutiny.

Consequently, there appears to exist a sound argument to evaluate these smaller-scale or incremental developments in the context of adherence to business requirements, technology readiness/maturity, and impact—with a view to evaluating whether a trajectory of uptake is evident, or not, and where this alternate implementation approach may be leading.

3. Materials and Methods

Building from the findings in the previous section that—(i) institutional constraints mean whole-of-sector blockchain transformations of the land administration sector will not be realised in the short term; (ii) the technical readiness level (TRL) of smart contracts, underpinned by blockchain, is considered to be at best, at the level of proof-of-concept (2–3); and (iii) that specific land dealings appear highly suited to smart contract application with respect to maximising the benefits, whilst mitigating the current limitations of blockchain technology—a methodology was developed to support the examination of the possibility and benefits of hybrid land dealings solutions, combining land registry processes with smart-contract/blockchain technology.

The developed methodology was fundamentally built from the pragmatist research paradigm and can be considered "design research", or at least design evaluation: the methodology sought to assess a solution "that works" in a given context and application area, rather than seeking any absolute truth [48]. Justification for this design approach can be found in other land administration research and development work, recently including [49–54]. Building from these works, the methodology can be said to be inspired by, although not a direct application, of the living labs approach [55], reflexivity and action research [56]; these specific methods already finding justification and application in land administration studies. On this, it should be noted that some of the co-authors were involved in the proofs-of-concept work in the jurisdictions, however, these were considered separate undertakings to the analyses informing this work.

In this vein, the methodology primarily utilised publicly available data and findings from rapid prototyping development work completed by ChromaWay, between 2016 and 2020 in three jurisdictions. This data includes reports, presentations, published code, technology descriptions, amongst other artefacts. ChromaWay is a software and blockchain solutions provider operating globally, but is primarily based in Sweden. ChromaWay has developed a generic suite of blockchain tools, customising them for multiple jurisdictions, across multiple sectors, including the land administration sector. A brief overview of the ChromaWay design and development process, and the resultant key components of the solution, are explained in Section 4.1. The publicly available ChromaWay data was supplemented by the pre-existing expert knowledge and contextual awareness of the authors.

The results of three specific proof-of-concept studies are drawn upon for this specific work, namely undertakings in Sweden (2016–2018), Australia (State of New South Wales) (2018), and Canada (Province of British Columbia) (2018), and further follow-up activities in 2019–2020. For each case, smart contracts were applied, and proofs-of-concept developed for a land dealing, namely a portion of the land conveyance process, in those jurisdictions. Subsequent to those studies, a framework was developed to enable assessment of each; first individually, and then comparatively. The framework assessed the specific use case of smart contracts for land conveyance within the jurisdiction, in terms of lead sponsors, required partners, smart contract technology components and features (i.e., the hybrid approach), the specific land conveyance use case, process participants, pre-existing problems with the process, challenges with the specific study, and key benefits. The results of these analyses are presented in Section 4.2, Section 4.3, and Section 4.4. Additionally, the cases were comparatively assessed against theories, including the core business requirements of land dealings, technology readiness and maturity levels, and strategic grid analysis. These results are presented in Section 5.

The findings were then synthesised to make generalised determinations of the contemporary potential for more nuanced and incremental application of smart contracts, implemented using blockchain technology, for the land administration sector.

4. Results

4.1. The Hybrid Approach

First, an overview of the developed generalised hybrid approach is provided in terms of the developed business, application, information, and technology architecture.

In terms of the **business architecture**, the philosophy behind the hybrid proof of concepts was to move beyond the concept of "big bang" sector-wide blockchain transformation for land administration, which would necessarily involve comprehensive and substantial re-engineering of all the business processes, in terms of land dealings, actors, and tasks. Instead, the focus was placed on specific land administration tasks or transactions that would most immediately utilise and benefit from smart contract application, envisioning connection to the existing land registration technology infrastructure, with minimal disruption. This is referred to as the "hybrid solution". It should be noted that the proof-of-concept did not connect to the production-level technology infrastructure.

For smart contracts to take hold in land registration processes—at least those transactions taking place in formal, legal, and/or statutory systems—a level of reform to existing legislative, regulatory, and administrative processes would be required: transactions involving immovable property are subject to specific laws in each jurisdiction (i.e., beyond regular contract law). That said, it is possible to explore the potential role of smart contracts in the land registration and transfer process.

In terms of the **application architecture**, in each of the proofs-of-concept, the existing processes were mapped in terms of actors and tasks. From this, alternative conceptual workflows were developed that incorporated smart contract technology. The actual smart contracts were designed using code which drove automated tasking and workflows where the rules defined what "messages" were acceptable at every state of the contract. For example, the property purchase process code defined a "buyer" who must sign-off on a contract. Digital signatures were used to establish message authenticity, in order to prevent "attackers" impersonating one of the parties, posting invalid or malicious information, and to indicate that the signing party was responsible for message contents, similar to a signature on a paper document or contract.

Here, it is also necessary to introduce the concept of "boundary connectors", the opposite of barriers to entry. Boundary connectors—technology, data, and collaborative business arrangements—enable business processes to cross organisational and jurisdictional lines. A simple example is the strategy of introducing regional "smart pass" transponders to allow vehicles to quickly and securely cross local highway jurisdictions (a "smart pass" is simply a smart contract device; a fee is deducted in exchange for the right to pass through a toll gate). The processing of land registration and associated mortgage lending processes can be thought of in a similar fashion (see Figure 1). The distributed ledger serves to connect buyers, sellers, settlement agents, lenders, and land registries into a single network (i.e., the road network) and the smart contract acts as a sort of "transponder", guiding the property transfer (or other transaction) to move across the ledger network.

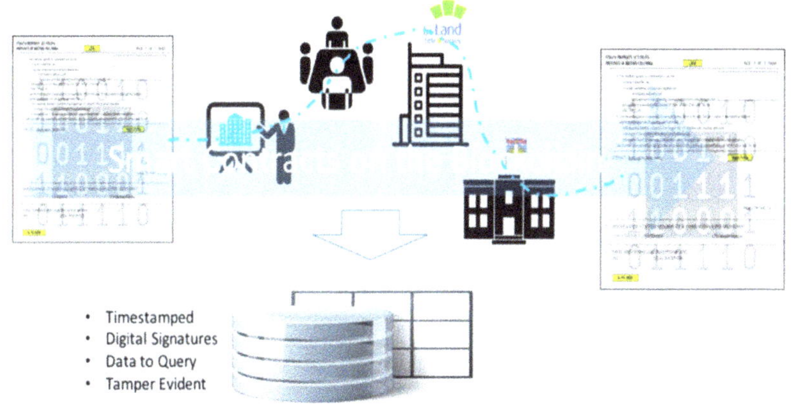

Figure 1. Simplified workflow schematic of a land dealing process enabled via smart contracts on the blockchain.

Information Architecture: Figure 2 illustrates the components of the hybrid conveyance solution proposed in this paper. There are three high level architectural components: (1) client-distributed applications (Dapps); (2) source of on/off chain data; and (3) the blockchain Backend. In the blockchain backend, smart contracts are written either with Esplix or using Postchain (explained below). The Conveyance Dapp is written like any other business application, except data is written to the blockchain nodes (BC Nodes) instead of a central database. Data in the Dapp is appended to/presented from the blockchain. The behaviour of the Dapp is driven by the codified smart contract (for each conveyance type) and the workflow which orchestrates the interaction of property ecosystem participants.

The smart contract in a land registration transaction employs digital signatures (typically through the use of cryptographic key pairs) that provides the signed transactions that are submitted to the blockchain ledger; specifies the data required by network partners to process/approve a transaction; enables automated processing of escrow payments or other types of actions based on predetermined rules; describes the definition of a completed task(s) (e.g., signatures, collected data, etc.) that permits the contract to proceed; enables participants with a user interface (e.g., as a smartphone) or systems (e.g., application servers) to complete the tasks required by the codified contract.

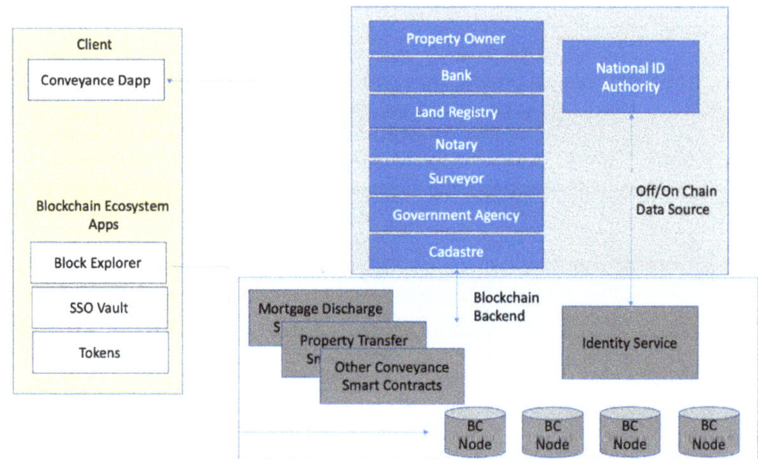

Figure 2. Hybrid smart contract/blockchain information/application architecture.

Technology Architecture: In each study, ChromaWay utilised a technology called Esplix. Esplix is non-Turing complete (not a fully operational program to avoid loops which enabled the DAO (decentralised autonomous organisation) hack [57]) and operates as a system for exchanging signed messages. The ChromaWay Esplix solution, like other similar frameworks, allowed the smart contracts to operate in a more settled legal space defined by laws like the US's Uniform Electronic Signatures Act ("UETA") and Electronic Signatures in Global and National Commerce Act ("ESIGN") that already recognise, enable, and validate the use of electronic signatures and electronic records. The Postchain module for Esplix allows one to utilise the Postchain consortium database as a consensus (witness) component of an Esplix system. Postchain provides a reliable message store and guarantees that once a message is confirmed its position relative to other messages is certain and that the message chain is unambiguous. Note that Postchain can also be used as the consensus-based blockchain data repository for a distributed application (Dapp) client.

4.2. The Swedish Case—Property Sales

The Swedish proof-of-concept was completed between June 2016 and June 2018 and was primarily sponsored by Lantmäteriet—the Swedish mapping, cadastral and land registration authority. Lantmäteriet is a government agency primarily providing information on Swedish geography and property. The project was sponsored by a consortium, including business consultants, technology providers, and financial institutions. Respectively, these were Kairos Future (business), Telia (ID Provider), ChromaWay (technology provider), SBAB Bank, and Landshypotek. SBAB Bank is a state-owned bank that borrows funds to support the Swedish mortgage market. It provides loans to private individuals, tenant-owner associations, and real estate companies. Landshypotek is a bank owned by farmers and foresters, almost forty thousand, and reinvests profits back into those enterprises.

The proof-of-concept focused on all of the phases of the property sales process including property transfer (or land conveyance). In the Swedish system, this includes a buyer, seller, real estate agent, buyer bank, seller bank, and the land registry. Perceived problems with the existing process are generally related to complexity, duration, duplication, and documents being physical. That is, the existing process was found to: include thirty-four (34) steps; take weeks or months to complete; only involved the land registry very late in the process; make use of only limited data re-use between steps; still be largely paper based, with signed documents sent by regular email; and to require manual identity checking.

The technology suite applied by ChromaWay included the Esplix Smart Contract, and Postchain Blockchain solutions. An excerpt of the smart contract developed for the real estate consortium formed in Sweden to process property transactions is shown in Figure 3. The developed smart contract dictates that the buyer's bank must sign-off (with its key pair) that it has (1) received the purchase sum and (2) sent the contract to the land registry. The land registry, in turn, must sign that it approves the purchase contract received from the buyer's bank. In this way, the smart contract defines and orchestrates and enforces the actions they must take to advance the contract and associated processes towards completion. The demonstrator illustrates the potential to use the smart contract in the context of established land transfers in a developed economy context.

```
actions
    (offer ((property-id string :description "Official ID of the property"))
        "Offer the property on the market (pending description of the property by the broker)"
        (guard
                (signatures seller)
                (eql state nil)
        )
        (
                update property property-id
                state :register-broker
        )
    )

    (register-broker ((broker-pk pubkey))
        "Invitation of broker to the contract"
        (guard
                (signatures seller)
                (eql state :register-broker)
                (eql broker nil)
        )
        (
                update broker broker-pk
                state :describe
        )
    )

    (describe ((description-param string :description "Description of the property, including it's state"))
        "Describe the property, including its extent and state"
        (guard
                (signatures broker)
                (eql state :describe)
```

Figure 3. Smart Contract Excerpt.

Features of the solution included that only parties to the contract were privy to the data in the contract; that the contract would not fully execute without satisfaction of the data and signing requirements; and that the contract protocol could be distributed through third party vendors or directly through registry developed apps. Given the number of actors involved in the conveyance process, this last feature was considered highly beneficial in the Swedish case.

The results of the proof-of-concept revealed a significant reduction in the number of manual steps needed for a property transaction (down from 34 to 13); greater transparency into the process for all parties (including the banks, land registry, etc.), in terms of being able to view the status of a transaction at any time during completion; and a simpler, less expensive distribution of the standard property transfer protocol using a smart contract. A major challenge identified during the proof-of-concept assessment is that Swedish law does not allow for the use of electronic signatures for property transactions, obviously a major constraint in terms of scaling the project to production level.

In terms of current status, the blockchain network and smart contract proof-of-concept protocol were trialled and externally tested. Further progress can start once the digital signature restrictions are addressed.

4.3. The Australian Case—Mortgage Discharge

The Australian proof-of-concept was completed from January 2018–October 2018 and led by New South Wales Land Registry Services (NSW LRS), supported by ChromaWay Asia Pacific and ChromaWay AB. NSW LRS operates the NSW land registry for the NSW State Government. It is a private company and has a 35-year concession to run the registry, commencing in July 2017. Part of the concession involves ensuring improvement and upgrade of the technology infrastructure underpinning the registry, in terms of service and security. In this vein, the 2018 proof-of-concept provided an ideal opportunity to trial the use and potential integration of blockchain enabled smart contracts.

Unlike the Swedish case, focused on the complete case of land conveyance, the Australian case focused on an even more specific land administration process. The core land administration process focused upon was "Discharge of Mortgage Lien". The Discharge of Mortgage Instrument is used to remove the recording of a mortgage from a land title. It usually applies to a whole parcel. Mortgage discharges are the most common transaction supported by NSW LRS, with between 20 K to 25 K completed per month. In comparison, there are usually from 13–17 K land transfer transactions (or land conveyance). The core participants in the process are the mortgagor (usually an owner), mortgagee (usually a financial institution), and the land registry (NSW LRS).

In the NSW context, the existing lien removal process was considered to be overly complicated and included more steps than were seen to be necessary, particularly given the possibilities provided by digital lodgement and processes. Due to the complexity, it has been found that, in some cases, even when debts are settled, liens removals have neither been appropriately lodged nor processed. In these cases, mortgage holders may not even be aware that the lien still exists on the property.

Similar to the Swedish technology solution, the NSW LRS Discharge of Mortgage case utilised Esplix Smart Contract and the Postchain Blockchain. The latter enabled the hybrid solution—an interaction between the existing NSW LRS technology infrastructure and the Explix Smart Contract: the smart contract could automatically call the NSW LRS system to return the title data into the smart contract, without demanding radical changes to the NSW LRS land registry databases. In addition, the new approach would enable the mortgagor to initiate the lien release, and not be dependent on the mortgagee, via automated enforcement. This contracting protocol could easily be distributed through third party vendors or directly through other NSW LRS apps, again without disrupting the underpinning and existing technology infrastructure.

The Australian NSW LRS case also revealed implementation and scaling challenges. Australia is a federation, divided into 6 States and 2 territories, and land administration responsibilities reside with those federated jurisdictions. This creates challenges for developing a national standard for e-conveyance and land transactions. For example, PEXA, the national e-conveyancing platform, not based on blockchain, took more than a decade to develop, and even still, many transactions in many States are completed outside PEXA. In law and regulations, it is considered an ELNO (electronic lodgement network operator) and could be open to competition from other ELNO. Whilst this creates competition, it also created inefficiencies, potential duplication of effort, and disaggregated market data. In such a small market (e.g., Australia has a population of ~25 M people), it is generally desirable to build a consensus around standards and processes between States/Territories; otherwise, getting buy-in and interest from private sector operators (e.g., software vendors; financial institutions), is more difficult. Therefore, any blockchain based solution, even if just for a limited number of land transactions, would need to address this issue.

That said, numerous future benefits were evident from the proof-of-concept work. In 2016–2017, NSW LRS processed 930,809 conveyance transactions, of which 25% (237,964) were mortgage-related. At the time of the study, less than 20% of mortgage lien releases were submitted fully electronically. A better "uptake" could be possible through decentralised smart contracts.

In terms of current status, the Discharge of Mortgage Lien process prototype was completed and approved as addressing technical requirements by NSW LRS.

4.4. The Canadian Case—Re-Assignment Reporting in British Columbia

The Canadian proof-of-concept was completed between June 2018 and October 2018. Like the Swedish work, a consortium approach was used, with the Land Title and Survey Authority of British Columbia (LTSA) the sponsoring organisation. LTSA was setup as a statutory corporation under the Land Title and Survey Authority Act 2005, giving it delegated authority over land title and survey systems in British Columbia. This allowed LTSA to focus on efficiency using a digital first approach. In this vein, it was a leader in developing online electronic filing, search, and parcel map services for land sector stakeholders in the 2000s. Other partners for the proof-of-concept included ChromaWay AB and Landsure Systems Ltd. Landsure is a wholly owned subsidiary of LTSA, and primarily supports the continued improvement of the LTSA via the development and management of LTSA's core technology infrastructure.

Like the other cases, rather than seeking a big bang whole-of-sector transformation solution, the LTSA focused upon a new land administration process that could benefit from a smart contract approach, whilst also providing for minimal disruption of the existing technology infrastructure. In this regard, the transaction focused upon was "Re-Assignment Reporting". This activity addresses the reporting of a re-sale of previously assigned condominium properties (primarily) prior to sale. An assignment is a right (and commitment) to purchase a property in the future. Typically, this occurs when a new condominium property is being built and the builder needs to presell a percentage of the properties before banks release funds for the formal build activity to commence. The new process involves numerous stakeholder bodies, including the assignee, assignor (new buyer), and realtor, property developers, LTSA (land registry), and the government planning branches.

The Re-Sales Assignment Reporting process is a new business function of LTSA and the smart contract alternative approach was evaluated in parallel to the development of the "traditional" approach using a central database. Note that the overall goal of the business function was to inject more transparency into condominium re-assignment for tax and planning purposes.

In terms of the developed technology infrastructure, use was made of the Esplix Smart Contract solution and Postchain Blockchain solution. The developed solution considered the planning agency (OSRE) to "push" a pre-sale filing number to LTSA for database storage. Moreover, when an assignor requested assignment of the property, the platform utilised the filing number in the smart contract. Like the NSW case, the contract protocol can be distributed through third party vendors or directly through registry apps.

In terms of the proof-of-concept results, the project experienced no significant challenges. The LTSA project team was primarily comprised of their technology and business analysis organisation—LandSure Systems. This greatly facilitated the technology knowledge transfer and development process.

Further key benefits, as against the more conventional database prototype, were identified as the property taxation branch (PTB) being able to query the smart contract data ledger at any time to view the state of transfers. The solution envisioned the property developers reporting these transfers, as they are in a better position to provide that information. Asking the buyers and sellers to report was considered as well.

In terms of the current status, the prototype project was completed, but due to scaling and change management constraints with all the stakeholders and various agencies, the prototype approach was not deployed, and a more traditional approach was used. LTSA is now evaluating other opportunities for the use of smart contract/blockchain technology.

5. Discussion

Moving beyond individual case examinations, this section undertakes comparative analysis against the core principles of land dealings, technology readiness/maturity, and strategic grid analysis. The accompanying interpretations help to shed light on the relative merits of the hybrid approach at both an organisational and sector level, update the status of smart contract application in the land sector, and enable hypothesis development for future development trajectories.

5.1. Business Requirements Adherence

The core business requirements for land dealings, outlined in Section 2.1, regardless of system antecedence being deeds or title, included the principles of: (i) registration; (ii) publicity; (iii) consent; and (iv) specialty. For all three (3) proofs-of-concept examined, it was shown that each of these principles can be met. That is, the smart contract approach, underpinned by blockchain, and offering integration with existing technology infrastructure (e.g., via APIs), enabled the registration of transaction details (i.e., parties, dates, transference details), and could be configured for wider public reading/viewing. Moreover, the consent principle could be realised through the developed Esplix code and subsequent automation. The principle of speciality, that is, the unambiguous identification of parties and parcels, can also be observed via the integration with pre-existing land registry databases, enabled with Postchain. In this way, achievement of the speciality principle is reliant on how land parcels and parties are identified within the jurisdiction.

Other core business requirements, also outlined in Section 2.1, those associated with titling systems—including curtain, mirror, and insurance—can also be shown to be supported. However, these principles demonstrate the limitations of only undertaking a technical assessment: the hybrid approach can certainly be shown to support obedience towards the principles, but it does not guarantee it. That is, adherence to the land dealing business requirements is not only dependent on technology, but broader socio-technical arrangements (e.g., specifics in legislation). Moreover, it could be argued that with a smart contract approach, the insurance principle becomes redundant altogether: the idea being that the technology confounds the possibility of land dealing fraud altogether. In the hybrid approach, the fraud protection benefits of blockchain exist within the transaction system, but, dependent on existing controls, unauthorised changes could still be made to the land registry itself. The use of hybrid approach provides an immutable record of transactions against which changes in the land registry can be checked. This will be more effective where all transactions are processed through the blockchain transaction system.

Looking beyond core business requirements, Section 2.1 also outlined limitations of existing systems for land dealings. These included perceived excesses in time, cost, duplication, complexity, and fraud with regards to land dealings. Across the three (3) cases, within the controlled hybrid test environments, it was illustrated that the integration and automation, enabled by the Esplix and Postchain solution, could result in reductions in time, complexity, and duplication. Less manual handling of a single dealing should also result in lower costs (i.e., less actors involved) for responsible agencies. However, this does not necessary equate to reduced costs of transacting parties: costs are often associated with set fees or duties, and these are not necessarily determined in simple cost recovery terms. Finally, as already mentioned, a key tenant of the smart contract and blockchain approach, is the reduction of fraud, via publicity, and in this regard, the hybrid solutions provide for this.

Putting aside the limitations of a technology-centric assessment, each of the hybrid proofs-of-concept were shown to support adherence to the core business principles of land dealings. Moreover, the hybrid approach also appears to deal with some of the limitations of existing technological approaches. In this regard, for the unit of analysis of "land dealing", evidence of the benefits of smart contract and blockchain approaches is apparent, beyond earlier theoretical espousals.

The positive results again raise the question as to why uptake has not been more apparent in the land sector? An immediate answer, in the framework of busines requirements, is that whilst the existing technology solutions may not be perfect, they largely already enable adherence to the same core principles. Why take on the risks of a new and potentially immature technology implementation, when existing systems work? This question invites analysis of the actual maturity level of the hybrid approach (e.g., TRL), and whether anecdotal perceptions that the technology is not yet matured, are valid.

5.2. Technology Readiness and Maturity Levels

Against the TRL adapted framework from [38], all three (3) proofs-of-concept reviewed appear to reflect relatively early phases of readiness and maturity with respect to and in terms of blockchain and smart contract maturity. That is, having moved beyond conceptualisation and early experimentation, the Swedish, Australian, and Canadian projects all appear to be at the TRL 4 (Component validation in a laboratory environment—or technical "Proof of Concept") levels. The cases have progressed further towards technological readiness than previously documented proposed applications of blockchain and smart contracts (e.g., see [1]). These earlier efforts were more worthy of TRL 2 designation (Conceptualisation of a new potential application for the technology) or TRL 3 (i.e., Analytical "Proof of Concept" without progressing to demonstrating technology proof of concept [1]). The technical proof of concept projects demonstrate that the hybrid solutions proposed can meet the technical functional requirements of land transaction applications explored.

Technical proofs-of-concept are important to understand the potential benefits and implications for processes and roles [46]. They are also important for communication between technologists, academics and practitioners (registry operators and intermediaries) (adapted [58]). From a systems supplier perspective, proofs-of-concept enable IT experts to highlight issues to be solved; potential clients to verify IT supplier capabilities and supports pre-sales process of IT providers [58]. However, while the technical proofs-of-concept are a step forward and are important to potential adoption of blockchain technology at a jurisdiction and industry level, there is significant further work required before this technology is likely to be adopted as a dominant approach to administering land records.

Going further, from a blockchain maturity perspective [46], see Table 2, all three (3) proofs-of-concept are assessed to be moving from Phase 1 (initial) into Phase 2 (structured). The land registry operators used the projects to explore how blockchain and smart contracts could be used to improve processes and governance structures. The more nuanced approaches of examining how a hybrid approach could utilise blockchain and smart contracts integrated into the existing land registry database technology reflects an advancement in maturity. While the proofs-of-concept successfully showed the technical feasibility of such solutions, none of the operators have yet moved to a formal technology assessment process to select and finalise an appropriate platform or have finalised partner network and governance frameworks. While the potential roles and actions of intermediaries were explored with differing levels of stakeholder engagement across the projects, substantial institutional work is required to move towards implementation of automated smart contracts (maturity phase 3) and business collaboration (i.e., DAOs) (maturity phase 4) in land registry and land transaction space. Government responsibility for land registry functions may limit the viability of moving to the DAO phase for this application of Blockchain.

The proofs-of-concept considered here also provide some insights on TOE for the adoption of smart contracts and blockchain hybrid solutions in land dealing applications. From a technology perspective, the examples indicate that hybrid solutions can meet technology requirements of different transactions that interact with land registries. The proofs-of-concepts did not progress to prove full integration of the technology used by existing registries. However, such integration is relatively straightforward with the analytical design, including interface to existing registry APIs. The cases suggest substantial potential benefits in the hybrid technical solutions explored compared to the existing manual, paper-based transaction systems. Potential benefits in terms of time, cost, duplication, complexity and potential fraud were identified. Many of these benefits flow from digitalisation and the cases did not compare the potential benefits and costs of the prototype blockchain solutions versus other forms of digital technology. This last one was demonstrated by the NSW case with the national PEXA solution (non-blockchain), already providing many of the digitalisation benefits identified and the Canadian case, resulting in the client selecting a more traditional solution.

Furthermore, the cases provided insights into organisational and environmental considerations of the TOE framework for the adoption of hybrid blockchain solutions for land registry transactions. From an organisational perspective, the proofs-of-concept increase organisational knowledge of the technology and how it can be integrated into existing strategies and processes [39]. From an environmental perspective, the cases show the complexity of land registry environments with Sweden halting their project due to legislation requiring wet signatures for land transactions; NSW being part of a broad national environment for which a national solution was available and the Canadian case did not proceed, partially due to change management issues with stakeholders.

In summary, the proofs-of-concept analysed show that in terms of readiness and maturity, blockchain and smart contract technologies, applied to land dealings and land administration more generally, have progressed towards more structured and analytical proofs-of-concept than previously observed. Finally, for further confirmation of these findings, or otherwise, we briefly consider where the case applications lie on the IT strategic grid.

5.3. Strategic Grid Positioning

The IT strategic grid [59] was previously used to assess blockchain adoption, amongst other database technologies, in the land sector [1,60]. The approach considers the impact a specific technology has on an organisation or its business processes from two perspectives: operational and strategic. In this vein, for any technology adoption, four (4) categories (represented graphically as quadrants) of impact can be identified: support; turn-around; factory; and strategic [59].

For the three (3) proofs-of-concept assessed, it is seen that despite the furthered maturity of the technology application, like [1], the adoption still remains within the turn-around quadrant. That is, the technology is being experimented with, with a view to understanding longer-term strategic impacts, but is not yet significantly impacting on day-to-day operations, production, and service delivery of land agencies.

Given the coarseness of the strategic grid analytical categories, it is not unsurprising that the three cases have not moved into the strategic categories in the intervening period since the work in [1] was undertaken: progress towards the strategic or operational quadrants most likely will require more lead time.

Nonetheless, despite the development trajectory shown in Section 5.2, it can be argued that there are still very real barriers to more scaled implementation of blockchain and smart contracts within the land sector. In the final part of this discussion, we further hypothesise the nature of these barriers, and the necessity for alternative adoption approaches, if not techniques for assessment.

5.4. Necessity for Sector-Wide Approaches

The three (3) comparative approaches illustrate that: (i) blockchain and smart contracts are viable solutions for delivering on the business requirements of land administration processes; (ii) whilst blockchain and smart contract uptake has progressed in the land administrations sector beyond mere conceptual work, it remains very much at the level of structured proof-of-concept work; and (iii) implementation and assessment tools focused on specific technologies and organisations, whilst able to reveal levels of socio-technical alignment and uptake, cannot explain the full context in terms barriers (if not benefits) to adoption. The first two points directly respond to the aims of the paper; however, the last point has the most significant implications for furthering work—and in this regard, several points are made.

First, it should be explored whether the barriers are discrete in nature, being able to be responded to with isolated interventions. Several hypotheses can be made for such discrete barriers, but each would require its own independent validation work. As one example, is uptake merely an issue of economics? In more developed contexts, although transaction fees may be the equivalent of hundreds or thousands of USD, comparatively, these amounts are still often very small against the total cost of a property[5]. As another example, the technology risk may be too high for conventional land sector players: new technological approaches bring new risks, and for mandated government monopolies, risk control and business continuity often trumps innovation or efficiency gains as an organisational driving force. The decentralised approach to data management is also a step away from the controlled centralised data approach of conventional land agencies. That is, the risk versus reward ratio is considered too great[6]. Legislative and regulatory barriers are other examples of discrete blockers. If, for example, barriers like these are shown to be highly influential blockers, then furthering the implementation pathway appears to be more straightforward (i.e., revisit the business model; modify laws; or examine the risk appetite or an individual agency).

Second, if the above is not the case, it could be that the discrete examples are mere fragments of a broader sector-level or even societal level resistance. If this is the case, having proven the technical and local validity via the proofs-of-concept, it becomes necessary to return to sector-wide perspectives, or even broader analysis of societal trust. That is, "sector-aware" approaches for organising and assessing smart contract implementations appear more relevant than ever. Whilst a consortium approach, with a sector-wide mindset, was appropriately taken in each of the three proofs-of-concept, it seems any furthering or expanding of those cases will require more structured attention to sector-wide awareness raising, communications, and partnership building activities: the major benefits of the blockchain solution are likely found from a sector level analysis, rather than firm level analysis. Likewise, greater attention to policy, legal, financial, cross-institutional, capacity, and educational aspects appears necessary. Interestingly, it is these aspects and activities, alongside more technical aspects such as "data" and "standards", that the recently endorsed Framework for Effective Land Administration (FELA) [61], developed by UN-GGIM (United National Global Geospatial Information Management) (August 2020), argues as being essential for effective land administration in member countries. Indeed, the framework may act as a guide or blueprint for the land sector with regards to further scaling blockchain and smart contract implementation.

Third, in the same vein, with regards to assessment of implementations, as was already outlined in Section 2.3, mere consideration of business requirements adherence, technology readiness, or strategy grid, at the firm level, is not enough. More useful assessment of blockchain and smart contract implementations demands sector-wide (if not society-wide) tools and techniques. Here, the TOE framework and equivalents were already shown to have previous utility [39,42,43], and even in this study, with regards to blockchain application, however, arguably a greater focus on "processes" over "states" (or entities) is needed, as is an easily and simply applied analytical tool. This is no small challenge in the context of complex industry settings.

6. Conclusions

This article commenced by arguing that the emergence of "blockchain" technology spawned conceptual and design work across multiple sectors aimed at realising the earlier "smart contract" concept. These developments were also occurring in the land admin-

---

[5] It should be noted that the transaction cost to property price ratio, in developing contexts, may make the technology more economically viable. This helps explain the numerous blockchain property starts-ups observed in those contexts.

[6] Again, in this regard, land sector smart contract solutions may come from outside the existing institutional frameworks, as demonstrated in [1], with start-ups offering alternative registration approaches.

istration domain: researchers had actively investigated conceptual and logical designs, with sector-wide digital transformation often driving the thinking. It was also shown that less reporting of actual implementations of land sector blockchain solutions was evident, particularly those undertaken in collaboration with practicing land sector actors.

Building on these assumptions, this paper continued the discourse, giving an updated and more nuanced view of example applications, opportunities, and emerging blockers. In contrast to the earlier sector-wide transformative visions, this work focused on examining emerging hybrid solutions—those that mix the use of smart contract technologies with more conventional, and pre-existing, database and internet technology infrastructure. The approach appeared to offer a way to overcome blockers by minimising disruptions, whilst maximising the benefits of the new technological approach.

Through the examination of multiple multi-actor industry proofs-of-concept case studies from Sweden, Australia (State of New South Wales), and Canada (Province of British-Columbia), and subsequent comparative analysis, against the core principles of land dealings, strategic grid analysis, and technology readiness/maturity levels, the hybrid approach was shown to be technically feasible, whilst also ensuring adherence to the core business requirements of land sector actors. The tangible artefacts of the proofs-of-concept, including code development, and resultant data and document outputs, served as stronger forms of evidence for the capability smart contract approach, for system stakeholders, as opposed to mere conceptual descriptions. In this regard, the hybrid proof of concept solutions can be understood as an important and necessary step in any scaling process.

In terms of strategic grid analysis, the hybrid approach within land applications can be said to sit within the "turn-around" quadrant: the hybrid approach offers the pathway to move towards more scaled operational and production level implementation. This aligned with the technology readiness and maturation analysis, where the hybrid solutions suggest the technology readiness has moved firmly into more considered proof-of-concept stages (e.g., level 3 or 4), and blockchain maturation to be at the level of "structured" inquiry (e.g., level 1 or 2). However, cross-cutting issues still requiring research attention with regards to scaled implementation and continuation of the development trajectory will depend on reverting and broadening to a whole-of-sector (if not societal-wide) perspective, and re-examining concepts of institutional trust, legal and policy issues, sustainable business models, stakeholder awareness, partnership building, and data decentralisation and security, in the light of these findings.

Author Contributions: Conceptualisation, R.B., T.M., A.-K.K. and M.P.; methodology, R.B., T.M. and M.P.; software, T.M., A.-K.K.; validation, T.M., A.-K.K., M.P. and R.B.; formal analysis, R.B. and M.P.; investigation, T.M. and A.-K.K.; resources, A.-K.K. and T.M.; data curation, T.M. and A.-K.K.; writing—original draft preparation, R.B. and T.M.; writing—review and editing, A.-K.K. and M.P.; visualisation, T.M. and R.B. All authors have read and agreed to the published version of the manuscript.

Funding: This research received no external funding.

Data Availability Statement: No new data were created in this study. Data sharing is not applicable to this article.

Acknowledgments: The authors wish to acknowledge the support of colleagues at ChromaWay (Sweden/USA), Land Title and Survey Authority of British Columbia (Canada), Kadaster International (Netherlands), and Swinburne Business School during the research and writing phases of this article. They also acknowledge that an abstracted version and less developed version of this work was previously presented at the 2019 Annual World Bank Conference on Land and Poverty.

Conflicts of Interest: The authors declare no conflict of interest, but acknowledge and declare members of the authorship team maintain affiliation and connection to several of the organisations included in the study. The potential for these affiliations to influence the representation or interpretation of reported research results is acknowledged.

References

1. Bennett, R.M.; Pickering, M.; Sargent, J. Transformations, transitions, or tall tales? A global review of the uptake and impact of NoSQL, blockchain, and big data analytics on the land administration sector. *Land Use Policy* **2019**, *83*, 435–448. [CrossRef]
2. Olsen, B.L. Beyond the Hype: Exploring Blockchain Technology in Land Administration—A Case Study of Ghana and Property Rights. Masters's Thesis, Copenhagen Business School, Copenhagen, Denmark, 17 September 2020.
3. Pisa, M.; Juden, M. *Blockchain and Economic Development: Hype vs. Reality*; Center for Global Development: Washington, DC, USA, 2017; Volume 107, p. 150.
4. Thomas, R.; Huang, C. Blockchain, the Borg collective and digitalisation of land registries. *Convey. Prop. Lawyer* **2017**, *81*. [CrossRef]
5. Martyn, A. The Concept of Land Plot as a Combination of Smart Contracts: A Vision for Creating Blockchain Cadastre. *Balt. Surv.* **2018**, *8*, 68–73. [CrossRef]
6. Karamitsos, I.; Papadaki, M.; Al Barghuthi, N.B. Design of the blockchain smart contract: A use case for real estate. *J. Inf. Secur.* **2018**, *9*, 177–190. [CrossRef]
7. Ameyaw, P.D.; de Vries, W.T. Transparency of Land Administration and the Role of Blockchain Technology, a Four-Dimensional Framework Analysis from the Ghanaian Land Perspective. *Land* **2020**, *9*, 491. [CrossRef]
8. Zevenbergen, J.; Frank, A.; Stubjkaer, E. *Real Property Transactions. Procedures, Transaction Costs and Models*; IOS Press: Amsterdam, The Netherlands, 2008.
9. Szabo, N. Formalizing and Securing Relationships on Public Networks. *First Monday* **1997**, *2*. [CrossRef]
10. Lauslahti, K.; Mattila, J.; Seppala, T. Smart Contracts How Will Blockchain Technology Affect Contractual Practices? *SSRN Electron. J.* **2017**. [CrossRef]
11. Vos, J.A.; Lemmen, C.H.; Beentjes, B.E. Blockchain based land administration feasible, illusory or a panacea. In Proceedings of the 2017 World Bank Conference on Land and Poverty, Washington, DC, USA, 20–24 March 2017.
12. Lemmen, C.; Vos, J.; Beentjes, B. Ongoing Development of Land Administration Standards: Blockchain in Transaction Management. *Eur. Prop. Law J.* **2017**, *6*, 478–502.
13. Larsson, G. *Land Registration and Cadastral Systems; Tools for Land Information and Management*; Longman Scientific and Technical: Norfolk, UK, 1991.
14. Zevenbergen, J. *Systems of Land Registration Aspects and Effects*; Publications on Geodesy: Delft, The Netherlands, 2002.
15. Simpson, S.R. *Land Law and Registration*; Cambridge University Press: Cambridge, UK, 1976.
16. Williamson, I.; Ting, L. Land administration and cadastral trends—A framework for re-engineering. *Comput. Environ. Urban Syst.* **2001**, *25*, 339–366. [CrossRef]
17. Griffith-Charles, C. The application of the social tenure domain model (STDM) to family land in Trinidad and Tobago. *Land Use Policy* **2011**, *28*, 514–522. [CrossRef]
18. Henssen, J. *Land Registration and Cadastre Systems: Principles and Related Issues*; Technische Universität München: Munich, Germany, 2010.
19. Barry, M.; Augustinus, C. Property theory, metaphors and the continuum of land rights. In *Securing Land and Property Rights for All*; UN-Habitat/Global Land Tools Network Nairobi: Nairobi, Kenya, 2015.
20. Lemmen, C.; Van Oosterom, P.; Bennett, R. The land administration domain model. *Land Use Policy* **2015**, *49*, 535–545. [CrossRef]
21. Zevenbergen, J.; Augustinus, C.; Antonio, D.; Bennett, R. Pro-poor land administration: Principles for recording the land rights of the underrepresented. *Land Use Policy* **2013**, *31*, 595–604. [CrossRef]
22. Zakout, W.; Wehrmann, B.; Torhonen, M.P. *Good Governance in Land Administration*; World Bank: Washington, DC, USA, 2006.
23. Dale, P.; McLaughlin, J. *Land Administration*; Oxford University Press: Oxford, UK, 2000.
24. Burns, T.; Grant, C.; Nettle, K.; Brits, A.M.; Dalrymple, K. Land Administration Reform: Indicators of Success and Future Challenges. *Agric. Rural Dev. Discuss. Pap.* **2007**, *37*, 1–227.
25. Bennett, R.; Wallace, J.; Williamson, I. Organising land information for sustainable land administration. *Land Use Policy* **2008**, *25*, 126–138. [CrossRef]
26. Griggs, L.; Thomas, R.; Low, R.; Scheibner, J. Blockchains, Trust and Land Administration—The Return of Historical Provenance. *Prop. Law Rev.* **2017**, *6*, 180.
27. Zevenbergen, J.; De Vries, W.; Bennett, R.M. (Eds.) *Advances in Responsible Land Administration*; CRC Press: Boca Raton, FL, USA, 2015.
28. The Law Handbook 2020. Available online: https://www.lawhandbook.org.au/2020_00_00_05_glossary/ (accessed on 20 January 2020).
29. Ryan, F. *Round Hall Nutshells Contract Law*; Thomson Round Hall: Dublin, Ireland, 2006.
30. Hemmo, M.; Sopimusoikeus, I. *Uudistettu Painos*; Talentum Media Oy: Helsinki, Finland, 2003.
31. Hemmo, M. *Sopimusoikeuden Oppikirja*; Talentum Media Oy: Helsinki, Finland, 2006.
32. Nakamoto, S. Bitcoin: A Peer-to-Peer Electronic Cash System. Manubot. (From 2008). Available online: https://nakamotoinstitute.org/bitcoin/ (accessed on 20 November 2019).
33. Buterin, V. A Next-Generation Smart Contract and Decentralized Application Platform. White Paper. Available online: https://cryptorating.eu/whitepapers/Ethereum/Ethereum_white_paper.pdf (accessed on 3 January 2014).

34. Christidis, K.; Devetsikiotis, M. Blockchains and smart contracts for the internet of things. *IEEE Access* **2016**, *4*, 2292–2303. [CrossRef]
35. Wang, S.; Ouyang, L.; Yuan, Y.; Ni, X.; Han, X.; Wang, F.Y. Blockchain-enabled smart contracts: Architecture, applications, and future trends. *IEEE Trans. Syst. Man Cybern. Syst.* **2019**, *49*, 2266–2277. [CrossRef]
36. Webster, A.; Gardner, J. Aligning technology and institutional readiness: The adoption of innovation. *Technol. Anal. Strateg. Manag.* **2019**, *31*, 1229–1241. [CrossRef]
37. Sauser, B.J.; Ramirez-Marquez, J.E.; Henry, D.; Di Marzio, D. A system maturity index for the systems engineering life cycle. *Int. J. Ind. Syst. Eng.* **2008**, *3*, 673–691.
38. Hicks, B.; Larsson, A.; Culley, S.; Larsson, T. A methodology for evaluating technology readiness during product development. In Proceedings of the 17th International Conference on Engineering Design (ICED'09) Design Has Never Been This Cool, Stanford University, Stanford, CA, USA, 24–27 August 2009.
39. Janssen, M.; Weerakkody, V.; Ismagilova, E.; Sivarajah, U.; Irani, Z. A framework for analysing blockchain technology adoption: Integrating institutional, market and technical factors. *Int. J. Inf. Manag.* **2020**, *50*, 302–309. [CrossRef]
40. Ammous, S.H. Blockchain Technology: What is it Good for? *SSRN Electron. J.* **2016**, 2832751. [CrossRef]
41. Ølnes, S.; Ubacht, J.; Janssen, M. Blockchain in government: Benefits and implications of distributed ledger technology for information sharing. *Gov. Inf. Q.* **2017**, *34*, 355–364. [CrossRef]
42. Depietro, R.; Wiarda, E.; Fleischer, M. The context for change: Organization, technology and environment. *Process. Technol. Innov.* **1990**, *199*, 151–175.
43. Clohessy, T.; Treiblmaier, H.; Acton, T.; Rogers, N. Antecedents of blockchain adoption: An integrative framework. *Strateg. Chang.* **2020**, *29*, 501–515. [CrossRef]
44. Humphrey, W.S. *Managing the Software Process*; Addison-Wesley Longman Publishing Co., Inc.: Boston, MA, USA, 3 January 1989.
45. Paulk, M.C.; Curtis, B.; Chrissis, M.B.; Weber, C.V. *Capability Maturity Model for Software*, Version 1.1; Software Engineering Institute: Pittsburgh, PA, USA, February 1993; CMU/SEI-93-TR-24, DTIC Number ADA263403.
46. Osterland, T.; Rose, T. From a Use Case Categorization Scheme Towards a Maturity Model for Engineering Distributed Ledgers. In *Blockchain and Distributed Ledger Technology Use Cases*; Springer: Cham, Switzerland, 2020; pp. 33–50.
47. Kaczorowska, M. Blockchain-based Land Registration: Possibilities and Challenges. *Masaryk. Univ. J. Law Technol.* **2019**, *13*, 339–360. [CrossRef]
48. Tashakkori, A.; Teddlie, C.; Teddlie, C.B. *Mixed Methodology: Combining Qualitative and Quantitative Approaches*; Sage: Thousand Oaks, CA, USA, 24 June 1998.
49. Steudler, D.; Rajabifard, A.; Williamson, I.P. Evaluation of land administration systems. *Land Use Policy* **2004**, *21*, 371–380. [CrossRef]
50. Çağdas, V.; Stubkjaer, E. Design Research for Cadastral Systems. *Comput. Environ. Urban Syst.* **2011**, *35*, 77–87. [CrossRef]
51. Mitchell, D.; Clarke, M.; Baxter, J. Evaluating land administration projects in developing countries. *Land Use Policy* **2008**, *25*, 464–473. [CrossRef]
52. Cete, M.; Yomralioglu, T. Re-engineering of Turkish land administration. *Surv. Rev.* **2013**, *45*, 197–205. [CrossRef]
53. Ramadhani, S.A.; Bennett, R.M.; Nex, F.C. Exploring UAV in Indonesian cadastral boundary data acquisition. *Earth Sci. Inform.* **2018**, *11*, 129–146. [CrossRef]
54. Polat, Z.A.; Alkan, M. Design and Implementation of A LADM-Based External Archive Data Model for Land Registry and Cadastre Transactions in Turkey: A Case Study of Municipality. *Land Use Policy* **2018**, *77*, 249–266. [CrossRef]
55. Asiama, K.O.; Bennett, R.M.; Zevenbergen, J.A. Participatory Land Administration on Customary Lands: A Practical VGI Experiment in Nanton, Ghana. *ISPRS Int. J. Geo-Inf.* **2017**, *6*, 186. [CrossRef]
56. Unger, E.M.; Bennett, R.M.; Lemmen, C.; de Zeeuw, K.; Zevenbergen, J.; Teo, C.; Crompvoets, J. Global policy transfer for land administration and disaster risk management. *Land Use Policy* **2020**, *99*, 104834. [CrossRef]
57. Mehar, M.I.; Shier, C.L.; Giambattista, A.; Gong, E.; Fletcher, G.; Sanayhie, R.; Kim, H.M.; Laskowski, M. Understanding a revolutionary and flawed grand experiment in blockchain: The DAO attack. *J. Cases Inf. Technol.* **2019**, *21*, 19–32. [CrossRef]
58. Jobin, C.; Le Masson, P.; Hooge, S. What does the proof-of-concept (POC) really prove? A historical perspective and a cross-domain analytical study. In Proceedings of the XXIXth Conference of the International Association of Strategic Management (AIMS), Baku, Azerbaijan, 11–13 June 2020.
59. Applegate, L.M.; Austin, R.D.; McFarlan, F.W. *Corporate Information Strategy and Management: Txct and Cases*; McGraw-Hill: New York, NY, USA; Irwin Custom Publishing: New York, NY, USA, 2006.
60. Bennett, R.M.; Wallace, J.; Williamson, I. Integrated Land Administration in Australia: The need to align ICT strategies and operations. In Proceedings of the Annual SSSI Conference, Melbourne, Australia, 25 August–1 September 2005.
61. UNGGIM. The Framework for Effective Land Administration. In Proceedings of the United Nations Global Geospatial Information Management—Expert Committee, New York, NY, USA, 26–27 August 2020.

 *land*

Article

Gender, Land and Food Access in Ghana's Suburban Cities: A Case of the Adenta Municipality

Kwaku Owusu Twum [1,2], **Kwabena Asiama** [3,*], **John Ayer** [4] and **Cosmas Yaw Asante** [4]

[1] Department of Spatial Innovation, Huts and Cities Limited, Accra GD0028407, Ghana;
    kwaku@koa-impact.com
[2] Department of Project Development, Gold Coast Sustainability and Governance Institute,
    Accra GD0889430, Ghana
[3] Geodetic Institute, Faculty of Civil Engineering and Geodetic Science, Leibniz University Hannover,
    30167 Hannover, Germany
[4] Department of Geomatic Engineering, Kwame Nkrumah University of Science and Technology,
    Kumasi AK000-AK911, Ghana; jayer.soe@knust.edu.gh (J.A.); ycasante.coe@knust.edu.gh (C.Y.A.)
*   Correspondence: asiama@gih.uni-hannover.de; Tel.: +49-(0)511-762-2406

Received: 21 August 2020; Accepted: 29 October 2020; Published: 31 October 2020

Abstract: The disparity in land and food access in Ghana often overlooks the possibility of an underlying gender disparity. This paper explores and interrogates the disparity between land and food access with respect to gender and the evolution of this relationship over the years as a result of the settlement expansion and urban growth within the Adenta Municipality in Ghana. Adopting a mixed pairwise approach of combining spatial analytical tools, vulnerability indexing and resilient indicators, the paper examines the levels and rates of land accessibilities within the stream of modern cities. It assesses the land market system complexities within developing economies and attempts to address the potential threats of gender-land access gaps. The paper finally assigns weights of ranks to model the phenomenon and recommends trends that can facilitate predictions and early cautionary systems for effective urban land governance in Ghana. The paper concludes that though it is noticed that women engage in power structures on a daily basis, this both benefits and burdens them, depending on their socio-cultural status and other factors in terms of access to land and food.

Keywords: gender; land tenure security; suburban competition; innovative spatial governance; food security

---

1. Introduction: Land and Competitive Urban Transformation

Though the Sustainable Development Goals (2015) and the New Urban Agenda (2016) point to the role of cities in promoting equity and sustainability, the rapid urbanisation in Ghana along with the changing land governance dynamics significantly influences these advances towards equity in household livelihoods [1–6]. Urban land demands have risen from 2% in the 1900 to 50% in 2017 with an estimated rise to 67% by 2050 [7]. Ghana's inter-censual regional data has consequently indicated a similar trend. The Greater Accra Metropolitan Area's (GAMA) urban population overflow of outside the city of Accra (Ghana's capital, referred from hereon as Accra) shows that Accra's urban share of GAMA's overall population growth had significantly declined from 70% to 40% within 1960–1970 and 2000–2010 inter-censual periods, respectively [7]. The speed of population growth on the fringes of Accra has resulted in a surge in the competition for residential land, infrastructure and commercial centres, as well as sufficient food-causing. This has influenced the urban change pattern of Ghana's capital, which is overwhelmingly moving towards suburban districts such as Adenta. Subsequently the land distribution pattern of peri urban and suburban spaces within the capital region is experiencing rapid growth. These rapid changes, in terms of land use and tenure in Ghana, has gained attention in Ghana since land is an important asset for both women and men [8]. Despite the attention, although studies have recognized the essential role and functions that women play in the development and management of natural resources such as lands in Africa, their involvement in ownerships, decision-making and economic developments on space, among others, has been poorly regarded [9–11]. In fact, Dery [12] confirms that women have faced different forms of tenure insecurity, both as wives and in their relations with wider kinsmen within the social and cultural settings of Ghanaian communities. In terms of women's contribution to farming and agriculture, though women make up 49.32% out of a population of 30.42 million, 52% of the farmers are women, with women producing 72% of Ghana's food stock [13].

Though deemed significant, studies on women's access to land in suburban cities have not been adequately explored. Women's access to land in the Global South has been discussed mostly in the context of rural space. However, land management arrangements in the suburban and periurban areas have been gradually transformed from the traditional liberal market in this urbanizing district to an almost exclusive male dominated market [14–16]. These suburban areas therefore sit at the nexus of traditional practices and practices influenced by colonial masters [17,18]. The mix of these systems has shown to be a disadvantage to women, especially with regards to land access [10]. As a result, the rise of land tenure insecurity and discrimination is high. Furthermore, single mothers by choice within these areas are heavily depended on by their children and in some cases, other relatives [10]. Characterized by low participation in the land market, the woman's abilities are severely curtailed, with their ownership rights restricted. This also accounts for the gross marginalization and poverty in the district especially among those without formal education. Consequently, this disparity in gender within the land management sector has weakened decision making and land governance in Ghana. Despite the seemingly widening gap between men and women in land access as well as the growing challenge of food limits, women within these areas still play significant roles as bread winners as well as food providers [10,11,19].

In response, this paper explores and interrogates the disparity between land and food access with respect to gender and the evolution of this relationship over the years as a result of the settlement expansion and urban growth within the Adenta Municipality. In the next section, the key concepts in this paper, land access, food access, gender and urban competition are explored, and the meeting points are identified. In the next section, the methodology adopted for the study, as well as a brief profile of the study area is given. The next section lays out the results of the study and further discusses the implications of the results shown. The fifth section concludes the paper with a summary of the findings.

2. Land and Food Access and Gender Imbalance within a Competing Urban Space—The Nexus

This section describes the relationship between land and food access on one hand, and gender on the other hand, within the urban setting. The section starts with an explanation of the relationship between urban competition and land access. This is followed by a description of the current dynamics of women's land rights with respect to access to land as well as food. The section concludes with a conceptual framework that depicts the land-gender-food nexus.

Cities are transforming close periurban areas into ad hoc cities, which Acheampong [20] describes as suburban areas or cities. This paper contextualizes suburban areas as suburban cities. Suburban areas are defined in contrast to central or inner-city areas, and are identified as commuting residential areas that possess a mixture of land uses [21]. Suburban areas or subcities are functions or outcomes of the overspills of the inner-city or core cities. They are "absorbers" of the population influx of typical cities. This is a phenomenon strongly seen within the Accra Metropolis (the core city) that has produced suburban cities like Adenta.

2.1. Urban Competition and Land Access

Land access in most of Sub-Saharan Africa is governed by the customary system comprising well-intentioned socio-cultural rules that have evolved over centuries to grant equal access to families and individuals who are members of the land-owning group. Land access in the customary setting is akin to property rights in the western styled or statutory land administration, in that the group holds the ownership rights, with individuals holding the use rights (see below). Hence, land access in this study refers to the mechanisms through which a person can be granted these rights to use the land. In Ghana, land ownership, land rights and tenures are administered in a plural legal environment with customary laws and norms operating alongside statutes [22]. Ghana recognizes two forms of land ownership: statutory and customary [23]. 80% of the total land area in Ghana is customarily owned by clans, stools, skins[1] and families. The state owns 18% while the remaining 2% is jointly owned by the state; the beneficiary interest being held by the community and the legal right being vested in the state [24,25]. Hence, land holding conditions are primarily determined by one's affiliation or membership to a land-owning group [26,27]. Many women in rural and urban spaces across Africa are restricted within the land market community due to the monetary barrier [4]. In fact, for the urban woman, whose economic activities are limited, this right of access to land seems to be a privilege granted by the male community [19]. In urban communities these cultural and discrimination against women is still prevalent except for those lands that are state-owned in which women may also participate in equally. This further aggravates the gender disparity to the disadvantage of women since they can only participate in the marginally owned portion of space in struggle with their majority male competitors. Urban competition is commonly influenced by population growth rates, hence making it a key indicator for evaluating the relative inequality in the distribution of any given resource whether land or food. This indicator compares the proportional variation of inequality visible in an urban area in relation to its residents and resource availability [28]. Consequently, rapid population growth without its corresponding structural developments and sufficient supply of services is conducive for urban inequalities. Thus, a spatially-balanced system through the monitoring of urban growth populations promises a better way of its management [29]. Urban population growth patterns need regular monitoring, as it is seen as a geospatial input that can be modelled for accurate assessments and predictions [30].

2.2. Women's Land Rights and Access

Urban population growth generally transforms the change pattern of the natural and built environment as well as the social fabric [31]. This growing change in the supply of physical resources, particularly land, is also seen in the functional and economic characteristics of the urban economy. Land is a highly valued asset across all regions in Sub-Saharan Africa, whose possession aids in wealth creation, and also improves a person's social standing and influence [32,33]. It is an essential base for food production and housing provision as well as economic undertakings which almost every human being relies (whether directly or indirectly) on it for their survival [8]. This philosophy positions the land governance literature as a multiple component system that embraces various aspects including land availability, equity, transparency and participation [19]. Within land governance, these rights are granted to all persons who are members of the group, male or female [12,34]. However, the foundation that establishes these rules relating to the ownership, use and entitlement of land narrows the inclusion of women [11]. Issues of land are complex and for it to be completely tackled, the following three conditions are to be met: the legal recognition, social recognition and enforceability by external authorities [19,22]. Failure to include any of these three elements renders the rights to land as incomplete. For instance, a land right that is legally recognizable but not socially recognized or

---

[1]   The stool (southern Ghana) and the skin (northern Ghana) is a body corporate (like the British crown) representing the people and headed by the chief.

enforceable is an incomplete right. On that basis, women are seen not to enjoy equal rights to land. Rather, they have secondary rights and entitlement to the use of land and can only gain access to land through the male members of their conjugal or uterine families [11,35,36]. Such access and rights are partial, in that they can be taken away anytime that the primary rights holder of the land feels appropriate [37]. This disconnect between legal and social backgrounds is one of the main challenges that most developing countries are facing [12]. Women's access to and control over land remains a contentious issue in political, social and economic discourses. Moreover, access to and control over land is not just an issue of academic and development considerations, but a question of fundamental human rights. Therefore, everyone, especially women, require an equal and equitable opportunity without any discrimination to have rights and access to land.

2.3. Women and Food Access

Gender issues pertaining to women within the land market and food access have become an important subject worth exploring especially in the urban space. Food access here is viewed as one of the four pillars of food security [38,39]. The others being food availability, food utilization, and food stability. Food access here further comprises three aspects—physical, economic and social aspects [38,40,41]. Though food is usually viewed as a monetized and abundant commodity in the urban setting, with urban dwellers having a high food accessibility, some have pointed to a "rural bias" in food studies that prioritize the rural areas [42,43].

Women are largely represented in the informal suburban sector and continue to dominate a major fraction of the informal workforces in cities; particularly as urban food producers and market retailers throughout the country [44]. Discussions concerning agricultural productivity are best explained through land rights and access analysis using a social systems methodology [27]. For instance, the women outlook approach facilitated the examination of growth and poverty within the Sub-Saharan region [32,45]. It recognized that women stand at the confluence of between economic growth and human development. Women and food access concerns in Ghana show that gender disparities persist, and that food inequality is costly to Ghana's economic and social development, as it limits the realization of growth and poverty reduction objectives [19].

The situation of inequality and poverty resulting from land access restrictions is prevalent among women with little formal education [11]. Thus, it is imperative to consider the existing inequitable, unequal and unbalanced (supremacy) relationships and practices confronting the rights and privileges of women in Africa and Ghana as far as land use conditions in the urban case is concerned. Consequently, the basis for sustainable land development based on legal and social justice is underscored by the provision of a defined framework that promotes urban land access and empowerment among women. This concept also provides room for a broader view of the urban social group [32]. It is recognized that women's poverty is directly related to the inadequate economic opportunities, and access to economic resources including land ownership and inheritance, credit, low access to education and support services resulting their minimal participation in the decision-making process [9]. Many households' heads in urban areas (largely women) are vulnerable to food insecurity due to an unsustainable labor wage and support from local policies [22]. Furthermore, areas characterized by low food production for home consumption heavily depend on food imports and others. The difficulties surrounding women's experiences in urban food and land access disparity stems from the cultural and gender shifts and dynamics of urban environments, which often combine women's individualities, duties, spaces and roles [12].

2.4. Conceptual Framework

The future food framework of the Adenta Municipality adopted a conceptual framework which serves also as the model builder.

This framework explains that the emergence of land holding restrictions among women is a cyclical flow particularly in the urban area (Figure 1). Following the constructs of the Ghana Statistical

Service [5] and the USAID [8] it is observed that urban growth transforms food producing areas into livable areas which comes with high economic costs. The economic functions and cost of prime lands or urban lands are still overwhelming, pushing the poor and marginalized in the society (especially women in the informal sector) out of the land market. These land access discriminations are further heightened by the cultural bias of land ownership in Ghana [46]. In effect, whilst land ownership and land rights are customarily defined and reduces women's access to land, the aforementioned conditions also cause it. Consequently, (potential) farm lands are lost to commercial areas and food supplies are left on the masses of women to confront and mitigate the resultant effect that comes with it, that is, food declines in these urban centers where informal single women are dominant.

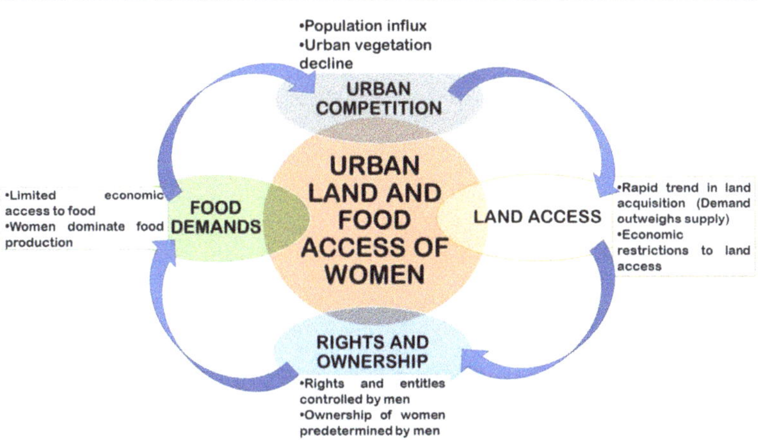

Figure 1. Conceptual framework of land-gender nexus in Ghana (author's construct).

3. Study Area and Methodology

This section provides an overview of the study area, with general information, as well as information relating to the study. The section further details the methodology used in the study.

3.1. The Study Area

73.3% of the employed urbanites in Greater Accra fall within the informal sector. Economically speaking, the women in Adenta Municipality dominate the informal sector [47]. They are mostly involved in economic ventures such as small-scale enterprises, trading, traditional hand jobs, food vending services and the swift up-and-coming commercial food retail centers popularly known as "provision shops". This sector is currently experiencing a major surge with a lot more women are becoming interested in this sector. This earns the municipality economic revenue through taxes. The agriculture sector employs a lot of women in the municipality with 80% involved in food production [48]. Nonetheless, with the recent urban changes (i.e., the farming land conversion to residential areas), the food vulnerability situation in the district has escalated. Thus, the women who used to farm on these lands for commercial purposes have resorted to subsistence farming, induced labor and other means of economic survival so that they can earn a living [49].

Possibly, with the urban sprawl from the capital (Accra) and the increase in housing units, agricultural areas in the Amrahia and Ashiyie for instance, have seen very little of the purpose for which they existed. Whilst the number of residents rises unusually, land sizes drastically diminish with land uses frequently altered. This has become a great concern for the future food index of the municipality since backyard agriculture is no longer feasible. Additionally, emerging literature, observing the growing urban food decline happening in peripheral spaces surrounding cities, are quick to suggest that farmers' livelihoods and roles are complicated by the hybridized situation of land access and gender disparities [50]. Demographically, Adenta municipality is among the fastest growing urban districts in the region. With a growth rate of 2.6%, and a regional growth rate of 4.4%, it is among the highest in the region and therefore reflects the fast-developing nature of the municipality [51]. The swift urban change patterns and local population influx have resulted in numerous urban challenges in the district. For instance, in 2015, there was a rainfall disaster in which 12 communities with 552 households and 2208 persons were affected [48]. This does not only question the security status but also creates an alert on the complex nature of this phenomenon that needs a multidimensional approach to solve its effects through geographic information systems (GIS). Local authorities have made it a priority to develop practical steps to combat the incidence. Accordingly, food inequality and land access disparities are soaring and this can only be collectively monitored and mitigated through scientific approach using geographic information systems (GIS). Over the years, the manual and poorly monitored urban changes have been a major contributor to the poor response of the local authorities and for that matter ineffective decisions. This sets a description for a consistent and replicative methodology that can holistically advise urban policy makers for pragmatic decisions [52].

3.2. Methodology

The study adopted a case study methodology that embraced a joined approach of connecting nonspatial data with spatial analytical tools. The research centered on a value-based perspective known as a "rights based approach" which espouses that access to and use of land is a basic human right issue and should be respected [5]. Thus, any barrier(s) to equal access concerning the use of land by both men and women is regarded as a breach of rights of affected members [22,53]. The adoption of the right-based approach helped to assess the level of land discrimination within the municipality. Furthermore, in order to appreciate the depth of issues, a convergent parallel mixed method approach was used with emphasis on 4 communities within the Adenta municipality. This featured critical perspectives in the research. Data were equally collected through focus group discussions (FGDs), interviews and administration of questionnaires through a purposive random sampling approach (see: [35]). In addition, satellite imagery was collected using a remote sensing approach; adopted to spatially classify areas of vulnerability based on population and municipal data. This enhanced the validity of the issues and climaxed the findings for strategic interventions [30].

Primary data from FGDs, direct interviews and questionnaires administration was complemented with satellite images and relevant secondary data obtained from divergent secondary sources, particularly, publications by government, policy think tanks and academicians (See Table 1). Additionally, some primary data were gathered from direct observations of the change of land use over time (in 1991 and 2018). Comparison was equally drawn to identify major matters that could be helpful to the selected communities. As part of the broader research design, a total sample of 60 households from the sampled areas in the municipality was estimated and interviewed. The focus of the study was to identify and interview households in urban and periurban areas in the Adenta Municipality. The Adenta Municipality is among the 151 urban districts growing within the country [7]. It was selected due to its proximity to the central business district (CBD) of Accra as well as the rapid rush for residential lands within the region. The district has similar characteristics with urban districts in the country [7]. Additionally, it has a unique blend of urban and periurban features. The assistance from the Adenta Planning Department was sought to give clear demarcations of the boundaries of the urban center. However, in order to effectively justify the case of food and land access disparity, purposive random sampling was used, where the zonal councils of the municipality were used as a basis for sampling for data collection. Again, respondents (being women household heads) were purposively sampled and interviewed. The selection of women was exclusively determined from local knowledge acquired from the local assembly's database. These women were selected based on their local knowledge on the topical issue and were willing to be interviewed.

Table 1. Details on respondents and sampling.

MUNICIPAL ZONES	Sample Communities	Number of Respondents *
KOOSE	Amrahia	15
Gbentanaa	Adenta Old Town and Housing Estate	15
Nii Ashale	Ashale Botwe	15
Sutsrunaa	Nmai Djorn	15

* Sample size for the five communities were 60 with 15 respondents for the various municipal zones[2] (Source: Author's construct).

It was possible to locate the communities since it was subdivided into 4 administrative zones. Thus, the municipality was put into 4 quadrants (subzones) according to the Adenta municipal report (2016) with the center "Adenta barrier" (a local neighbourhood) serving as the epicenter for economic and social activities. The boundaries of each quadrant were also identified with the communities. The goal was to have a relational representation of households throughout the municipality. In the second stage, subcommunities within each zone were identified and randomly selected. Respondents were later selected through purposive random sampling, where respondents were interviewed per their locations.

3.2.1. Weighting and Ranking of Data Indicators

Considering the land management conditions and subcategorizations of the district, a set of criteria was determined to indicate the presence of gender gaps and food inequalities within the land market domain. The validation and normalization of indicator values were primarily based on other scholarly methods (see: [36,37] In effect, the study used SPSS and Microsoft Excel to standardize and normalize indicator values as shown in the following steps and formula in reference to the approach of [38].

1. To normalize all the indicators to matching units and scale, standardized values were calculated using the standard deviation formula as shown in Equation (1):

$$\text{Standard Value } Z = (X_i - \mu)/\sigma \tag{1}$$

   where Z = standardized value, X_i = Indicator value (quantitative data collected on the land use change and access, through questionnaires and PHC data), μ = mean value and σ = standard deviation.

2. The next step was to normalize the standardized indicator values such that the values fell within 0 to 1 in order to facilitate the weighting of the indicators. The normalized standardized value as shown in Equation (2) is defined as

$$Y = (Z_i - a)/(e - a) \tag{2}$$

   where Zi lies between a to e, Z_i = standard score, a = minimum value, e = maximum value and Y ranges from 0 and 1.

3. Reverse indicators, such as incidence of vulnerability and gender were further standardized using the formula $(1 - Y)$ so that all values nearer or equal to 1 are those approaching food security, while those nearer zero (0) means land insecurity or within the food stress areas.

---

[2] Sample size was limited as a result of the time constraint factor of the research.

3.2.2. Weighting and Calculating Aggregate Scores

In this research, equal weight was assigned to each indicator based on the premise that they all possess equal significance to the land access equity of the city. Aggregate scores for each land and food dimension, defined as the Dimension Sustainability Score, were calculated using

$$f(x) = \sum_{i=1}^{n}\left(\frac{Yi \times Wi}{Wi} \right) \tag{3}$$

where, W = weighting of each indicator, Y = normalized value of each indicator, n = number of indicator and i = year of assessment.

Finally, the overall Urban Food and Land Access Index (UFLAI) was calculated by summing the sustainability of each dimension score year by year.

Again, equal weight (1/4) was assigned each dimension of Urban Growth (UG), Economic Access (EA), and Social Access (SA) as shown in Equation (4)

$$\text{Urban Food and Land Access Index (UFAI)} = \sum_{i=1}^{n}\left(\frac{(UG \times W)\ (EA \times W)\ (SA \times W)}{\sum W} \right) \tag{4}$$

3.2.3. Using Pairwise to Simulate Land and Food Access Disparities

The maps for the known indicators were derived from their various measurements, and were connected to the prevailing rate of land use, access and food disparity, based on the relative importance of every indicator that was evaluated. Ref. [54] categorize AHP into 3 stages: (I) Disintegration—where the urban situation is identified and structured into indicators (II) Relative judgment—this is done through pair wise comparison (III) Aggregating the priorities—calculate suitability index. Structuring of the indicators is fairly a subjective activity and somewhat relies on decision maker's expertise and experience. The indicators were therefore considered based on the importance of the field data acquired (explained above). Table 2 presents the pair-wise conditions developed in AHP for each adopted standard. Eigenvalues were used to designate the relative importance weight of each indicator according to the parameters of [55].

Table 2. Pairwise conditions.

	Pairwise Comparison—Criteria Comparison Matrix ©			
	Urban Growth	Economic Access	Social Access	Total
Urban Growth	1.000	0.140	3.000	4.140
Economic Access	7.000	1.000	5.000	13.000
Social Access	0.330	0.140	1.000	1.470
Sum columns	8.330	1.280	9.000	

Variables	Description
1	Equal importance
3	Moderate importance
5	Strong importance
7	Very strong importance
9	Extreme importance
2,4,6,8	Intermediate values between adjacent scale values

Source: Author's construct.

3.3. The Adoption of Spatial Frameworks and Multiple Evaluation Methods

More than millions of urban inhabitants currently face land and food challenges in Ghana precisely within its urban cities like Adenta. This has largely been associated with the threatened residual income size, food expenditure and the deficient capacity to produce food arising from the land access restriction of the urban majority [46]. Therefore, the processes involved in how land is acquired, how it is distributed and the actors involved have a direct bearing on the affordability and access to food [50]. This chain of variables could be well appreciated if closely related to space and analyzed using suitable spatial models and techniques. Subsequently, the ability to understand and monitor the change systems of urban areas and evaluate its patterns is a right direction to address land and food insecurities [56]. Attempts have been made by various national and international bodies to face urban land issues and food security the technological way. Like Hagai, [56], this research follows a pattern of coupling the diagnoses of land access disparities and their relationships with food systems and women in the district using GIS. These dynamics have been overlooked in many studies of urban land and food security, but recently even international bodies including World Bank, RUAF and FAO are all moving towards the direction of multi-evaluation of gender, land access and food declines. Against this background, this research has modelled the process of food and land security assessment particularly among women using GIS tools, based on data representations on urban transformation and land rights indicators. These were in the form of GIS layers and were well integrated using the multi-evaluation model approach [29]. Further, the application of GIS modelling approaches is relatively faster and convenient than statistical manual methods, where components of gender, land and food systems can be analyzed. Information ought to be executed readily, so as to facilitate evidence-based understanding of food imbalances, thereby guiding city authorities in decision making with regard to designing equitable methods of food frameworks and urban models for effective local decisions.

4. Results and Discussion

This section sets the paradigm of underscoring the multidimensional nature of land tenure security, gender disparity and the preposition of land use governance within the suburban cities of Ghana. From a baseline analysis of the Adenta municipality, the indicators were analyzed and placed in perspective to align critically with the system of measuring the land access disparity in the country. The section presents and discusses the results of the study from the social, economic and spatial aspects of land and food access as shown in Section 2.

4.1. Urban Competition and Land in Adenta Municipality

One of the key urban indicators is the population distribution and spatial emphasis of the municipality. This phenomenon monitors and tracks urban spills, population increases and deficiencies of an area. It further plays a key role in monitoring the rate of access and use of land. The population of the municipality reckons to be increasing at a significant rate with a sporadic urban growth of 4.87 currently. This growth correlates to the sporadic regional urban growth of 4.4 (Figure 2). This calls for pragmatic attention to be given to urban land monitoring and food inequalities within the municipality. The phenomenon of urban competition and land access restriction has contributed to a gender gap within the municipality. The study explored, according to the literature review, indicators to check the level of access. The urban land and food systems indicators used to assess inequalities were urban growth, economic access and social access. These parameters were adopted from [38] as a metric system for evaluating the food and land access situation of Adenta municipality.

Likewise, the urban growth of the municipality is projected to rise in the next 5 to 10 years and this is a determinant for land access and food disparity in the city. However, the results gathered revealed that the rate of urbanization in the Accra city (1.97%, [57]) gradually reduces, whereas that of the municipality is increasing rapidly at an annual growth rate of 4.87% per annum. This is as a result of the presence of sprawl and urban migration from the "choking" city center to the "free periurban" areas. Adenta municipality is one of the key population absorbers from the Accra central city of the Greater Accra region. This urban character has positioned the Adenta municipality in a critical view that requires a consistent approach to salvage its associated urban problems particularly towards land and food inequalities. For instance, the municipality records various levels of poverty, informality and vulnerabilities among women. In order to understand this change, the internal system of the municipality was considered. These internal systems include urban population growth, congestion and others which have led to continuous struggle for occupational and residential land in the subcity.

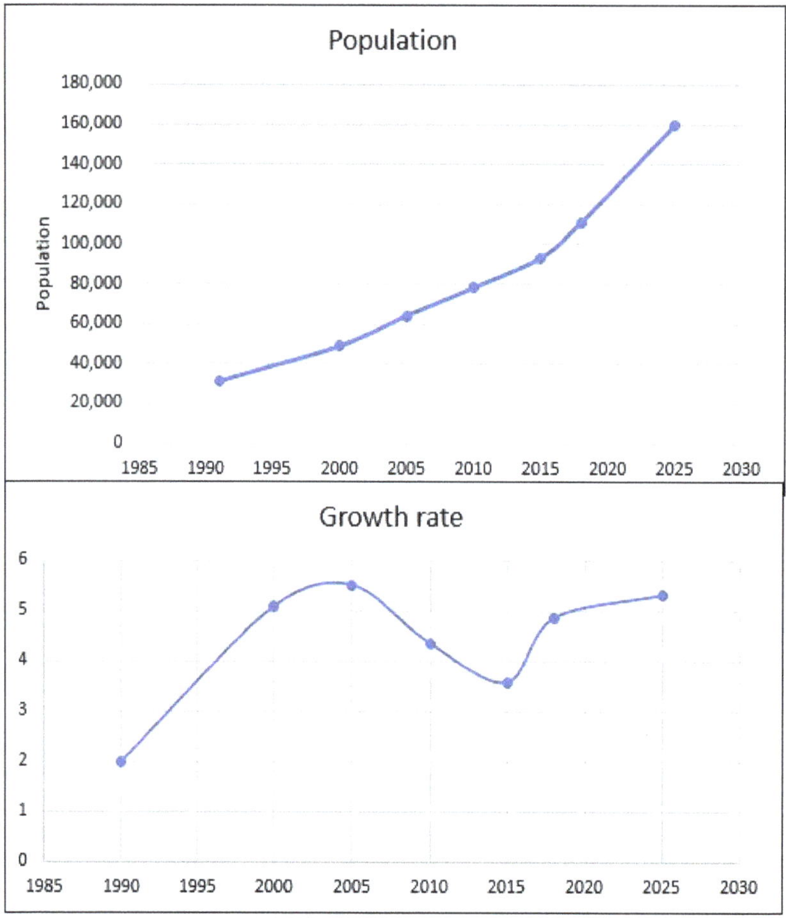

Figure 2. Population distribution of Adenta municipality from 1991 to 2025.

4.2. Urban Sprawl, Vegetation Loss and Food Declines

Africa continues to transition into an urban continent over the years and the rate of urban growth translates into the decline of agricultural lands for food. In Figure 3, it is identified that many countries including Nigeria, Cameroon, Benin and Togo (boundary nations of Ghana) have similar characteristics with the rate of change in their agricultural land loss. The rise is evidently noted in the early 1990s and 2000s and the trend has remained the same since then. However, projections from the Ghana Statistical Service [7] estimates that these rates could double in the next 10 years. Hence, the rapid change in urbanization with its direct impact on land use and access is a phenomenon not only in Ghana but throughout the continent.

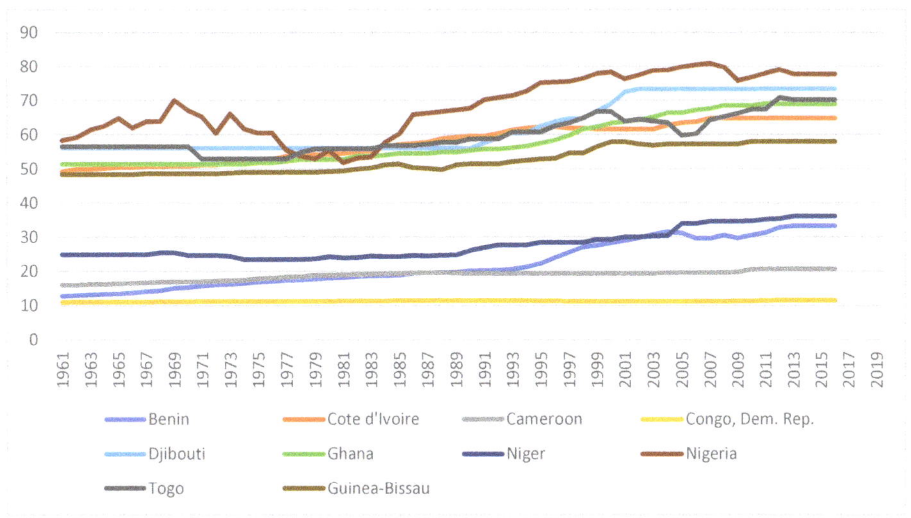

Figure 3. National distribution of agricultural land loss in selected countries in Africa. Source: World Bank Database, FAO Database, (2020).

Similarly, the struggle for lands for residential and commercial use particularly in African urban and suburban cities is pronounced. Population growth and urban change which causes the sprawl of cities is a major contributor towards urban food loss [4]. For instance, rapid population growth in Ghana is reflected in the rapid expansion of land cover class in settlements [45]. Urban areas grew from 1460 sq km in 1975 to 2560 sq km in 2000 and 3830 sq km in 2013, a rise of 161% over 38 years [7]. Among these regions, the greater Accra (shown in Figure 3) recorded high rates of change in agricultural or vegetational land loss. This shows that the more urbanized districts in this region, the more food availability and access is limited. There is the need for a cross-regional and cross-national collaboration towards sustainable land management policies that includes gender indices.

According to Table 3, the population change of the municipality was analyzed with respect to its zones which displays the internal change and distribution pattern. As indicated above, despite the gross change of population rise within the municipality, its level and rate of change vary from location to location. In other words, the population and growth pattern within the municipality is unevenly distributed (Figure 4). This is particularly due to the functional differences and socio-economic evolutions that characterize each zone (Figure 5). The physical concentration and socio-economic limitations of cities constantly affects the access to land and food systems in the municipality.

Table 3. Population distribution of Adenta municipality represented in their zones.

Municipal Zones	1991	2000	2005	2010	2015	2018
Koose	2686	4282	5511	6995	27,436	33,631
Gbentanaa	9274	14,432	18,974	23,009	8341	9938
Nii Ashale	9898	15,414	20,198	24,535	29,256	34,852
Sutsurunaa	9499	14,942	19,511	23,676	28,231	32,684
Total	31,357	49,070	64,194	78,215	93,264	111,105

Source: AdMA Medium Term Development Report, (2008 to 2018).

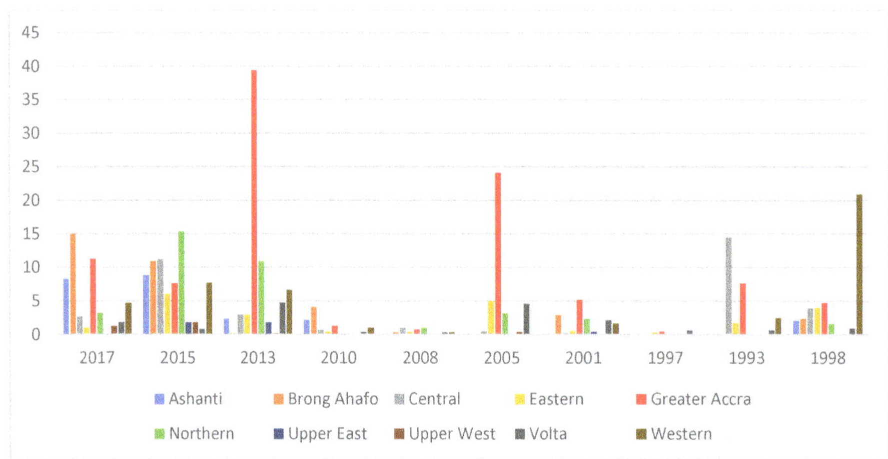

Figure 4. Regional distribution of urban agricultural land loss due to urbanization. Source: Agricultural Stress Index, FAO (2017).

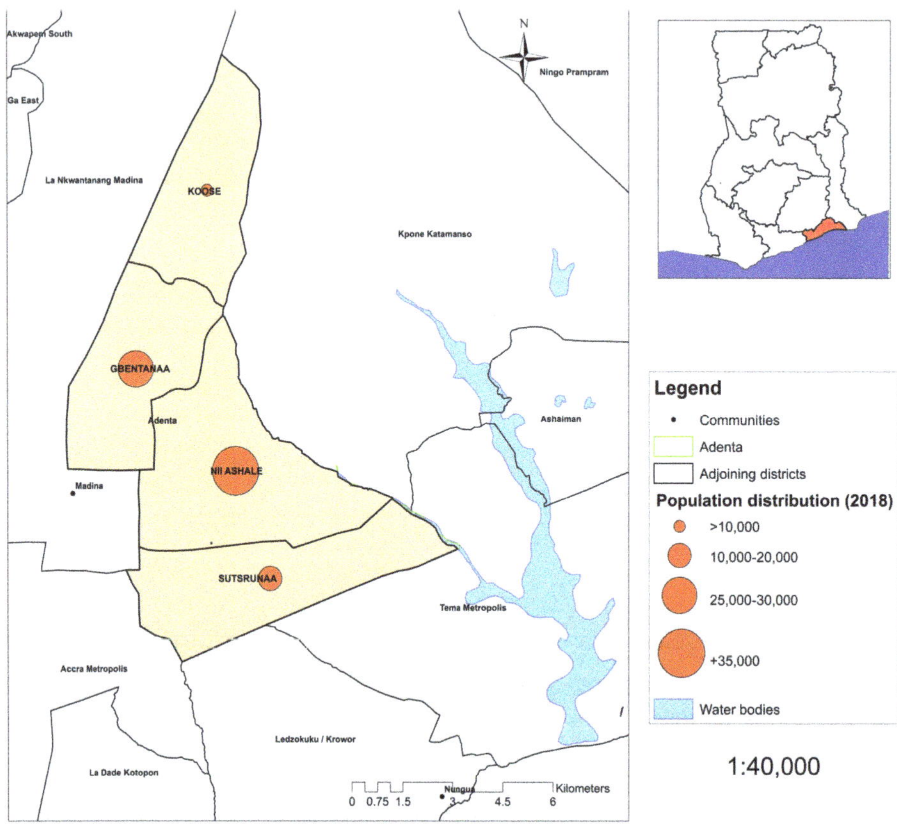

Figure 5. Population distribution of Adenta municipality in 2018 shown in zones (author's construct).

4.3. Economic Access

As a component of land and food access, economic access of the citizenry, particularly women, is an essential component for the development of every nation. The economic component of land and food access disparity discussed considers the income levels, land and food expenses in the municipality. Although the Adenta municipality is perceived as a middle-class city, the terrain has expressively transformed due to urban changes. Currently, the informal sector is catching up with almost 40% of the working class moving into the informal sector. Again, a significant portion (12.7%) of the population (mostly women) do not work and therefore depend on their working relatives for food access. This represents a typical characteristic of urban sprawl, as the city keeps expanding. In effect, the urban economy of the Adenta municipality has been polarized, with an average monthly income of Ghc 1200 (USD 220)[3]. Additionally, the economic access to food has become a great concern to both the formal and informal income groups. For instance, over 50% of the population spend close to 40% of their net income on food which is worrying. This situation is worsened in the case of low-income earners.

Another aspect of the economic access was the eating times of the urban dwellers. According to the definition for food security, one must be able to afford food in their required nutritional proportion at all times. The situation is somewhat different in the municipality. Almost 50% indicated that they only eat twice a day; whilst close to 15% eat three times a day. It was recorded that some even eat once a day. This implies that despite the income levels and social status of the urban residents, low food security is evident. In fact, it must be noted that this indicates traces of urban marginalization, inequality and poverty. Conversely, a major reason for their eating times was due to high costs of food for the majority and the high rental values on the part of others. Figure 6 indicates the notable areas that this urban phenomenon occurred.

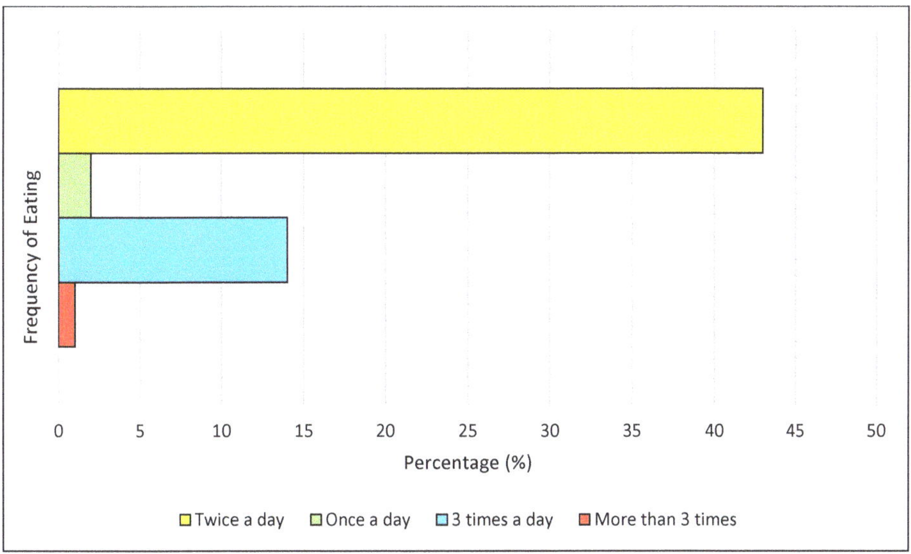

Figure 6. Frequency of eating for people in Adenta municipality.

The nature of land and food access disparities breed sensitive areas of vulnerability and insecurity patterns that emerge in the life of the urban woman. Darfour and Rosentrater [57], indicate that about

---

[3]   Exchange rate used $1 = Ghc 5.5.

2 million women are prone to food insecurities and are normally characterized as urban vulnerable groups as a result. This suggests an unanticipated natural or human-induced shock that can greatly affect the chain of food consumption in the district. From Figure 7, it can be observed that although traces of low access cuts across the municipality, urban residents in Koose and Gbentanaa record severe vulnerabilities with low access to food. This is largely due to the rate of urbanization and inability to access land for economic gains. This is because many (80%) of these female occupants are farmers and farm workers who depend on their farms for survival. Unfortunately, all these areas keep experiencing conversions from its agricultural areas into commercial and residential land uses.

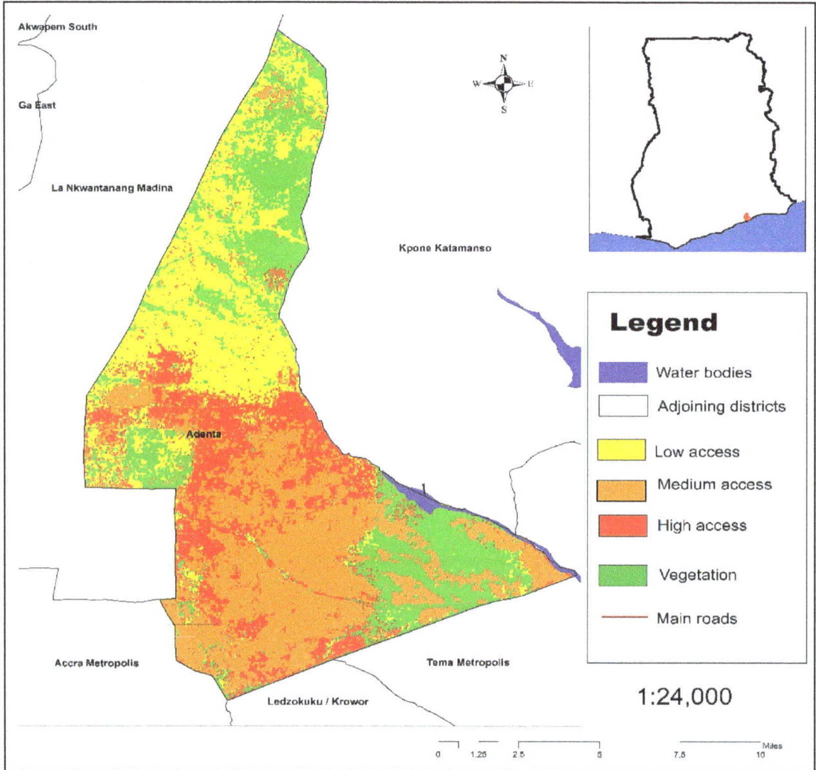

Figure 7. Economic access to food shown in zones (author's construct).

Again, food is arguably available in the country but it is not accessible to all sections of the population, especially the marginalized women and children in the society [57,58]. In Ghana, 1.2 million people suffer acute food insecurity and a further 2.07 million are vulnerable to poor diets in the country [21]. Incidentally, food insecurities are traced to poor land access conditions popular within rural and urban areas that are economically weak. Subsequently for urban areas, the informal communities are characterized by significantly less food production for home consumption since the major household heads are women who heavily depend on food imports due to the land access restrictions. Notably, 73.3% of the employed urban inhabitants in Greater Accra fall within the informal sector of employment with over 55% being women. It is therefore expensive to afford food at all times with the right nutrients by these urbanites since most resort to street foods or already-processed food. According to the food prices as at February, 2019 for instance, a 25kg bag of maize was sold at Ghc 146.26 (EUR 25.89) in Accra with a 2.49% increment whilst it was sold at Ghc 115 (EUR 27.44) in Bawku. Likewise, a bag of local rice gained a 1.24% increment from Ghc 326.71 (EUR 57.83) to Ghc 400

(EUR 70.80) in Accra whilst it was recorded at Ghc274 (EUR 48.50) at Tamale [59]. This clearly shows that due to the fast rate of urbanization and human concentration in the municipality, food is becoming more and more expensive as food demand keeps soaring whilst land keeps diminishing. Moreover, the lack of space and unavailability of food largely accounts for this shortage within these sub cities.

4.4. Social Access

Figure 8 indicates the social dimension of the disparities of women's access to food and land in the Adenta municipality. With an increasing growth of women in the informal sector, the economic groups of the urban food system confirm that despite the variations of income, over 50% of net income is spent on food. It can be realized that despite the presence of land and food, the link to access and utilization is broken. In fact, it is observed that the poor and marginalized who represent the low-income groups stand rather higher risks of inadequate access to food.

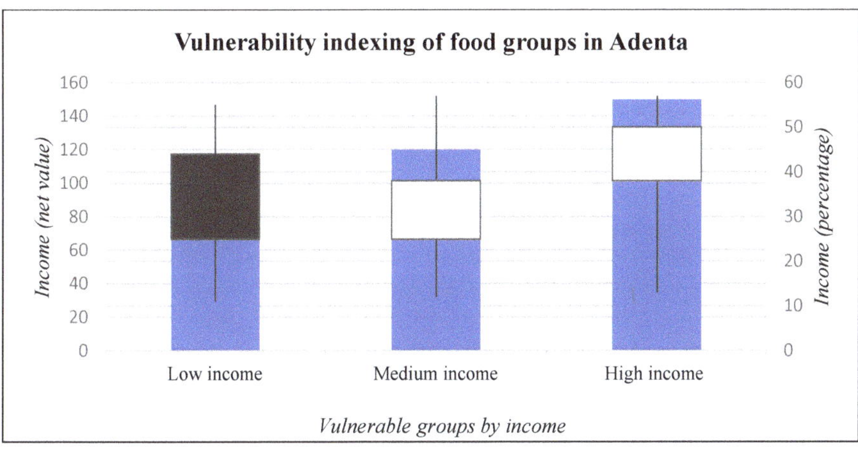

Figure 8. Vulnerability indexing of food groups in Adenta municipality (Source: Adenta Municipality medium term report, 2017).

The women of Adenta Municipality progressively dominate the informal sector [59]. Most household heads in the Adenta Municipality are women, particularly as a result of the high rate of single parenthood. Hence, they constitute a majority of household leaders who ensure household food security. This leaves a heavy burden of urban survival on the women and children despite all developments to make their lives better in the municipality [60]. In addition, most resort to informal job opportunities particularly in the food sector, an area that seems to be of great priority to the survival index of the Adenta Municipality and Ghana.

According to AdMA [48] the agriculture sector employs a lot of women in the municipality with 80% involved in food production (shown in Figure 9). Nonetheless, with the recent urban changes (i.e., the farming land conversion to residential areas), the food vulnerability situation in the district has escalated. Thus, the women who used to farm on these lands for commercial purposes have resorted to subsistence farming, induced labor and other means of survival so that they can earn a living. The women lack the necessary support such as access to and control over land which remains one of the fundamental sources of power defining women's status, identity and opportunity in many communities in the Adenta municipality.

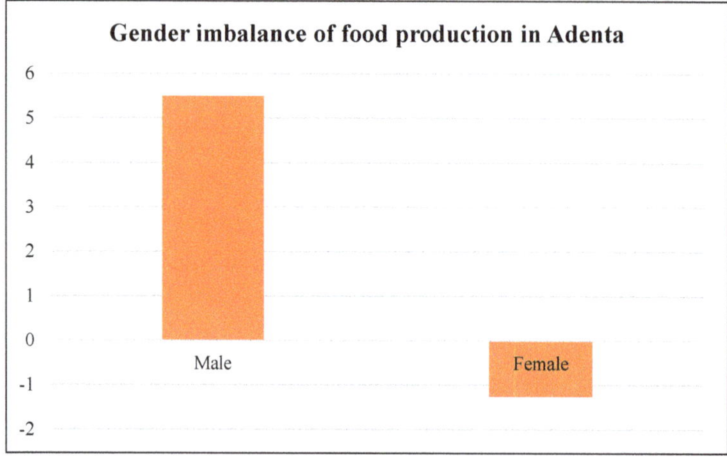

Figure 9. Gender imbalance of food groups in Adenta municipality [48].

4.5. The Role of Sustainability Food Frameworks in Urban and Suburban Districts

Recent research in the area of urban and regional sustainable food system community has shown that various strategies and plans devised by city authorities particularly in the Global North, but not in the Global South. For instance, FAO [61] points out that over 100 cities globally are setting examples for the first time in global urban food frameworks. As a result, the built-up cities are taking the lead in food strategies and seeking steps to reimagine food as an urban subsystem whose sustainability is firmly knitted with that of all other basic urban systems such as mobility, housing, utilities and waste management. The concept of livable cities clinches on sustainable food systems which profess a healthy, equal and ecologically balanced urban space [50]. This is progressively receiving recognition. As a matter of local policy, it is becoming a major responsibility for national and international government organizations to pursue sustainable food frameworks [3]. Accordingly, the urban food phase has evolved from the agrarian-dependent perspective to a complex modern theory. The projections of suburban cities and food security are not only concerned with its generation, but the reality of modern socio-ecological functions expressed by urban societies such as Adenta Municipality. A significant solution to these difficulties may be to develop a stronger analytical linkage between a series of methodologies and sources of information that have the potential to contribute to food and land insecurity assessment coupled with vulnerability monitoring. In this regard, the recognition and implementation of a multifunctional framework in land management influences its sustainable spatial use [62]. The need for a sustainable urban food and land framework is therefore, indispensable.

The major challenge has been the lack of a monitoring system as a result of the inconsistent and essential data. Addressing the unreliable, analogous and relevant data combined with the fragile information base therefore represents a prime priority and precondition for future work. Addressing this challenge, the strategy of a sustainable urban food system consisting of two steps is recommended: first, discovering and prioritizing urban data, and monitoring information needs in the local communities is essential; secondly, efforts to determine multiple, inventive and efficient ways of systematically collecting and analyzing this data can facilitate decision-making [63]. Hence, depictions of these urban areas derived from satellite data principally promises the most congruent measure of defining surface properties [61]. This can help set a rudimentary step of assessing the urban land and food patterns of the district and ensure a quantitative description of the urban space for predictions and framework designs. Frameworks assist city authorities and other stakeholders to better comprehend issues, distinguish problems, prioritize tasks or programs and facilitate policies. It also guides policy analysis and contributes to effective policy assessments. For instance, food systems and vulnerability analysis among others support the identification and targeting of urban classes such as women to help in easy intervention and implementation of plans [64]. Geospatial analysis provides a socio-spatial perspective to examine many urban food issues including food accessibility and urban agriculture land uses [55]. It should be noted that sustainable food and land frameworks are not intended to be the most comprehensive structure, rather, the most appropriate for contemporary food analysis, considering suburban change complexities [58].

4.6. The Complex Nature of the Land and Food Disparity

The Adenta Municipality experiences multiple dimensions of gender disparity as far as food and land access is concerned. From Table 2, the pairwise comparison tool was adopted as parameters for identifying the food and land access disparities within the municipality. The speed of population growth on the fringes of Accra has resulted in an equivalent surge in demand for land to reside, infrastructure and commercial centers to trade, as well as food to satisfy—causing a profound swelling of the suburban populace. This has successively influenced the urban growth pattern of Ghana's capital, which is overwhelmingly moving towards previously periurban districts (such as Adenta), that surround the Accra city. In effect, the express changes and condensations have altered the physical and socio-economic features of thes suburban communities [4,65]. A key component of this change is the urban food dynamics that follow. For instance, before the recent redemarcation of districts, the official boundaries of Accra covered only 300 sq. km, or 7.4% of Greater Accra region's total land area, with the rest dominated by agriculture [7]. However, due to the limited capacity to contain the rapidly growing urban population and economic activities, Accra's massive spill over into periurban settlements have drastically reduced their vegetative presence and affected farmlands. For instance, from 10, it is realized that the urban growth dimension which represented the physical composition of the municipality is grossly distributed but more evident in the zonal areas of Gbentanaa and Nii Ashale. This is because the city center is within the suburban region of the Accra city and it keeps absorbing the population influx of the main cities of Accra.

Again, the economic indicator represented the gross decline of the income levels of people particularly women. Lands and rental values in the municipality keep escalating at a competitive rate. This situation prices out individuals in the society who cannot afford habitable lands. In fact, most of these sections of the urban space are made up of women. It is observed from Figures 10 and 11 that people residing in the municipality have limited economic access to land especially those in Koose and Gbentanaa.

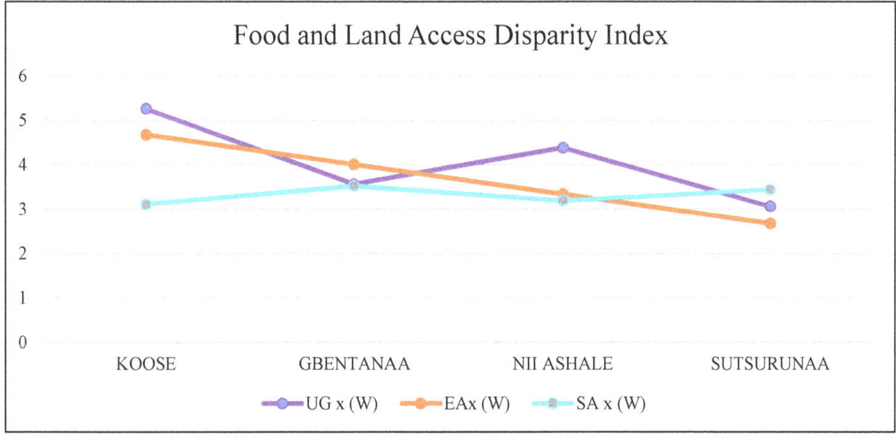

Figure 10. Food and land access disorder index in the municipality (source: field survey, 2019).

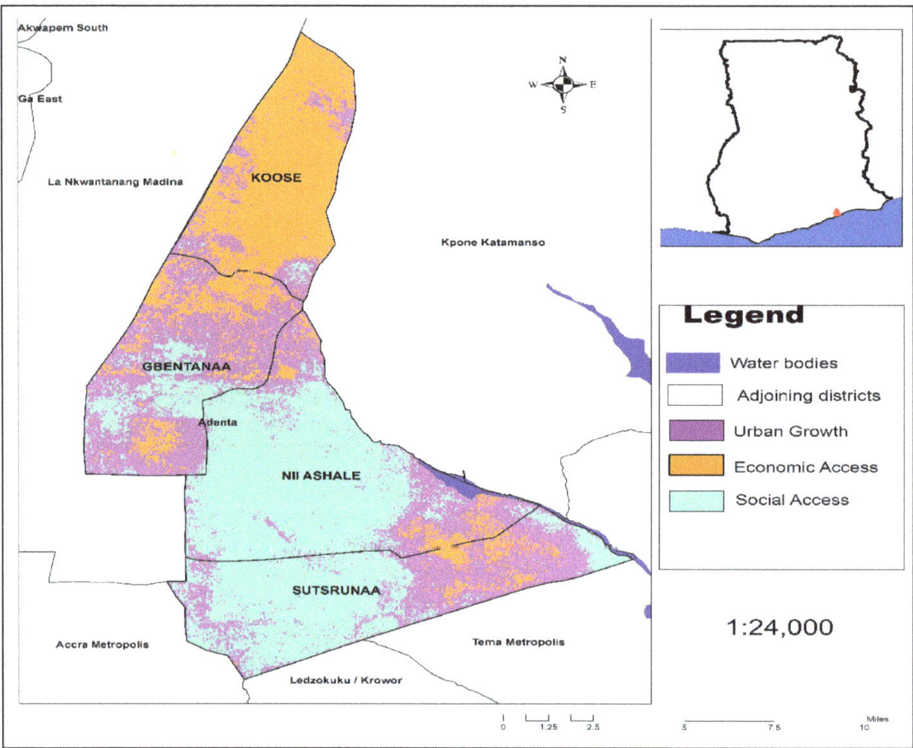

Figure 11. Food and land access disparity index in the municipality (source: field survey, 2019).

Furthermore, the risk of food is directly associated with forms of land vulnerability, and at some point, poverty. Therefore, the social dimension of land and food access disorder was predominantly associated with urban inequality, gender discrimination in buying land for occupational and commercial purposes such as farming and trading. The Adenta municipality was previously known to be a middle-income level district upon its creation. However, the change pattern of its urban domain has embraced various levels of income groups and therefore reflects different forms of food risks, vulnerability and inequality. In fact, the rate of urbanization is rapidly altering the so-called food and land equity state of the country and if the steps of multidiagnostics and criteria-based decision-making are not embraced in addition with the right technology of GIS; the country could be in a food crisis in the coming years, with women being the most significantly affected urban actors. Therefore, it is necessary for steps to be taken to reinforce the land and food subsector for pragmatic decisions to be made now and in future. These indicators could serve as strategic benchmarks for the municipal assembly to adequately track the rates and levels of urban change to facilitate decision making.

4.7. Land and Food System Dynamics in the Adenta Sub-City

Food and land access are increasingly becoming urban development issues in urban and suburban areas [61] for which the concept of urban land and food systems has gained considerable popularity from local and inter-regional levels [21,23]. The food systems of the growing sub city are prone to a range of socioeconomic and agro-climatic shocks and this could further aggravate if food systems measurements are trivially recognized. Suburban cities like Adenta in the Greater Accra region will always be dependent on cross food systems; that is, they will continue to outsource food, from further locations and global food chains as well as from nearby rural, periurban and urban producers. Although its urban residents are entitled to various food options such as processed foods, street foods, fresh foods, imports and others, its over dependence on global food supply and systems has increased vulnerabilities and risk with a rippling effect on local economies. The narrative surrounding its constituents contributes to the dynamic nature of land and food systems in the municipality, and has of late affected the social and economic landscape (as indicated in Figure 8 above). For example, compared to rural food systems in Ghana, suburban food systems are having more stress index in the urban communities than the rural. Particularly within the economic and social dimensions, many urban groups (especially women) within the country are at risk when it comes to food. For instance, the findings confirmed that about 50% of the urban groups were willing to give up their organic foods to go for genetically modified (GM) foods. Again, over 20% were willing to patronize genetically modified foods, whilst almost 30% greatly considered it (Figure 12). Many attributed this condition to their economic standings: "I feed 7 children every day; we sometimes eat once a day, because we cannot afford all food ingredients" (Street hawker and single mother, Adenta Commandos). This alarming finding indicates that despite local repulsion for GM foods, there is the likelihood on the rush for GM foods over organic foods due to economic limitations.

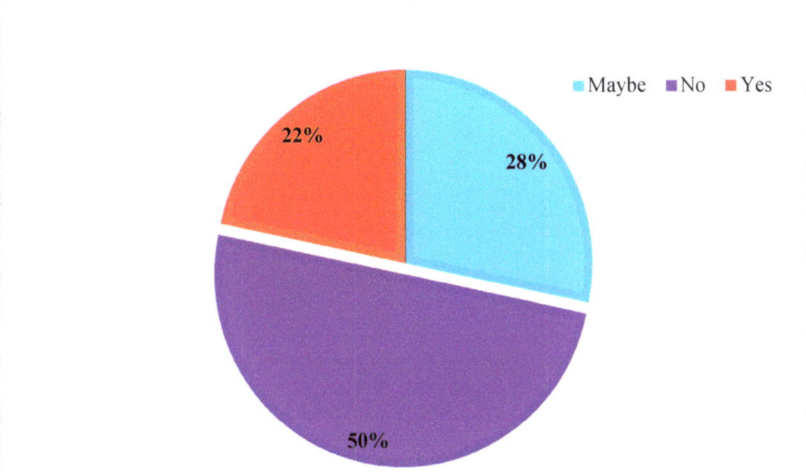

Figure 12. Preference for genetically modified foods (source: author's construct, 2019).

5. Conclusions

The complex change system of urban areas coupled with the poor resource distribution among women reflects a challenging way of monitoring the efficiency of land use and food access in communities. Agriculture or food access could be a luxury in urban communities in Ghana. Although most women in Accra originate from rural areas, it is unreliable to predict that their only motivation for farming and food trading stems from their inability to enter into other urban livelihoods. Rather, this research evidently suggests that women farmers and food traders have personal stakes in this urban phenomenon particularly due to the land access disparities. Despite their restrictions to land access and ownership, they have still managed to contribute immensely towards the availability and distribution of food. Consequently, GIS-based urban growth modelling can provide measurable and visualized methods for determining spatial and nonspatial information. It provides the leveraging tool of importing indicators for the systematic monitoring and predictions of issues in cities. Women in urban communities like Adenta have various social, economic and political limitations as far as acquiring land properties and supporting food efficiency is concerned. In this study, the local women interviewed had various educational backgrounds, formal and informal experience, and diverse limited processes for entering the land market. These processes evaluated previously held notions that construct women and informal workers to be poor, uneducated and unemployed. However, it was also noticed that women engaged in power structures on a daily basis that both benefit them and burden them, depending on their socio-economic status and other factors.

This study identified the plurality of land and food systems in urban districts. It is followed with the impact of urban changes on land and food demand and its accessibility in the urban regions of Ghana. It concluded on the importance of modelling land and food disparities in Adenta which is characterized by high population growth, destruction of vegetative and agricultural lands for development, among others. Ghana's urban districts need to embrace the package of tools and methodologies in numerical modelling and simulation capacities to review cities and municipalities as multidimensional, socio-ecological systems that have an evolving character in order to appreciate gender-based functionalities. Exploring these urban pointers was systematically and seamlessly allowed for ranking and prioritization of the land and food components that needed more attention, with women not being disregarded. Whilst urban growth, economic and social changes were influencers on urban land and food disparity index, the economic access indicator had weights closer to the urban food stress index. This meant that initiating policies to empower women economically could facilitate their chances of acquiring lands and becoming independent.

Author Contributions: Conceptualization, K.O.T., K.A., J.A., and C.Y.A.; Data curation, K.O.T.; Formal analysis, K.O.T. and K.A., Investigation, K.O.T. and K.A., Methodology, K.O.T., Supervision, K.A., J.A., and C.Y.A.; Validation, K.A., J.A., and C.Y.A.; Visualization, K.O.T., and K.A.; Writing—original draft, K.O.T. and K.A.; Writing—review & editing, K.O.T., K.A., J.A., and C.Y.A. All authors have read and agreed to the published version of the manuscript.

Funding: The publication of this article was funded by the Open Access Fund of Leibniz Universität Hannover.

Acknowledgments: The data used in this study were curated as part of the MPhil. studies of K.T. under a DAAD In-Country/In-Region Programme NELGA Research Fellowship (2019). Further analysis was solely based on the research interests of the authors without any funding.

Conflicts of Interest: The authors declare no conflict of interest.

References

1. Farvacque-Vitkovic, C.; Raghunath, M.; Eghoff, C.; Boakye, C. *Development of the Cities of Ghana: Challenges, Priorities and Tools*; Africa Region Working Paper Series; The World Bank: Washington, DC, USA, 2008.
2. Anand, S.; Jagadeesh, K.; Adelina, C.; Koduganti, J. Urban food insecurity and its determinants: A baseline study of Bengaluru. *Environ. Urban.* **2019**, *31*, 421–442. [CrossRef]
3. Tefft, J.; Jonasova, M.; Adjao, R.; Morgan, A. *Food Systems for an Urbanizing World—Knowledge Product*; The World Bank: Washington, DC, USA, 2017.
4. Oduro, C.Y.; Adamtey, R. The Vulnerability of Peri-Urban Farm Households with the Emergence of Land Markets In Accra. *J. Sci. Technol.* **2017**, *37*, 85–100.
5. United Nations UN. *Transforming our World: The 2030 Agenda for Sustainable Development*; Division for Sustainable Development Goals: New York, NY, USA, 2015.
6. New Urban Agenda—Habitat III. Available online: https://habitat3.org/the-new-urban-agenda/ (accessed on 15 October 2020).
7. Ghana Statistical Service (GSS). *Population and Housing Census*; Ghana Statistical Service (GSS): Accra, Ghana, 2012.
8. Kuusaana, E.D.; Kidido, J.K.; Halidu-Adam, E. Customary Land Ownership and Gender Disparity, Evidence from the Wa Municipality of Ghana. *Ghana J. Dev. Stud.* **2013**, *10*, 63–80. [CrossRef]
9. Rünger, M. Governance, Land Rights and Access to Land in Ghana—A Development Perspective on Gender Equity. In Proceedings of the 5th FIG Regional Conference, Marrakech, Morocco, 18–22 May 2011; FIG: Accra, Ghana, 2011.
10. Asiama, S.O. Crossing the Barrier of Time. The Asante Woman in Urban Land Development. *Afr. Riv. Trimest. di Stud. e Doc. Dell'istituto Ital. per l'Africa e l'Oriente* **1997**, *52*, 212–236.

11. Lambrecht, I.B. As a Husband I Will Love, Lead, and Provide. Gendered Access to Land in Ghana. *World Dev.* **2016**, *88*, 188–200. [CrossRef]

12. Dery, I. Access to and Control over Land as Gendered: Contextualising Women's Access and Ownership Rights of Land in Rural Ghana. *Africanus* **2015**, *45*, 28–48. [CrossRef]

13. USAID. *Gender Analysis Report*; USAID: Washington, DC, USA, 2020.

14. Arko-Adjei, A. Adapting Land Administration to the Institutional Framework of Customary Tenure. Ph.D. Thesis, Delft University of Technology, Delft, The Netherlands, 2011.

15. Kalabamu, F.T. Divergent paths: Customary land tenure changes in Greater Gaborone, Botswana. *Habitat Int.* **2014**, *44*, 474–481. [CrossRef]

16. Ubink, J.M.; Quan, J.F. How to combine tradition and modernity? Regulating customary land management in Ghana. *Land Use Policy* **2008**, *25*, 198–213. [CrossRef]

17. Abdulai, R.T.; Ndekugri, I.E. Customary landholding institutions and housing development in urban centres of Ghana: Case Studies of Kumasi and Wa. *Habitat Int.* **2007**, *31*, 257–267. [CrossRef]

18. Kuusaana, E.D.; Eledi, J.A. Customary land allocation, urbanization and land use planning in Ghana: Implications for food systems in the Wa Municipality. *Land Use Policy* **2015**, *48*, 454–466. [CrossRef]

19. Namubiru-Mwaura, E. *Land Tenure and Gender: Approaches and Challenges for Strengthening Rural Women's Land Rights*; Women's Voice, Agency, & Participation Research Series; World Bank Group: Washington, DC, USA, 2014.

20. Acheampong, R.A. Urbanization and Settlement Growth Management. 2019, pp. 171–203. Available online: https://link.springer.com/chapter/10.1007/978-3-030-02011-8_9 (accessed on 10 October 2020).

21. Twum-Baah, K.A. Population growth of Mega-Accra—Emerging issues. In *Visions of the City: Accra in the 21st Century*; Mills-Tettey, R., Adi-Dako, K., Eds.; Woeli Publishing Services: Accra, Ghana, 2000.

22. Higgins, T.; Fenrich, J. Legal Pluralism, Gender, and Access to Land in Ghana. *Forham Environ. Law Rev.* **2012**, *23*, 7–12.

23. Kasanga, K.R.; Kotey, N.A. *Land Management in Ghana: Building on Tradition and Modernity*; IIED: London, UK, 2001.

24. Mireku, K.O.; Kuusaana, E.D.; Kidido, J.K. Legal implications of allocation papers in land transactions in Ghana—A case study of the Kumasi traditional area. *Land Use Policy* **2016**, *50*, 148–155. [CrossRef]

25. Asiama, K.O.; Bennett, R.M.; Zevenbergen, J.A. Participatory Land Administration on Customary Lands: A Practical VGI Experiment in Nanton, Ghana. *ISPRS Int. J. Geo-Inf.* **2017**, *6*, 186. [CrossRef]

26. Alden Wily, L. Custom and commonage in Africa rethinking the orthodoxies. *Land Use Policy* **2008**, *25*, 43–52. [CrossRef]

27. Magigi, W.; Drescher, A.W. The dynamics of land use change and tenure systems in Sub-Saharan Africa cities; learning from Himo community protest, conflict and interest in urban planning practice in Tanzania. *Habitat Int.* **2010**, *34*, 154–164. [CrossRef]

28. Jat, M.K.; Garg, P.K.; Khare, D. Monitoring and modelling of urban sprawl using remote sensing and GIS techniques. *Int. J. Appl. Earth Obs. Geoinf.* **2008**, *10*, 26–43. [CrossRef]

29. Bhanjee, S.; Zhang, C.H. Mapping Latest Patterns of Urban Sprawl in Dar es Salaam, Tanzania. *Pap. Appl. Geogr.* **2018**, *4*, 292–304. [CrossRef]

30. Bowyer, D. Measuring Urban Growth, Urban Form and Accessibility as Indicators of Urban Sprawl in Hamilton, New Zealand. Master's Thesis, Lund University, Lund, Sweden, 2015.

31. Twum, K.O.; Ayer, J. Connecting the complex dots: A review of urban change complexities in Ghana. *Cogent Soc. Sci.* **2019**, *5*, 1677119. [CrossRef]

32. Appiah, J.A. Gender Differences in Access to, and Use of, Farmlands: A Case Study of Abokobi in the Ga East Municipality. Ph.D. Thesis, University of Ghana, Accra, Ghana, 2015.

33. Danso, G.; Cofie, O.; Annang, L.; Obuobie, E.; Keraita, B. *Gender and Urban Agriculture: The Case of Accra, Ghana*; Resource Centre on Urban Agriculture and Forestry: Accra, Ghana, 2004.

34. Asiama, K.O.; Lengoiboni, M.; van der Molen, P. In the Land of the Dammed: Assessing Governance in Resettlement of Ghana's Bui Dam Project. *Land* **2017**, *6*, 80. [CrossRef]

35. Kalabamu, F.T. Patriarchy and women's land rights in Botswana. *Land Use Policy* **2006**, *23*, 237–246. [CrossRef]

36. Yngstrom, I. Women, Wives and Land Rights in Africa: Situating Gender Beyond the Household in the Debate Over Land Policy and Changing Tenure Systems. *Oxf. Dev. Stud.* **2002**, *30*, 21–40.

37. Knox, A.; Duvvury, N.; Milici, N. *Connecting Rights to Reality: A Progressive Framework of Core Legal Protections for Women's Property Rights*; International Center for Research on Women (ICRW): Washington, DC, USA, 2007.

38. Pinstrup-Andersen, P. Food security: Definition and measurement. *Food Secur.* **2009**, *1*, 5–7.

39. van der Molen, P. Food security, land use and land surveyors. *Surv. Rev.* **2017**, *49*, 147–152.

40. WFP. *World Hunger Series: Hunger and Markets*; World Food Programme: Rome, Italy, 2009.

41. Rockson, G.; Bennett, R.; Groenendijk, L. Land administration for food security: A research synthesis. *Land Use Policy* **2013**, *32*, 337–342.

42. Crush, J.S.; Frayne, G.B. Urban food insecurity and the new international food security agenda. *Dev. South. Afr.* **2011**, *28*, 527–544.

43. Crush, J.; Caesar, M. City without choice: Urban food insecurity in Msunduzi, South Africa. *Urban Forum* **2014**, *26*, 165–175.

44. Chen, S.; Ravallion, M. *The Developing World Is Poorer Than We Thought, But No Less Successful in the Fight against Poverty*; The World Bank: Washington, DC, USA, 2008.

45. Lacey, E.J. Understanding The Livelihoods Of Women In The Local Foodscape: A Case Study of Accra, Ghana. Master' Thesis, University of Oregon, Eugene, Oregon, 2010.

46. Agana, C. Women's Land Rights and Access to Credit in a Predominantly Patrilineal System of Inheritance: Case Study of The Frafra Traditional Area, Upper East Region. Ph.D. Thesis, KNUST, Kumasi, Ghana, 2012.

47. Simon, D.; McGregor, D.; Thompson, D. The search for peri-urban resources sustainability. In *The Peri-Urban Interface: Approaches to Sustainable Natural and Human Resource Use*; Simon, D., McGregor, D., Thompson, D., Eds.; Sinica: London, UK, 2006; pp. 1–17.

48. Adentan Municipal Assembly. Medium Term Development Report. Adenta, Ghana. 2017. Available online: https://new-ndpc-static1.s3.amazonaws.com/CACHES/PUBLICATIONS/2016/04/04/GR_Adentan+Municipal_2014-2017+MTDP.pdf (accessed on 31 October 2020).

49. Acheampong, R.A. *Spatial Planning in Ghana*; The Urban Book Series; Springer: Cham, Switzerland, 2019; ISBN 978-3-030-02010-1.

50. COHRE. *Bringing Equality Home: Promoting and Protecting the Inheritance Rights of Women: A Survey of Law and Practice in Sub-Saharan Africa*; Centre on Housing Rights and Evictions: Geneva, Switzerland, 2004.

51. Capaldo, J.; Karfakis, P.; Knowles, M.; Smulders, M. A model of vulnerability to food insecurity. **2010**. [CrossRef]

52. FAO. *Regional Overview of Food Security and Nutrition in Africa 2017*; FAO: Rome, Italy, 2017.

53. Torvikey, G.D. Strengthening Women's Voices in the Context of Agricultural Investments: Lessons from Ghana. Accra/London. 2016. Available online: https://pubs.iied.org/pdfs/12591IIED.pdf (accessed on 10 October 2020).

54. Saaty, T.L. A scaling method for priorities in hierarchical structures. *J. Math. Psychol.* **1977**, *15*, 234–281.

55. Saran, S.; Ramana, K.V. Site Suitability Analysis for Industries Using Gis and Multi Criteria Decision Making. *ISPRS Ann. Photogramm. Remote Sens. Spat. Inf. Sci.* **2018**, *IV-5*, 447–454.

56. Hagai, M. Food Security Modeling Using Geographic Information Systems (GIS) Techniques: A Strategy Towards Reliable Food Security Information & Early Warning Systems (FSIEWS) for Tanzania. *J. L. Adm. East. Afr.* **2014**, *2*, 130945610.

57. Darfour, B.; Rosentrater, K.A. Agriculture and Food Security in Ghana. In Proceedings of the 2016 ASABE International Meeting, Orlando, FL, USA, 17–20 July 2016; American Society of Agricultural and Biological Engineers: St. Joseph, MI, USA, 2016.

58. FAO, R. and W.B. *Urban Food Systems Diagnostic and Metrics Framework, Roadmap for Future Geospatial and Big Data Analytics*; World Bank: Washington, DC, USA, 2017.

59. GSS. Ghana Living Standards Survey (GLSS6). 2012. Available online: https://www.ilo.org/surveyLib/index.php/catalog/466/stuady-description (accessed on 10 October 2020).

60. Dubbeling, M.; Renting, H.; Hoekstra, F.; Wiskerke, J.S.C.; Carey, J. City Region Food Systems. *Urban Agriculture Magazine*, 29 May 2015.

61. Potere, D.; Schneider, A. A critical look at representations of urban areas in global maps. *GeoJournal* **2007**, *69*, 55–80. [CrossRef]

62. The World Bank. *Urban Agriculture, Findings from Four City Case studies*; The World Bank: Washington, DC, USA, 2013.

63. FAO. *Rome Declaration on World Food Security*; FAO: Rome, Italy, 1996.
64. United Nations. *Open Working Group Proposal for Sustainable Development Goals*; United Nations: New York, NY, USA, 2015.
65. Yankson, P.W.K. *Urbanisation Industrialisation and National Development: Challenges and Prospects of Economic Reform and Globalisation*; Ghana Universities Press: Accra, Ghana, 2006.

Publisher's Note: MDPI stays neutral with regard to jurisdictional claims in published maps and institutional affiliations.

 *land*

Communication

History and Prospects for African Land Governance: Institutions, Technology and 'Land Rights for All'

Robert Home

Department of Business and Law, Anglia Ruskin University, Bishop Hall Lane, Chelmsford CM1 1SQ, UK; robhome47@gmail.com

Abstract: Issues relating to land are specifically referred to in five of the United Nations' (UN) 17 Sustainable Development Goals, and UN-Habitat's Global Land Tools Network views access to land and tenure security as key to achieving sustainable, inclusive and efficient cities. The African continent is growing in importance, with climate change and population pressure on land. This review explores an interdisciplinary approach, and identifies recent advances in geo-spatial technology relevant to land governance in sub-Saharan Africa (SSA). It discusses historical legacies of colonialism that affect the culture of its land administration institutions, through three levels of governance: international/regional, national and sub-national. Short narratives on land law are discussed for four Anglophone former British colonies of SSA. A wide range of sources are drawn upon: academic research across disciplines, and official publications of various actors, including land professions (particularly surveyors, lawyers and planners), government and wider society. The findings are that African countries have carried forward colonial land governance structures into the post-independence political settlement, and that a gulf exists between the institutions, language and cultures of land governance, and the mass of its peoples struggling with basic issues of survival. This gulf may be addressed by recent approaches to land administration and technological advances in geo-spatial technology, and by new knowledge networks and interactions.

Keywords: land governance; Fit-for-Purpose Land Administration; sub-Saharan Africa; legal history of land; distributed ledger technology; citizen participation

Citation: Home, R. History and Prospects for African Land Governance: Institutions, Technology and 'Land Rights for All'. *Land* **2021**, *10*, 292. https://doi.org/10.3390/land10030292

Academic Editor: Kwabena Asiama

Received: 12 February 2021
Accepted: 8 March 2021
Published: 12 March 2021

1. Introduction and Approach

'The land is the dearest thing that we have. Without the land there is no nation.'

(Armenian writer, Sero Khanzadyan (1915–1998))

'Land is the only thing in the world that amounts to anything.'

(Gerald O'Hara, fictional slave plantation owner in Georgia, USA, in Margaret Mitchell, *Gone with the Wind*)

The quotations above suggest the power that land can exert in society, a power that is being rediscovered in the 21st century. Five of the United Nations' (UN) 17 Sustainable Development Goals (SDGs) for 2015–2030 refer specifically to land; the 2018–2030 strategy for UN-Habitat's Global Land Tools Network (GLTN) has restated its pro-poor agenda with a strapline 'A world in which everyone enjoys secure land rights' [1,2].

Land administration needs at the start to be distinguished from land governance. Land administration comprises an extensive range of governmental systems, whose processes include: transferring rights from one party to another; regulating uses; gathering land-based public revenues; and resolving conflicts involving land. The concept of land governance is wider, and recognizes the importance of power and political relations, and multiple stake-holders and actors with their own cultures and specialist languages, for instance professions, academia, government and wider society.

The academic scholarship around governance explores processes of interaction and decision-making between the institutions by which authority is exercised, and is moving towards a more radical and de-centred view of the state that takes account of cultural traditions and practices [3]. Various concepts and theories about land governance have been advanced, including the following. Historical institutionalism investigates the influence of social, political, and economic change over time upon institutional and political structures and outcomes [4]. Political settlement theory explores the effects of power relations upon institutions and patterns of development [5,6]. Path dependence theory argues that decisions we face are limited by past decisions, even when past circumstances may no longer be relevant; critical junctures occur when existing political structures fail, and new dynamics and institutions emerge [7,8]. Actor-network theory examines networks of causation that are both material (between people and things) and semiotic (between concepts) [9]. Credibility theory, used by insurers to assess risk from historic claims, is being applied to institutional change and land administration [10,11]. The main academic discipline dealing with land is geography, and the sub-discipline of critical legal geography investigates law's effect upon physical landscapes and spatial boundaries, one of its more esoteric manifestations being the 'nomosphere' [12–14]. The professions of law, surveying and planning also have their own academic disciplines, and influence land governance agendas at all levels, particularly through their professional associations. All these professions, disciplines and sub-disciplines have their own 'academic tribes' and 'silo mentalities' that may impede mutual understanding [15,16].

This review explores how institutional histories and cultures around land governance operate. It outlines some relevant developments in global geo-spatial technology and land administration, and investigates land governance in sub-Saharan Africa (SSA) at three levels: supranational institutions (specifically the World Bank, land professions, UN/Habitat/GLTN, and the African Union (AU)), national land law and institutions, focusing on four countries of Anglophone SSA, and sub-national activities of local authorities and civil society. The review draws from a range of sources: recent academic research and scholarship across disciplines, and official publications of agencies, governments and professional associations [17]. The conclusions are that African countries have largely carried forward colonial land governance structures since independence; this has created a gulf between the institutions, language and cultures of land governance on the one hand, and, on the other, the mass of its peoples struggling with basic issues of survival. Recent developments in citizen participation and community-based action offer better prospects for the future.

2. Surveyors in Land Governance: The Power of Technology

Among the professions involved with land, surveyors can claim the closest direct connection. The International Federation of Surveyors (FIG) has strongly influenced the evolving international approach to land administration and governance, and its publications offer a timeline of those changes, engaging with wider concepts of land governance, the recognition of a continuum of land tenure types, and more democratic geospatial technologies. FIG's Young Surveyors Network pursues a theme of rapid response to change by the surveyor of tomorrow, and has an African network applying new approaches through survey technologies.

National governments create and maintain land administration systems: surveying and mapping land ownership and rights, and providing information for users. One country may have deeds registration, another title registration; some systems are centralised, others decentralized; some are based on a general boundaries approach, others on fixed boundaries; some may prioritise state interests, others private interests. This is 'the institutional memory of the map and the archive' [18], which is increasingly being transferred to digital form through schemas for describing spatial characteristics, combining information from multiple sources. In 2011 the UN brought together member states' national mapping agencies in a Committee of Experts on Global Geospatial Information Management (UN-GGIM), which adopted a FIG proposal for the concept of Fit-for-Purpose Land Administration (FFPLA), from which evolved a Framework for Effective Land Administration

(FELA), as an enabling environment for developing policies and standards [19,20]. A more flexible approach suited to regions of the 'global South' now accepts the concept of a continuum or spectrum of land rights, from informal to formal, and applying general rather than fixed boundaries to land, aerial images rather than field surveys, accuracy related to purpose rather than technical standards, and continuous improvements.

An international standard on the Land Administration Domain Model (LADM) was adopted in 2012, and structures the LADM into conceptual packages [21,22]. The party package contains classes representing information about a person or organisation with a relationship to land; this could be an individual, company or other legal entity, or a group of parties. The administrative package records rights, restrictions, responsibilities and transactions associated with a spatial unit or group of units. Spatial units (or land parcels) are defined by geographical extent, and may be aggregated from sub-units or subdivisions (e.g., all land parcels within a local government administration), and legal spaces of buildings and utility networks. The surveying and representation package identifies the spatial sources, perhaps a registered survey plan or orthophotos, with point, line and surface representations of spatial units through terrestrial surveys, global positioning satellites or field sketches [23].

New technology for land administration is being embraced by surveyors, policy-makers, governments and communities. The technical terms and acronyms may be impenetrable to the uninitiated: GDAL (Geospatial Data Abstraction Library), OSGeo (Open Source Geospatial Foundation), QGIS (Free and Open Source Geographic Information System), GML (Geography Markup Language), REST-API (Representational State Transfer Application Programming Interface), LIDAR (Light Detection and Ranging). Such complex technical language may impede understanding by those making major decisions about adopting these technologies. Three examples illustrate the recent advances in geo-spatial technology and applications: drones (or unmanned aerial vehicles, UAVs), automatic feature extraction (AFE), and blockchain (or distributed ledger technology, DLT).

UAVs are supplementing or even replacing terrestrial survey methods in many situations, and remote sensing from drones can be more cost-effective than from satellites. They allow land surveyors to accomplish more in less time, especially in hazardous or physically hard-to-reach areas, offering centimetre-level accuracy with fewer control points. UAV uses high-resolution imagery with optical sensor data to delineate boundaries, and generate accurate, real-world 3D models from 2D imagery [24].

AFE uses machine learning, pattern recognition and image processing to derive values or features from a large data set, and can record land parcel boundaries, both visible and invisible. The morphology of cadastral boundaries can be complex; they may be defined socially, perhaps covered by thick vegetation canopy, and not visible through remote imagery. AFE can help in both initial data capture and updating/maintenance, with initial working-draft land records or updating a cadastral map. One such application is Smart Sketchmaps, a set of sub-tools to align sketched information with base-map data and existing geo-referenced datasets [25].

Blockchain/DLT in landed property developed together with crypto-currencies, and has been ambitiously claimed as 'the next industrial revolution'. It can offer a decentralised, secure database in a transparent network which allows peer-to-peer transactions without an intermediary. It requires an architecture of overlaid technologies to support changes of data, confirm digital identity and privacy, ensure legal compliance and enforceability of smart contracts. Data can be updated without loss of historic data, through remote imagery and verified by smartphone. It has the potential to bring speed, certainty and clarity to property decisions, but its technical language can lead to confusion over key principles, misinterpretation by non-technical people, and inappropriate products, while the network may be vulnerable to systemic shock [26–28].

Such technological advances are challenging data collection by qualified and authorized survey professionals ('top-down') with more democratic approaches ('bottom-up'). Users can now create open-source geo-spatial data with cost-effective measurement, and combine data from multiple sensors and techniques, allowing local communities to become involved in data collection and management. The big survey companies like Leica and Trimble now offer easy-to-use field data collection tools, with scalable accuracy for initial registration and documentation of land rights, and potential for a 'vertical' spectrum of land administration services [29].

3. The African Dimension

These developments are ready for application in Africa, where land governance is increasingly seen as a major challenge. Africa is the largest continent by land area, and has the largest number of countries proportionate to land area, and human population of over a billion. The 2020 population estimates for the AU's 55 member states range from the most numerous (Nigeria 206 million, Ethiopia 116 million) to some 20 with less than five million, and several off-shore island states [30]. Africa's five-fold population growth in the last half-century has created a so-called 'youth bulge', which has been seen as a predictor for social unrest, leading to war and terrorism from stress factors of poverty, mass unemployment, unmanaged urban growth, food and water shortages, and disease; this youth bulge can also be seen as a strength in future, but requires education and livelihood opportunities for young people [31]. Hundreds of languages are spoken in Africa, and create communication problems between people and their governments. The legal systems of most African countries were imposed and imported by past European colonial powers, and the political settlement at independence brought to power new indigenous elites, while maintaining largely intact colonial laws and institutions that facilitated great inequalities in wealth and land ownership [32,33]. Former British colonies or protectorates in SSA cover a larger combined land area than any other former colonial power (see Table 1), and this article explores the land law histories of one country in each sub-region (South Africa, Nigeria, Kenya, and Zambia), drawing from the author's research.

Table 1. Anglophone countries in sub-Saharan Africa (SSA, by region). Source: https://www. worldometers.info/population/countries-in-africa-by-population/ (accessed on 10 December 2020).

Region	Countries	Population	p/km^2	Urban Pop %
West	Nigeria	206	226	52
	Ghana	33.1	103	43
	Sierra Leone	8	111	53
	Gambia	5.1	239	59
	Liberia	5.1	53	43
East	Kenya	53.8	94	28
	Uganda	45.7	225	26
	Tanzania	59.7	67	43
Central	Zambia	18.4	203	45
	Zimbabwe	14.9	25	38
	Malawi	19.1	38	18
South	South Africa	59.3	49	67
	Namibia	2.5	3	55
	Botswana	2.4	3	73
	Lesotho	2.1	71	45
	eSwatini	1.2	67	41

Policies of indirect rule and the dual mandate, associated with Lord Lugard, the influential sometime governor of Nigeria, were transferred to other British colonies and protectorates. He claimed with lofty superiority a hundred years ago that:

> It is still a matter of indifference to the people whether Government takes up a few square miles, here for a township, or there for a railway, or elsewhere as leases to commercial, mining, agricultural or ranching companies. Even if occupiers are

expropriated in the neighbourhood of a large town, there is as a rule abundant land elsewhere in the great unoccupied spaces of this vast country. [34] (p. 29)

Population pressure on land has increased dramatically since them, challenging the systems of land tenure. Recent research with satellite mapping data has shown that cities of Anglophone colonial origin now have less intense land use than Francophone ones, more irregular layouts, and poorer electricity and piped water connections at the informal urban edge [35,36]. British colonial policy outside the towns and white settler farmlands designated 'native reserves' where customary tenure was maintained, but it allowed such land to be taken (or 'set aside') without compensation if required by the state for a 'public interest', such as for mining, forestry or township creation. African states still use such inherited laws to allow large-scale land-based investments ('land grabbing'), and in 2000–2012 leased an estimated six million hectares of customary land for biofuel and food production, largely ignoring the needs for land and livelihoods of those displaced [37,38]. Survey findings on global tenure security suggest that about a fifth of households risk losing their homes within the next five years, uncertainty which holds back investment and sustainable development. The World Bank supports mass land titling, yet SSA land is mostly legally undocumented, and accounts for a third of global forced evictions. The political will to reduce insecure tenure is often not only lacking, but empowers the evictors [39,40].

The so-called 'colonial masters' favoured an evolutionary theory of land rights, under which customary or communal tenure would be extinguished over time by an inevitable progress towards individual property rights [41,42]. The GLTN's land rights continuum now identifies a spectrum of rights, from customary, occupancy, anti-evictions, adverse possession, group tenure, and leases, ending with registered freehold. Private property is seen as the highest form of land right, guaranteed in the constitutions of post-independence African states, yet an estimated two-thirds of the continent's usable land remains under communal or customary land tenure (the highest proportion in the world). British colonial officials opposed granting private property rights for Africans:

The Land Officer and myself are of the opinion that it would be a disaster to allow the African to slide into possession of what would, to all intents and purposes, be an absolute freehold over land which the African occupies under native law and custom. [43]

Theories of property have often been reluctant to recognize plural property relations, particularly 'non-owner' interests that may be collective or communal, yet such land relations can defend people and communities against the penetrative forces of globalisation and capitalism, fulfilling an important welfare function, and serving as a reservoir of cheap, un-serviced resource in peri-urban areas. This review next explores how SSA deals with tensions between private and customary land rights, distinguishing between three levels of land governance: supra-national, national and sub-national actors. States have sovereignty and control over their land laws and institutions, but the other two levels—above and below the nation state—also have power and influence [44].

3.1. Supranational, Regional and Human Rights Actors

This section considers the role in SSA land governance of the World Bank, UN-Habitat and GLTN, professional associations, and the AU. Each has a particular institutional narrative, and operates through its own institutions and actor networks, with differing degrees of power and influence.

The World Bank is the largest single lender for development in Africa, approving some $20 billion in 2020. Its core mission has changed since its origins in the 1944 Bretton Woods conference, from post-war reconstruction to ending extreme poverty, and it has been involved in framing the SDGs, especially SDG1 ('no poverty') and SDG16 ('peace, justice and strong institutions'). It embraces Hernando de Soto's argument that a framework of secure, transparent and enforceable property rights is a critical precondition for reducing poverty, and its annual Land and Poverty conference presents research and good practice from its many projects around the world. Its Worldwide Governance Index which ranked SSA lower than other regions in all six key variables: voice and accountability, political stability, regulatory quality, rule of law and control of corruption, and government effectiveness [45]. Its Land Governance Assessment Framework has been applied in 40 participating countries (seven in Africa, and four of those Anglophone), and identified 116 'dimensions' in such areas as land tenure recognition, institutional arrangements, urban planning, and dispute resolution. The four Anglophone SSA countries scored poorly on such dimensions as dealing with longstanding land disputes, and transfer of land from public to private ownership; worst scoring was Sierra Leone (good on 22, bad on 45), and the best was Rwanda (good on 47, bad on 9) [46].

Within the UN system, rural land governance is mainly the preserve of the Food and Agriculture Organization (FAO), and urban land governance that of UN-Habitat with its mission of 'a better quality of life for all in an urbanizing world'. UN-Habitat's headquarters in Nairobi houses the GLTN (founded in 2006), which works through a 'dynamic and multisectoral alliance of international partners committed to increasing access to land and tenure security for all, with a particular focus on the poor, women and youth' [47]. The GLTN reviewed the international frameworks for land governance, and tracks tenure security through its Global Land Indicator Initiative. With the 2020s declared the 'decade of action' for the SDGs, the GLTN has enlisted 80 partner organizations in four cluster groups: international civil societies (both urban and rural), training/research institutions, and professional bodies. The UN-GGIM also promotes a partnership approach by establishing an Academic Network and a Private Sector Network. Another aspect of growing global and regional co-operation is the emergence of regular conferences and forums, with UN agencies as partners/participants [48,49]. The GLTN has some 20 land tools at different stages of development, among which are the Social Tenure Domain Model (STDM) and Participatory and Inclusive Land Readjustment (PILaR). STDM provides a universal standard for representing people–land relationships independent of levels of formality, legality and technical accuracy [50]. PiLAR seeks to expand the existing land readjustment model by adding more inclusive negotiation processes, so that costs and benefits may be better shared among landowners and other stakeholders, in a less confrontational approach than compulsory expropriation [51,52].

Law, surveying and planning are three professions particularly involved with land governance. All were closely associated with past colonial power structures in SSA, and have continued that role in post-colonial political settlements. They are part of the system that Hernando de Soto compared to a laboratory bell jar, sealed off from the rest of society:

> Inside the bell jar are elites who hold property using codified law borrowed from the West . . . The bell jar makes capitalism a private club, open only to a privileged few, and enrages the billions standing outside looking in. [53] (p. 66)

The late Patrick McAuslan, a pre-eminent land, law and development academic, entitled one of his books 'Bringing the law back in' to express his concern that development agendas gave insufficient attention to law and its institutions, allowing anti-poor policies from the colonial period to continue [54]. Lawyers' work is mostly paper-based and typically happens in court-room or bureaucratic settings, while surveyors and planners are more likely to operate closer to the land and people [55]. Colonial surveyors facilitated land-taking from indigenous peoples, mapping boundaries by 'systematic survey' methods, and introducing government registration of title. In east, central and southern Africa they divided land into 'sections' (meaning a cutting) and transferred often huge tracts of land to white settlers and companies [56]. Town planning, claimed as an apolitical, technical and modern approach to colonial management, was responsible for applying on the ground racial segregation policies, and has continued to shape African urban landscapes [57–60].

The AU in 2015 (the same year that launched the SDGs) adopted its own Agenda 2063 as its 'collective vision and roadmap'. This aspired to 'people-centered development, gender equality and youth empowerment', 'access to affordable and decent housing to all in sustainable human settlements', 'effective and territorial planning and land tenure, use and management systems' and 'improving the livelihoods of the great percentage of the people working and living in slums and informal settlements'. The SDGs claim to rest 'on a set of universal principles, values and standards, such as human rights, that are applicable in all countries, in all contexts and circumstances and at all times' [61,62], this approach can be in tension with that of the AU ('African solutions to African problems'), and the AU still depends upon external financial support from the European Union and other core donors for most of its budget [63,64]. Over half of its budget goes on 'peace support operations', with conflicts within and between states that are often related to competition for land and resources; these include displacements for foreign investments, ethnic antagonisms, urban evictions, clashes between farmers and pastoralists, resistance to natural resource exploitation, and tensions between indigenes and 'strangers' [65,66].

The AU adopted a Declaration on Land in 2009, and afterwards created an African Land Policy Centre as a joint programme of the AU Commission, the African Development Bank and the United Nations Economic Commission for Africa (UNECA) [67,68]. The centre has produced land policy guidelines, which recommends reducing the 'overwhelming presence of the state in land matters', but depends upon member states themselves being willing to apply them. The guidelines also assert equal legal status for customary and 'modern' property rights, and an AU Forum of African Traditional Authorities was created, but the two tenure types are often in competition [69,70].

The AU has also organized a Network of Excellence in Land Governance for Africa (NELGA) to build capacity in higher education because member states lack the human and institutional capacity required to implement sustainable land policies [71]. Linguistic legacies of colonialism, and the AU's origins in anti-colonial struggles, have contributed to the location of its regional 'nodes'. Kwame Nkrumah University of Science and Technology (KNUST) in Ghana for West Africa, Ardhi in Tanzania for East Africa, Namibia University of Science and Technology (NUST) and University of Western Cape (UWC) for Southern Africa, all are located in Anglophone countries with socialist backgrounds and significant levels of customary tenure, while Francophone countries have separate nodes in Dakar (West Africa), Cameroon (Central Africa) and Morocco (North Africa).

The AU's judicial organs, the African Commission and Court of Human and Peoples Rights, which are mandated to apply the Banjul Charter (1981), have considered several land and property cases, attracting significant attention. The charter, signed in a time of decolonization, frequently refers to the rights of 'peoples' (in the plural) to development, and to hold natural resources and property, but this commitment has been complicated by the UN Declaration on the Rights of Indigenous Peoples (2007). In three key cases affecting Kenya—two of them brought by indigenous peoples, the third by Nubian descendants of the colonial military—the African Court found for the appellants multiple breaches of their charter rights [72,73]. The Mbiankeu case confirmed that a valid land certificate was proof of property ownership guaranteed by the state, and required the Cameroonian government to annul a fraudulent title and compensate the victim. Banjul Charter rights have also been upheld in cases concerning peaceful enjoyment of property, family rights, and forced displacement without due process [74,75]. Such pro-poor judicial activism has, however, been met with a lack of implementation by African governments, and the AU's enforcement mechanisms are weak [76,77].

The tension between universal human rights and 'African solutions for African problems' is encountered in particular in the treatment of women, who are often disinherited and impoverished by patriarchal customary authorities. The proportion of female-headed households in Africa is growing because of male migration, male partner deaths from disease and conflicts, unpartnered adolescent fertility and family disruption. In a social landscape of many female-headed households, more women are establishing a home without men's involvement, as a 'domain of autonomy' where they can reproduce persons for whom they provide a home, and which they can let or sell. The 2020 pandemic is now leaving newly widowed women without family support, often denied inheritance rights, facing social stigma as perceived carriers of the disease, and at risk of destitution, especially for older women without pensions or bank accounts. African women are increasingly pressing for recognition of their land rights, encouraged by the Maputo Protocol on the Rights of Women in Africa (2003), and by support for gender equality in SDG5, many national constitutions and GLTN's 'gendered land rights' [78–81]. In Zambia a draft national land policy proposed in 2002 to make 30% of the land available for women was rejected by traditional authorities, but a Supreme Court case granted a greater property share to the wife after divorce, after considering both customary law and principles of equity or fairness [82,83].

3.2. National Land Laws and Institutions: Colonial Legacies and New Technologies

African governments after independence have jealously protected their national sovereignty under the AU Constitutive Act and following the Westphalia state model. Boundaries between colonies, often created arbitrarily, became mostly fixed at independence under the uti possedetis principle (paraphrased as 'what you have you keep'). Notwithstanding the many international agendas, declarations and goals that they may sign up to, national governments decide their own land laws and policies under political settlements largely shaped around the time of independence.

The GLTN aspires to a 'world in which everyone enjoys secure land rights'. International law protects private property rights, but does not include an explicit right of access to land, although it does recognize rights to housing, private property and an adequate standard of living [84,85]. After independence many African governments reformed their land and planning laws in attempts to redress injustices from the colonial past, while keeping control in the hands of the state. Such reforms are complex and highly political, and have not always helped to achieve broad-based socio-economic development [86–89]. Systems of control and exclusion inherited from colonial rule have allowed powerful vested interests to benefit from an environment of insecure land rights, and the institutions of land administration and planning may appear to follow international norms, but often not function effectively [90–92]. Corrupt and fraudulent land allocations are common, but are becoming less acceptable: an AU anti-corruption convention exists (2003, ratified by 44 AU member states as at 2020), some lands ministers and officials have been convicted in recent years, and the African Land Policy Centre's third conference (2019) had the theme: 'Winning the fight against Corruption in the Land Sector: Sustainable Pathways for Africa's transformation' [93–95].

This review next discusses the colonial roots of SSA land laws and administration, which have contributed to the continued dominance of the state and of powerful vested interests in land. Four short narratives of land laws draw upon the author's research in former British colonies, one in each of the SSA regions: South Africa, Nigeria (west), Kenya (east) and Zambia (central). The section then explores attempts to introduce digital technologies in SSA land administration, especially blockchain/DLT.

South Africa was one of the oldest European colonies in Africa, settled by the Dutch and other immigrants in the 17th century, then under British colonial rule from the 1790s. The 1913 Natives Land Act (subsequently renamed the Bantu Land Act and the Black Land Act) was the cornerstone of *apartheid* until abolished in 1994. The majority Africans population had been excluded from land ownership, and their land reserves, called tribal homelands or Bantustans, comprised only 7% of the area of South Africa; whites and other racial groups had the rest and best. Customary land tenure in the reserves was maintained, but often misinterpreted and undermined by the judiciary, manipulated by administrators, and overlooked in legislation. As late as 1986 a South African judge (white of course) could state in his court that: 'Whites own land by law, whether they are industrious or not, while non-whites must demonstrate their worthiness to own land through their labor' [96]. The government since 1994 has ended legal racial segregation and made other land law reforms. Land redistribution offered those prejudiced under the old regime (the urban and rural poor, farm workers, labour tenants and emergent farmers) to acquire land with state assistance, but only from willing sellers. The Communal Land Tenure Act (2004) allowed community ownership (community defined as a single juristic person, often a tribal authority), in a move intended to address the chaotic land administration in the former homelands. New land registration procedures allowed rights to be upgraded from 'initial ownership' under the 1995 Development Facilitation Act. In spite of such law reforms, progress has been limited in reducing racial inequalities in land ownership, or improving access to land and housing for the rapidly growing population. The South African government adopted a target in 1994 of transferring 30% of commercial farming land to 600,000 smallholders, but after a decade only 3% had been transferred, reflecting institutional weaknesses and the reluctance of owners to offer property voluntarily. The issue of land reform remains understandably highly contentious and complex [97–100].

West Africa had little white settlement and a different land governance story, as exemplified by Nigeria, now the most populous SSA country and a federation with many ethnic groups and languages [101]. When Southern Nigeria was created in 1900, the first act of the new administration was a Native Lands Acquisition Proclamation, which required the governor's permission for all state acquisitions, and the Northern Nigerian Land and Native Rights Proclamation (1910) conferred similar powers on its governor, a measure 'designed to define and secure the rights of the natives to the use of the land whilst providing opportunities for development on modern lines'. The British colonial authorities in West Africa had less power over land than in South Africa:

> The alienation of tribal lands first to Europeans for mining purposes and later to stranger Africans for cocoa farming has been one of the major problems of the Gold Coast [now Ghana]. The Colonial Government attempted to deal with this problem by securing control over land generally in a manner similar to that now in operation in Northern Nigeria and Tanganyika. Native resistance to such a measure was so intense that it had to be dropped. Local native feeling has always been extremely sensitive on land matters and any suggestion of Government interference has been represented as an attempt to dispossess the people of their lands. [43,102]

After Nigerian independence in 1960, its Land Use Act (1978) ended any residual private ownership, and allocated land rights through certificates of occupancy (equivalent to 99-year leases); minerals remained a federal matter. These certificates, originating in previous colonial policy, were controlled by state governors (19 of them in 1976, by 1996 grown to 36, plus a Federal Capital Territory). Complex procedures and often corrupt bureaucracies mean that most land transactions by-pass official consent, and take place as private contracts between the parties; less than 3% of land is thought to be formally registered with federal, state or local authorities, which can take up to two years because of complex chains of title and poor documentation. Nigeria's transfer fees are among the most expensive in the world, yet land is still seen as a reliable store of value and best hedge against inflation. The 1978 Act has not been reformed, and is much criticised: 'politically undemocratic, economically unproductive, but also socially segregative, particularly in its urban and non-urban dichotomy' [103–107].

In East Africa a tiny white settler community dominated Kenyan land law in the colonial period, strongly influenced by South African experience and placing strict controls over African residence and movement. In the early days of the East African Protectorate (later Kenya), a Crown Lands Ordinance (1902) vested in the crown 'all public lands in the Protectorate … including all lands occupied by native tribes', and a subsequent ordinance (1915) further empowered the colonial administration to grant land on behalf of the crown to individuals (not Africans) on 999-year leases, or 99-year leases in township surveyed lots. A deeds registry system for these grants was introduced, following South African practice, and afterwards a Torrens-style official registry of titles, following the South Australian example. In 1921 the Kikuyu (the tribe most affected by white settler land acquisition) took their case to the High Court of the newly constituted Kenya colony, but the judge determined that under the 1915 Ordinance all native title to the land disappeared, and their rights were confined to occupation, cultivation and grazing, in accordance with Privy Council case law at the time. A few years later the Chief Native Commissioner (a British official) was reporting 'the acute anxiety on the part of natives of every tribe with regard to the present insecurity of their land tenure' [108]. The Native Lands Trust Ordinance (1930) later empowered declared the governor to preside over a trust board to administer the native reserves that were created on the South African model, and he could 'set aside' (in effect confiscate without compensation) trust lands for 'public interest' purposes such as townships or mining. Later, in an attempt to counter the Mau-Mau insurgency by creating an African land-owning class, the Registered Lands Act (1963) provided for registration of African interests in trust land by a process of systematic adjudication, similar to that for enclosure claims in nineteenth-century Britain. President Kenyatta's post-independence government did not fundamentally reform the legal framework, and created a network of powerful new beneficiaries through the establishment of private land-buying companies, often headed by prominent politicians, who took over white settler lands, subdivided them or sold on to public corporations at inflated values. The Trust Land Act (1970) transferred former tribal trust land to local authorities (called county councils), which could be converted into registered private land or 'set apart' for public purposes as under the colonial regime. Kenya's new constitution (2010) vested all land in the people collectively as a nation, and provided for three separate land tenure systems of apparently equal status (public, private and community). The subsequent Community Land Act 2016 gave potentially extensive powers to communities, but without the political will to make such a tenure regime effective [109–111].

In the fourth case, Zambia (the former Northern Rhodesia) had been Britain's richest African colony in the 1950s from its copper ore exports (excluding South Africa and diamond-producing Sierra Leone). After it transferred in 1924 from chartered company rule to become a protectorate of the British crown, new ordinances followed British colonial policy in East and South Africa. Upon independence in 1964, Zambia inherited four categories of land: state land (formerly crown land), freehold land, reserves and trust land. After it became a one-party state, legislation in 1975 vested all land in the President on behalf of the people: freehold land became leasehold, and all land transactions required the President's consent. A subsequent Lands Act (1995), which is still in effect, allowed the Lands Commissioner to convert customary land to 99 year leases, thus restoring value to land. This reform worsened economic inequality by concentrating land titles in Zambian elites and foreigners, while existing occupiers could be deemed squatters and evicted [112,113].

The above brief land law histories of four SSA countries show the continuities of colonial systems with the political settlements after independence. The land administration systems have remained overwhelmingly paper-based, bureaucratic and vulnerable to exploitation and corruption, and now have to struggle with greatly increased populations and demand for land. International moves towards FFPLA, which include World Bank support for 'accelerating digital transformation in Africa', have meant experimentation with digital land titling and transactions, offering a potentially promising application for blockchain/DLT linked to the tokenization of property assets through crypto-currencies. DLT in real estate has been seen as an opportunity to replace paper-based public registers and prevent such abuses as double-selling, impersonation of buyers and sellers, and forged signatures. Instead computer authentication of users, time-stamped smart contracts, and tamper-proof records, even zero-visit online transactions, could support a transparent and publicly verifiable distributed ledger for property [114,115].

American corporations linked with crypto-currencies have experimented with blockchain/DLT projects in several SSA countries since 2018. Medici Land Governance, a subsidiary of a blockchain accelerator backed by controversial entrepreneur Patrick Byrne, partnered with three capital cities [116]. It made a memorandum of understanding with Zambia's Lands Ministry and Lusaka city council to test digital land titling in regularizing informal settlements. In Liberia, a Medici project set out to record digital rights for

1000 residential plots around the capital Monrovia, using high-resolution imagery, a street address system, and community education and data collection. In Rwanda, whose political leadership initiated mass land titling and aspired to transform the capital Kigale through real estate redevelopment, Medici piloted digitization of the Land Registry; Rwanda now claims to be second in the world for speed of registering new property [117,118].

Other experimental DLT projects have occurred in Ghana and South Africa. Bitland, a non-profit subsidiary of American bitcoin corporation Ethereum, partnered with the Ghana Land Commission to record 5000 properties in the capital Accra through a new system of GPS coordinates, digital proof of identity, and transfer of funds via smart contracts [119]. In South Africa, the African National Congress (ANC) built some 3 million houses under its Reconstruction and Development Programme after 1994, but less than 2 million were registered on hand-over, and after more than 20 years original beneficiaries had died or moved away, or let to tenants. In 2018, the Centre for Affordable Housing Finance in Africa undertook a pilot project in Khayelitsha township outside Cape Town, surveying about 1000 properties and occupiers, cross-checking results against original subsidy data, and transferring documentation to a digital record [120].

These experimental DLT projects in SSA seem to have struggled with scaling up to national level because of basic infrastructure problems (power outages, poor internet connectivity, low computer ownership and computer literacy), and the technology may be poorly understood by over-eager politicians. The 'placelessness' of technologies that allow digital transactions and money transfers over the heads of the actual occupiers create risks of occupiers losing their homes for the lure of quick money, and predatory lenders forcing distress sales [121].

3.3. Below the State Level: Local Government, Public Awareness and Participation for Meeting Basic Needs

Land governance not only operates at national level, but also through sub-national institutions: local government, non-governmental and citizen-based organizations (NGOs, CBOs). This review now addresses past and current local governance arrangements, and new approaches that are changing the relations between citizens, communities and government.

Past colonial regimes In SSA ran local administration through centrally appointed (usually white) 'district officers', and local authorities were generally weak and under-resourced. Racially based master–servant legislation, requiring employers to house their workers, but not where suitable housing was available nearby, allowed for African housing to be neglected. Across much of SSA services for the African population were financed through the so-called 'Durban system', under which labourers paid a registration fee for their own policing and accommodation, and profits from a municipal monopoly over beer halls were paid into a 'Native Revenue' account, kept separate from other municipal finances to pay for community services. Africans soon recognized the injustice of the system: 'How much beer must I drink before my children can drink water? Do other countries make poor people drink beer to collect money for water?' [122,123]. Now, sub-national authorities are increasingly asserting themselves. The global umbrella organisation, United Cities and Local Governments, founded in 2004, is the main NGO promoting democratic, effective and innovative local government, encouraged by GLTN initiatives on land-based local finance. Structural reforms of local government powers and finances, however, move slowly, even more slowly than land law reform. Citizen perceptions of local government are generally negative, but city-to-city learning and mentorship programmes are improving the transfer of knowledge and best practice [124,125].

With rule of law measures supported by the World Bank and SDG16 ('peace, justice and strong institutions'), ways are being developed to improve citizen awareness and access to law, including protection of land rights. Over 20 years ago pioneering American researchers Ewick and Silbey identified three common narratives that ordinary people tell about law: it is either magisterial and remote, or a game whose rules can be manipulated to one's advantage, or an arbitrary power to be actively resisted [126] SSA experience confirms these research findings, law being seen as arbitrary or remote from the realities of people's daily lives, and as a tool manipulated by elites. The many linguistic and ethnic communities within African states create conditions of 'legal pluralism'—the existence of multiple sources of law within a single geographical area. Courts use official languages of European origin, which may be poorly understood by people who speak different languages in their every-day lives. Attempts to codify traditional land practices have largely come from foreign lawyers remote from the people [127,128]. Tackling legal problems is harder if there is a shortage of lawyers, whether practising or in universities; land and human rights lawyers are even fewer in number, although Pretoria University's Centre for Human Rights has been actively training them [129]. Externally-funded NGOs sometimes represent communities in court, but governments SSA accuse them of being 'busy bodies' or 'meddlesome interlopers', and in 2017 the Institute for Human Rights and Development in Africa petitioned the African Commission to express its 'concern and alarm at the shrinking of the civic society space' in many countries [130]. The media, social media and NGOs can help improve public awareness, but women in particular still seem unaware of their legal property rights, which allows patriarchal attitudes over land to continue [131].

Democratization through data-gathering and communication technologies is opening new possibilities to close the gap between policies at international and local level and practical implementation on the ground. In Liberia the Amplio Talking Book, a battery-powered audio device, is reaching people in remote communities with poor literacy levels and without electricity or the internet, who can now learn about recent land law reforms in their local language, and record their feedback [132–134]. In Cape Verde open-source technologies created a mobile geographic information system (GIS)-web mapping application for informal areas of the capital at high flood risk to collect and transmit data at household level for evaluation and action [135]. International charity MapAction works with Oxfam and civil society partners to map quickly from satellite imagery and improve food security, livelihoods, and access to water, sanitation and hygiene services. Participatory mapping, sometimes called 'counter-mapping' or 'cadastral politics', uses local oral history and traditions so that local communities can record land uses previously unrecognised by state institutions, as evidence to assert their occupancy claims and engage with land governance institutions [136–139].

Local surveying and mapping can also help with settlement upgrading. After decades of government evictions and demolition, going back to colonial slum clearance measures, communities are increasingly demanding greater accountability and transparency from government, and empowerment of their improvement efforts. The 'informal' areas may lack basic services, and be difficult to navigate physically, with inadequate street and property addressing, and poor road and path networks, and are still seen as inferior compared with 'formal' developments. New attitudes towards tenure security have softened official hostility to slums or squatter settlements, as governments recognize the political costs of eviction, and tenure regularisation can be linked to physical upgrading measures. Surveying and title registration may be complex and expensive, but upgrading can evolve gradually over time with local political leadership. Kenya (where UN-Habitat has its headquarters) has various community-based organizations that have resisted government evictions and promoted access to land, shelter and basic services for the urban poor, including free conversion of title deeds [140–143].

Recent research identified three stages in the development of one informal settlement (Mindolo North, Kitwe, Zambia). In the initial occupation stage, marginalised groups of society, mostly young and unemployed, occupied vacant land, and struggled against council hostility that demolished about 600 homes in 2014. The second or consolidation stage saw a rapid expansion of land coverage and buildings, as the undeterred settlers adapted through a mix of social norms and borrowed statutory rules. The settlement ultimately received official approval, and the third stage (maturity) saw intensified construction and house completions, an informal local governance structure, documentation of property rights, and the beginnings of health and other social facilities [144,145].

A shared sense of citizenship, trust and reciprocity creates social capital, with its own formal and informal rules. The local community can become the curator of collective memory as a basis for education and empowerment. Cape Town's District Six Museum, celebrating the multi-racial neighbourhood that racial segregation policy demolished, displays a large street map, embellished by handwritten notes from former residents showing where they lived before removal, to create a sense of neighbourhood community and instil a pride in heritage. In the Kalingalinga poor neighbourhood of Lusaka, Zambia, an exhibition of work by local photographers and visual artists which toured internationally, helped to empower the community [146–148]. Citizen frustration with basic infrastructure shortages of water, electricity and sanitation pushes them to learn tactics to negotiate improvements [149–153]. Networks of local community actors, for instance, are recycling waste materials into energy briquettes, as alternative cooking energy solutions using locally available technologies [154–156].

4. Conclusions

Africa's land governance challenges can seem daunting, even before the recent impact of the coronavirus disease 2019 (COVD-19) pandemic. This review has shown the complex technical languages of the different stakeholders, and the constraints upon resources. The problems are 'wicked' (in the sense of resisting solutions rather than evil), where their complexity means that efforts to solve one aspect may reveal or create other problems. The political settlement at independence for African countries, which often continued colonial land inequalities and governance structures, is being increasingly challenged half a century later by population pressures and climate change, while the expectations of the global community from the SDGs have grown. Dysfunctional national land laws and administration are increasingly seen as a major economic obstacle to African development. African land governance may now be at a critical juncture (in path-dependency terms), because the demographic youth bulge puts new demands upon the governing class, while existential threats grow [157,158].

There is no easy route to improve land administration and governance in SSA; political leadership, law reform and investment in systems and technologies are all needed. There are, however, reasons for optimism, and much has been achieved within a few years to better understand the problems. The traditional skills of the professions of land and built environment are being supplemented by more political skills of mediation, dispute resolution and local coalition-building, while academic scholarship is being enriched by interdisciplinary approaches. Theory and practice is being rethought to change negative colonial legacies by reform of laws and regulations, improve public space both physical and figurative, stimulate entrepreneurship and innovation, and build stronger civic society. New knowledge and actor networks—academic and professional, global and local—are developing new thinking and connections, such as FIG, the GLTN partner group, UN-GGIM, NELGA, and the Right to Development network. Advances in survey technology are adding more democratic data-capture techniques, championed by younger surveyors, and digital land administration offers the potential of better registration and protection of property rights. Land governance at a local everyday level means communities and neighbourhoods negotiating their own formal and informal rules, which seems to be occurring across SSA, and new legal structures for land management are developing, such as co-operatives or community land development trusts. Traditional authorities are returning as part of the modern political landscape, as much as any constitution, legislature or local council, and their resilience is contributing to better community participation in development efforts.

Funding: This research received no external funding.

Data Availability Statement: Not applicable.

References

1. UN. *Transforming Our World: The 2030 Agenda for Sustainable Development*; UN: New York, NY, USA, 2015.
2. UN. *New Urban Agenda*; Habitat III, United Nations; A/RES/71/256; UN: New York, NY, USA, 2017.
3. Bevir, M.; Rhodes, R. *The State as Cultural Practice*; Oxford University Press: Oxford, UK, 2010.
4. Mahoney, J.; Thelen, K. (Eds.) *Explaining Institutional Change: Ambiguity, Agency and Power*; Cambridge University Press: Cambridge, UK, 2010.
5. Khan, M. Political settlements and the analysis of institutions. *Afr. Aff.* **2018**, *117*, 636–655. [CrossRef]
6. Goodfellow, T. Seeing Political Settlements through the City. *Dev. Chang.* **2017**, 1–24. [CrossRef]
7. Sorensen, A. Taking path dependency seriously: An historical institutionalist research agenda in planning history. *Plan. Perspect.* **2015**, *30*, 17–38. [CrossRef]
8. Prado, M.; Trebilcock, M. Path Dependence, Development, and the Dynamics of Institutional Reform. *Univ. Tor. Law J.* **2009**, *59*, 341–380. [CrossRef]
9. Farias, I.; Bender, T. (Eds.) *Urban Assemblages: How Actor-Network Theory Changes Urban Studies*; Routledge: London, UK, 2010.
10. Special issue on Credibility of informality. *Cities.* 2020, *97*. Available online: https://www.sciencedirect.com/journal/cities/vol/97/suppl/C#article-44 (accessed on 10 December 2020).
11. Ho, P. *Unmaking China's Development: The Function and Credibility of Institutions*; Cambridge University Press: Cambridge, UK, 2017. [CrossRef]
12. Braverman, I.; Blomley, N.; Delaney, D.; Kedar, A. *The Expanding Spaces of Law: A Timely Legal Geography*; Stanford University Press: Stanford, CA, USA, 2014.
13. Delaney, D. *The Spatial, the Legal and the Pragmatics of World-Making: Nomospheric Investigations*; Routledge/Glasshouse Press: Abingdon, UK, 2010.
14. Parnell, S.; Oldfield, S. (Eds.) *Routledge Handbook on Cities of the Global South*; Routledge: London, UK, 2014.
15. Becher, T.; Trowler, P. *Academic Tribes and Territories: Intellectual Enquiry and the Culture of Disciplines*, 2nd ed.; Open University Press: New York, NY, USA, 2001.
16. De Waal, A.; Weaver, M.; Day, T.; van der Heijden, B. Silo-Busting: Overcoming the Greatest Threat to Organizational Performance. *Sustainability* **2019**, *11*, 6860. [CrossRef]
17. Home, R. (Ed.) *Land Issues for Urban Governance in Sub-Saharan Africa*; Springer: Geneva, Switzerland, 2021; Subsequent references to chapters in this book are cited as 'in Home, *op.cit.*; 2021'.
18. Pottage, A. The Measure of Land. *Mod. Law Rev.* **1994**, *57*, 361–384. [CrossRef]
19. FIG. *Fit-for-Purpose Land Administration*; FIG Publication 60; Annex to UN Economic and Social Council Resolution 2011/24; FIG: Eilat, Israel, 2015.
20. UN-GGIM. *Framework for Effective Land Administration.* E/C.20/2020/29/Add. UN-GGIM, Adopted Decision 6/101. Available online: https://ggim.un.org/UN-EG-LAM/ (accessed on 10 December 2020).
21. ISO 19152:2012. *Geographic Information—Land Administration Domain Model*, 1st ed.; ISO: Geneva, Switzerland, 2012.
22. Van Oosterom, P.; Lemmen, C.H.J. The Land Administration Domain Model: Motivation, standardisation, application and further development. *Land Use Policy* **2015**, *49*, 527–534. [CrossRef]
23. Lemmen, C.; Unger, E.-M.; Bennett, R. How Geospatial Surveying Is Driving Land Administration: Latest Innovations. *GIM Int. Newsl.* **2020**, *34*, 25–29.
24. Calafate, C.T.; Tropea, M. Unmanned Aerial Vehicles—Platforms, Applications, Security and Services. *Electronics* **2020**, *9*, 975. [CrossRef]
25. Is4land. Available online: http://platform.its4land.com (accessed on 10 December 2020).
26. Konashevych, O. Constraints and Benefits of the Blockchain Use for Real Estate and Property Rights. *J. Prop. Plan. Environ. Law* **2019**. [CrossRef]
27. Thomas, R. Blockchain's incompatibility for use as a land registry. *Eur. Prop. Law J.* **2017**, *6*, 361–390.
28. Lemieux, V.L. Evaluating the Use of Blockchain in Land transactions: An Archival Science Perspective. *Eur. Prop. Law J.* **2017**, *6*, 392–440. [CrossRef]
29. FIG. *The Land Surveyors Role in the Era of Crowdsourcing and VGI*; FIG Publication 73; FIG: Eilat, Israel, 2019.
30. Sankoh, O.; Dickson, K.E.; Faniran, S.; Lahai, J.I.; Forna, F.; Liyosi, E.; Kamara, M.K.; Bu-Buakei Jabbi, S.-M.; Johnny, A.B.; Conteh-Khali, N.; et al. Births and deaths must be registered in Africa. *Lancet* **2020**, *8*, 33–34. [CrossRef]
31. Cincotta, R. *The Security Demographic*; Youth Bulge Defined: When over 40% of the Adult Population Are Young, Particularly Young Men; Population Action: New York, NY, USA, 2003.
32. Mamdani, M. *Citizen and Subject: Contemporary Africa and the Legacy of Late Colonialism*; Princeton University Press: Princeton, NJ, USA, 1996.
33. Taiwo, O. *How Colonialism Pre-Empted Modernity in Africa*; Indiana University Press: Bloomington, IN, USA, 2010.
34. Lugard, F.D. *Political Memoranda*; Waterlow: London, UK, 1919; p. 29.
35. Baruah, N.G.; Henderson, J.V.; Peng, C. Colonial Legacies: Shaping African Cities. *J. Econ. Geogr.* **2021**, *21*, 29–65. [CrossRef]
36. Cobbinah, P.B.; Aboagye, H.N. A Ghanaian twist to urban sprawl. *Land Use Policy* **2017**, *61*, 231–241. [CrossRef]
37. Cotula, L. *The Great African Land Grab? Agricultural Investments and the Global Food System*; Zed Books: London, UK; New York, NY, USA, 2013.

38. Hufe, P.; Heuermann, D.F. The local impacts of large-scale land acquisitions. *J. Contemp. Afr. Stud.* **2017**, *35*, 168–189. [CrossRef]
39. Gallup. *Global Property Rights Index: Perceptions of Land Tenure Security in Nine Countries*; Gallup: Washington, DC, USA, 2017.
40. African Union Commission. *Guiding Principles on Large Scale Land Based Investments in Africa*; African Union Commission: Addis Ababa, Ethiopia, 2014.
41. Platteau, J.-P. The evolutionary theory of land rights as applied to Sub-Saharan Africa. *Dev. Chang.* **1996**, *27*, 29. [CrossRef]
42. Allen, T. *The Right to Property in Commonwealth Jurisdictions*; Cambridge University Press: Cambridge, UK, 2000.
43. Home, R.K. Are Africans culturally unsuited to property rights? *J. Law Soc.* **2013**, *40*, 403–419. [CrossRef]
44. Davy, B. Polyrational property: Rules for the many uses of land. *Int. J. Commons* **2014**, *8*, 472–492. [CrossRef]
45. Kaufmann, D.; Kraay, A.; Mastruzzi, M. *Worldwide Governance Indicators Project*; World Bank Policy Research Working Paper 4149; World Bank: Washington, DC, USA, 2007.
46. World Bank. *Land Governance Assessment Framework: Implementation Manual*; World Bank: Washington, DC, USA, 2013; Available online: https://www.worldbank.org/en/programs/land-governance-assessment-framework (accessed on 30 January 2021).
47. FAO. *Voluntary Guidelines on the Responsible Governance of Tenure of Land, Fisheries and Forests in the Context of National Food Security*; FAO: Rome, Italy, 2012; Available online: https://gltn.net/about-gltn/ (accessed on 2 February 2021).
48. Wehrmann, B. *Land Governance: A Review and Analysis of Key International Frameworks*; HS/072/17E; UN-Habitat: Nairobi, Kenya, 2017.
49. *GLTN Strategy 2018–2030: A World in Which Everyone Enjoys Secure Land Rights*; Nairobi, Kenya, 2018; Available online: https://gltn.net/download/gltn-strategy-2018-2030/ (accessed on 30 January 2021).
50. Lemmen, C. *The Social Tenure Domain Model: A Pro-Poor Land Tool*; FIG Publication 52; FIG: Eilat, Israel, 2013.
51. UN-Habitat. *Remaking the Urban Mosaic: Participatory and Inclusive Land Readjustment*; UN-Habitat: Nairobi, Kenya, 2016.
52. Chavunduka, C. Stocktaking Participatory and Inclusive Land Readjustment in Africa. In *Land Issues for Urban Governance in Sub-Saharan Africa*; Springer: Cham, Switzerland, 2021; pp. 137–154.
53. De Soto, H. *The Mystery of Capital: Why Capitalism Triumphs in the West and Fails Everywhere Else*; Black Swan: London, UK, 2000.
54. McAuslan, P. Bringing the Law Back. In *Essays in Land, Law and Development*; Ashgate: Aldershot, UK, 2003.
55. Van Wagner, E. Seeing the Place Makes it Real. *Environ. Plan. Law J.* **2017**, *34*, 522.
56. Home, R. Scientific survey and land settlement in British colonialism. *Plan. Perspect.* **2006**, *21*, 1–22. [CrossRef]
57. Njoh, A. *Planning Power: Town Planning and Social Control in Colonial Africa*; UCL Press: London, UK, 2007.
58. Robinson, J.B. *The Power of Apartheid: State, Power and Space in South African Cities*; Butterworth-Heinemann: London, UK, 1996.
59. Home, R.K. Colonial Township Laws and Urban Governance in Kenya. *J. Afr. Law* **2012**, *56*, 175–193. [CrossRef]
60. Porter, L. *Unlearning the Colonial Cultures of Planning*; Ashgate: Burlington, MA, USA, 2010.
61. UNDG Sustainable Development Working Group. Universality and the 2030 Agenda for Sustainable Development from a UNDG Lens. Available online: https://www.un.org/ecosoc/sites/www.un.org.ecosoc/files/files/en/qcpr/un (accessed on 9 November 2020).
62. Huneeus, A.; Madsen, M. Between universalism and regional law and politics: A comparative history of the American, European, and African human rights systems. *Int. J. Const. Law* **2018**, *16*, 136–160. [CrossRef]
63. AU Commission. *Agenda 2063: The Africa We Want*; AU Commission: Addis Ababa, Ethiopia, 2015.
64. Pharatlhatlhe, K.; Vanheukelom, J. *Financing the African Union: On Mindsets and Money*; ECDPM Discussion Paper 240; ECDPM: Maastricht, The Netherlands, 2019.
65. Wehrmann, B. Land Conflicts—A practical guide to dealing with land disputes. In *Scoping and Status Study on Land and Conflict*; HS/050/16E; UN-Habitat: Nairobi, Kenya, 2019.
66. UN-Habitat. *Scoping and Status Study on Land and Conflict*; HS/050/16E; UN-Habitat: Nairobi, Kenya, 2019.
67. AU Commission. *Declaration on Land Issues and Challenges in Africa*; AU Assembly/AU/Decl.l (XIII), Sec 2.1; AU Commission: Addis Ababa, Ethiopia, 2011.
68. Amao, O. *African Union Law: The Emergence of a Sui Generis Legal Order*; Routledge: London, UK; New York, NY, USA, 2018.
69. GLTN. *Customary Land Tenure Security: Tools and Approaches in Sub-Saharan Africa*. HS/048/19E. 2019. Available online: https://gltn.net/download/customary-land-tenure-security-tools-and-approaches-in-sub-saharan-africa-a-synthesis-report/ (accessed on 9 November 2020).
70. Ubink, J. Customary Legal Empowerment in Namibia and Ghana? *Dev. Chang.* **2018**, *49*, 930–950. [CrossRef] [PubMed]
71. The Network of Excellence on Land Governance in Africa. Available online: https://nelga.org/ (accessed on 2 February 2021).
72. UN. *Declaration on the Rights of Indigenous Peoples*; 61/295; UN: New York, NY, USA, 2008.
73. Home, R.; Kabata, F. Turning fish soup back into fish: The wicked problem of African community land rights. *J. Sustain. Dev. Law Policy* **2018**, *9*, 1–22. [CrossRef]
74. *Mbiankeu Genevieve v Cameroon*. Commission Communication 389/10. 2015. Available online: https://ihrda.uwazi.io/en/document/ir5ahz00dat?page=1 (accessed on 2 February 2021).
75. AU. *Convention for the Protection and Assistance of Internally Displaced Persons in Africa (Kampala Convention)*; AU: Addis Ababa, Ethiopia, 2009.
76. Okoloise, C. Circumventing obstacles to the implementation of recommendations by the African Commission on Human and Peoples' Rights. *Afr. Hum. Rights Law J.* **2018**, *18*, 27–57. [CrossRef]

77. Ayeni, O.U. *The Impact of the African Charter and the Maputo Protocol in Selected African States*; Pretoria University Law Press: Pretoria, South Africa, 2016.
78. Maputo Protocol on the Rights of Women in Africa. 2003. Available online: https://au.int/en/treaties/protocol-african-charter-human-and-peoples-rights-rights-women-africa (accessed on 2 February 2021).
79. Munyoni, L.M. (Ed.) *Women and Land in Africa*; Zed Books: London, UK, 2003.
80. Onyango, L. Gender Perspectives of Property Rights in Rural Kenya'. In *Essays in African Land Law*; Home, R., Ed.; Pretoria University Law Press: Pretoria, South Africa, 2011; pp. 135–154.
81. Bhatasara, S. Women, land and urban governance in colonial and post-colonial Zimbabwe. In *Land Issues for Urban Governance in Sub-Saharan Africa*; Springer: Cham, Switzerland, 2021; pp. 207–224.
82. *Chibwe v Chibwe*. Supreme Court of Zambia Judgment 38/2000. Available online: https://zambialii.org/node/2468 (accessed on 2 February 2021).
83. Spichiger, R.; Kabala, E. *Gender Equality and Land Administration: The Case of Zambia*; DIIS Working Paper 2014:04; Danish Institute for International Studies: Copenhagen, Denmark, 2014.
84. UN. *International Covenant on Economic and Cultural Rights*; UN: New York, NY, USA, 1966.
85. UN-Habitat. *The Rights to Adequate Housing*; UN-Habitat: Nairobi, Kenya, 2009.
86. CAPRi/CGIAR. *Land Rights for African Development: From Knowledge to Action*; CAPRi/CGIAR: Washington, DC, USA, 2006.
87. Berrisford, S.; McAuslan, P. *Reforming Urban Laws in Africa: A Practical Guide*; African Centre for Cities: Cape Town, South Africa, 2017.
88. Berrisford, S. Revising Spatial Planning Legislation in Zambia: A Case Study. *Urban Forum* **2011**, *22*, 229–245. [CrossRef]
89. UN-Habitat. *Effectiveness of Planning Law in Sub-Saharan Africa*; UN-Habitat: Nairobi, Kenya, 2019.
90. Pritchett, L.; de Weijer, F. *Fragile States Stuck in a Capability Trap?* World Development Report Background Paper. Available online: https://openknowledge.worldbank.org/handle/10986/9109 (accessed on 2 February 2021).
91. Boone, C. *Property and Political Order in Africa*; Cambridge University Press: Cambridge, UK, 2014.
92. Onoma, A.K. *The Politics of Property Rights Institutions in Africa*; Cambridge University Press: Cambridge, UK, 2017.
93. *Convention on Preventing and Combating Corruption*; AU: Addis Ababa, Ethiopia, 2003.
94. Chiweshe, M. Urban land governance and corruption in Africa. In *Land Issues for Urban Governance in Sub-Saharan Africa*; Springer: Cham, Switzerland, 2021; pp. 225–236.
95. Ocheje, P.D. Creating an anti-corruption norm in Africa. *Law Dev. Rev.* **2017**, *10*, 477–496.
96. Bennett, T.W. Historic Land Claims in South Africa. In *Property Law on the Threshold of the 21st Century*; Maanen, G.E., van der Walt, A.J., Eds.; Maklu: Apeldoorn, The Netherlands, 1996.
97. Beinart, W.; Delius, P. The Historical Context and Legacy of the Natives Land Act of 1913. *J. South. Afr. Stud.* **2014**, *40*, 667–688. [CrossRef]
98. McCusker, B.; Moseley, W.G.; Ramutsindela, M. *Land Reform in South Africa: An Uneven Transformation*; Rowman & Littlefield: Lanham, MD, USA, 2015.
99. Cousins, B.; Walker, S. (Eds.) *Land Divided, Land Restored*; Jacana Media: Johannesburg, South Africa, 2015.
100. Pooe, T.K. Developmental State No Birth Right: South Africa's Post-1994 Economic Development Story. *Law Dev. Rev.* **2017**, *10*, 361–387. [CrossRef]
101. All Population Figures for Nigeria Show High Error Rates. The Census Results Are Disputed. Available online: www.citypopulation.de/Nigeria-Cities.html (accessed on 2 February 2021).
102. Nti, K. This Is Our Land: Land, Policy, Resistance, and Everyday Life in Colonial Southern Ghana, 1894–1897. *J. Asian Afr. Stud.* **2013**, *48*, 3–15. [CrossRef]
103. Otubu, A. The Land Use Act and land administration in 21st century Nigeria: Need for reforms. *J. Sustain. Dev. Law Policy* **2018**, *9*, 80–108. [CrossRef]
104. Amokaye. The impact of the Land Use Act upon land rights in Nigeria. In *Local Case Studies in African Land Law*; Home, R., Ed.; Pretoria University Law Press: Pretoria, South Africa, 2011; pp. 59–78.
105. Africa Check. Who Owns the Land in Nigeria? Factsheet: Oak Ridge, TN, USA, 2016.
106. Ekemode, B.G.; Adegoke, O.J.; Aderibigbe, A. Factors influencing land title registration practice in Osun State, Nigeria. *Int. J. Law Built Environ.* **2017**, *9*, 240–255. [CrossRef]
107. Nwuba, C.C.; Nuhu, S.R. Challenges to Land Registration in Kaduna State, Nigeria. *J. Afr. Real Estate Res.* **2018**, *1*, 141–172. [CrossRef]
108. Memo by, G.V. Maxwell, 18 May 1926, file CO 533/711 'Native Lands in Kenya', in UK National Archives. Available online: https://discovery.nationalarchives.gov.uk/details/a/A13530124 (accessed on 2 February 2021).
109. Klopp, J.M.; Lumumba, O. Reform and counter-reform in Kenya's land governance. *Rev. Afr. Political Econ.* **2017**, *44*, 577–594. [CrossRef]
110. Coldham, S. The effect of registration of title upon customary land rights in Kenya. *J. Afr. Law* **1978**, *22*, 91–110. [CrossRef]
111. Manji, A. Whose land is it anyway? The failure of land law reform in Kenya. *Afr. Res. Inst. Counterpoint* **2015**, *1*, 1–13.
112. Munshifwa, E.K. *Rural Land Management and Productivity in Zambia: The Need for Institutional and Land Tenure Reforms*; Surveyors Institute of Zambia, 2002; Available online: http://mokoro.co.uk/land-rights-article/rural-land-management-and-productivity-in-zambia-the-need-for-institutional-and-land-tenure-reforms/ (accessed on 2 February 2021).

113. Ali, D.A.; Deininger, K.; Hilhorst, T.; Kakungu, F.; Yi, Y. *Making Secure Land Tenure Count for Global Development Goals and National Policy: Evidence from Zambia*; World Bank Policy Research Working Paper 8912; World Bank: Washington, DC, USA, 2019.
114. UNECA. *Blockchain Technology in Africa*; UNECA: Addis Ababa, Ethiopia, 2017.
115. Dankani, I.M.; Mahmud, R.H.; Saadu, I. Digitalization: An inevitable tool in urban administration and management in Nigeria. *Sokoto J. Soc. Sci. Conf.* **2019**, *1*, 2384–7654.
116. Medici Land Governance. Available online: info@mediciland.com (accessed on 12 December 2020).
117. Goodfellow, T. Urban Fortunes and Skeleton Cityscapes: Real Estate and Late Urbanization in Kigali and Addis Ababa. *Int. J. Urban Reg. Res.* **2017**, *41*, 786–803. [CrossRef]
118. Payne, G. Land Issues in Rwanda's Post conflict law reform. In *Local Case Studies in African Land Law*; Home, R., Ed.; Pretoria University Law Press: Pretoria, South Africa, 2010; pp. 21–38.
119. Ghana's Land Administration at a Crossroads, Focus on Land in Africa. 2011. Available online: https://gatesopenresearch.org/documents/3-633 (accessed on 12 December 2020).
120. South Africa Pilots Blockchain for Property Registry. Available online: https://www.ledgerinsights.com/south-africa-pilots-blockchain-property-registry/ (accessed on 12 December 2020).
121. Odendaal, N. Everyday urbanisms and the importance of place: Exploring the elements of the emancipatory smart city. *Urban Stud.* **2020**. [CrossRef]
122. Hall, B. *Tell Me, Josephine*; Andre Deutsch: London, UK, 1965; p. 135.
123. Swanson, M.W. The Durban System: Roots of Urban Apartheid in Colonial Natal. *Afr. Stud.* **1976**, *2*, 159–176. [CrossRef]
124. Cirolia, L. The Fiscal City: Financing Africa's Urban Areas and Local Governments. In *Land Issues for Urban Governance in Sub-Saharan Africa*; Springer: Cham, Switzerland, 2021; pp. 35–52.
125. Kuenzi, M.; Lambright, G. Decentralization, Executive Selection, and Citizen Views on the Quality of Local Governance in African Countries. *Publius J. Fed.* **2018**. [CrossRef]
126. Ewick, P.; Silbey, S.S. *The Common Place of Law: Stories from Everyday Life*; University of Chicago Press: Chicago, IL, USA, 1998.
127. Woodman, G.R. *Customary Land Law in the Ghanaian Courts*; Ghana Universities Press: Accra, Ghana, 1997.
128. Cotran, E. *Casebook on Kenyan Customary Law*; Professional Books: Abingdon, UK, 1987.
129. Kahn-Fogel, N. The troubling shortage of African lawyers: Examination of a continental crisis using Zambia as a case study. *Univ. Pa. J. Int. Econ. Law* **2012**, *33*, 719–789.
130. IHRDA Statement to African Commission, Banjul. 5 November 2017. Available online: https://www.ihrda.org/2017/11/ (accessed on 12 December 2020).
131. Mutolo, N. Women's Access to Property Legal Information. Master's Thesis, Onati Institute for the Sociology of Law, Onati, Spain, 2019.
132. Amplio. Available online: www.amplio.org (accessed on 30 November 2020).
133. Siddiqi, B. *Law Without Lawyers: Assessing A Community-Based Mobile Paralegal Program in Liberia*; International Development Law Organization: Rome, Italy, 2012; Available online: https://namati.org/resources/law-without-lawyers-assessing-a-community-based-mobile-paralegal-program-in-liberia/ (accessed on 30 November 2020).
134. Park, A. Consolidating peace: Rule of law institutions and local justice practices in Sierra Leone. *S. Afr. J. Hum. Rights* **2008**, *24*, 536–564. [CrossRef]
135. Correia, R. Un système d'information foncière pour gérer le risque d'inondation: Expérimentation à Praia (Cap Vert). *Etudes de l'environnement. Université d'Avignon* **2019**, 54–69. [CrossRef]
136. Panek, J. How participatory mapping can drive community empowerment—A case study of Koffiekraal, South Africa. *S. Afr. Geogr. J.* **2015**, *97*, 18–30. [CrossRef]
137. Cook, S.B. Searching Through Silos: Assessing the Landscape of Participatory Mapping Research. *Int. J. E-Plan. Res.* **2020**, *9*, 23–30.
138. Ragan, D.; Tindall, D.; Muldoon, M. *Community Mapping Guide Volume 3: A Youth Community Toolkit for East Africa*; UN-Habitat: Nairobi, Kenya, 2011.
139. McCall, M.K. *Participatory Mapping and Participatory Cartography in the Urban Context Utilizing Local Spatial Knowledge: A Bibliography*; CIGA, UNAM: Morelia, Mexico, 2019.
140. Muchadenyika, D.; Waiswa, J. Policy, politics and leadership in slum upgrading: A comparative analysis of Harare and Kampala. *Cities* **2018**, *82*, 58–67. [CrossRef]
141. Lindell, I.; Ampaire, C. The untamed politics of informality: 'Gray space' and struggles for recognition in an African city. *Theor. Inq. Law* **2016**, *17*, 257–282. [CrossRef]
142. Meredith, T.; MacDonald, M.; Kwach, H.; Waikuru, E.; Alabaster, G. Partnerships for successes in slum upgrading: Governance and social change in Kibera, Nairobi. In *Land Issues for Urban Governance in Sub-Saharan Africa*; Springer: Cham, Switzerland, 2021; pp. 237–256.
143. Sait, M.S. Should Monrovian Communities Agree to Voluntary Slum Relocations. In *Land Issues for Urban Governance in Sub-Saharan Africa*; Springer: Cham, Switzerland, 2021; pp. 339–354.
144. Munshifwa, E.K.; Mooya, M.M. Property rights and the production of the urban built environment—Evidence from a Zambian city. *Habitat Int.* **2016**, *51*, 133–140. [CrossRef]

145. Munshifwa, E.K. Adaptive resistance amidst planning and administrative failure: The story of an informal settlement in the city of Kitwe, Zambia. *Town Reg. Plan.* **2019**, *75*, 66–76. [CrossRef]
146. Jessa, S. Cultural Heritage Regeneration of District Six. Master's Thesis, Cape Peninsula University of Technology, Cape Town, South Africa, 2015.
147. Hacker, K. Generation Z: Visual Self-Governance through Photography. In *Personas and Places*; Lam, C., Raphael, J., Eds.; Waterhill Publishing: New York, NY, USA, 2018.
148. Dodman, D.; Mitlin, D. Challenges for community-based adaptation: Discovering the potential for transformation. *J. Int. Dev.* **2013**, *25*, 640–659. [CrossRef]
149. Lemanski, C. (Ed.) *Citizenship and Infrastructure*; Routledge: London, UK, 2019.
150. Gaisie, E.; Poku-Boansi, M.; Adarkwa, K.K. An analysis of the costs and quality of infrastructure facilities in informal settlements in Kumasi, Ghana. *Int. Plan. Stud.* **2018**, *23*, 391–407. [CrossRef]
151. Acey, C.S. Silence and Voice in Nigeria's Hybrid Urban Water Markets: Implications for Local Governance of Public Goods. *Int. J. Urban Reg. Res.* **2018**. [CrossRef]
152. Nastar, M.; Abbas, S.; Rivero, C.A.; Jenkins, S.; Kooy, M. The emancipatory promise of participatory water governance for the urban poor. *Afr. Stud.* **2018**, *77*, 504–525. [CrossRef]
153. Nganyanyuka, K.; Martinez, J.; Lungo, J.; Georgiadou, Y. If citizens protest, do water providers listen? Water woes in a Tanzanian town. *Environ. Urban.* **2018**, *30*, 613–630. [CrossRef]
154. Borie, M.; others. Mapping narratives of urban resilience in the Global South. *Glob. Environ. Chang.* **2019**, *54*, 203–213. [CrossRef]
155. UN-Habitat. *Land Tenure and Climate Vulnerability*; UN-Habitat: Nairobi, Kenya, 2019.
156. Buyana, K.; Byarugaba, D.; Sseviiri, H.; Nsangi, G.; Kasaija, P. Experimentation in an African Neighborhood: Reflections for Transitions to Sustainable Energy in Cities. *Urban Forum* **2018**. [CrossRef]
157. Cotula, L.; Anseeuw, W.; Baldinelli, C.M. Between Promising Advances and Deepening Concerns: A Bottom-Up Review of Trends in Land Governance 2015–2018. *Land* **2019**, *8*, 106. [CrossRef]
158. Conklin, J. Wicked Problems and Social Complexity. Available online: www.cognexus.org (accessed on 10 January 2021).

 *land*

Review

Farmland Fragmentation, Farmland Consolidation and Food Security: Relationships, Research Lapses and Future Perspectives

Pierre Damien Ntihinyurwa * and Walter Timo de Vries

Chair of Land Management, Department of Aerospace and Geodesy, Technische Universitaet Muenchen (TUM), Arcisstrasse 21, 80333 Munich, Germany; wt.de-vries@tum.de
* Correspondence: pdamien.ntihinyurwa@tum.de

Abstract: Farmland fragmentation and farmland consolidation are two sides of the same coin paradoxically viewed as farmland management tools. While there is a vast body of literature addressing the connections between farmland fragmentation and farmland consolidation on the one hand and agriculture production and crops diversification on the other hand, their relationship with variations in food security is still under-explored. This challenges policy makers about whether and how to devise policies in favor of fragmentation conservation or defragmentation. Therefore, drawing on the multiple secondary data and the deductive logical reasoning through an integrative concept-centric qualitative approach following the rationalist theory, this study critically reviews and analyses the existing body of literature to identify how farmland fragmentation versus defragmentation approaches relate to food security. The goal is to develop and derive an explicit model indicating when, where, how and why farmland fragmentation can be conserved or prevented and controlled for food security motives as a novel alternative comprehensive scientific knowledge generation, which could guide and inform the design of future research and policies about farmland fragmentation management. The findings show that both fragmentation and consolidation variously (positively and negatively) impact on food security at different (macro, meso and micro) levels. While farmland fragmentation is highly linked with food diversification (food quality), acceptability, accessibility, and sovereignty at the local (household and individual) levels, farmland consolidation is often associated with the quantity and availability of food production at the community, regional and national levels. Theoretically, the best management of farmland fragmentation for food security purposes can be achieved by minimizing the problems associated with physical and tenure aspects of farmland fragmentation along with the optimization of its potential benefits. In this regard, farmland consolidation, voluntary parcel exchange and on-field harvest sales, farmland realignment, and farmland use (crop) consolidation can be suitable for the control of physical fragmentation problems under various local conditions. Similarly, farmland banking and off-farm employment, restrictions about the minimum parcel sizes subdivision and absentee owners, joint ownership, cooperative farming, farmland use (crop) consolidation, agricultural land protection policies, and family planning measures can be suitable to prevent and minimize farmland tenure fragmentation problems. On the other hand, various agriculture intensification programs, agroecogical approaches, and land saving technologies can be the most suitable strategies to maximize the income from agriculture on fragmented plots under the circumstances of beneficial fragmentation. Moreover, in areas where both rational and defective fragmentation scenarios coexist, different specific strategies like localized and multicropping based land consolidation approaches in combination with or without agriculture intensification programs, can provide better and more balanced optimal solutions. These could simultaneously minimize the defective effects of fragmentation thereby optimizing or without jeopardizing its potential benefits with regard to food security under specific local conditions.

Keywords: farmland fragmentation; farmland consolidation; food security; food sovereignty; agroecology; integrative review

Citation: Ntihinyurwa, P.D.; de Vries, W.T. Farmland Fragmentation, Farmland Consolidation and Food Security: Relationships, Research Lapses and Future Perspectives. *Land* **2021**, *10*, 129. https://doi.org/10.3390/land10020129

Academic Editors: Kwabena Asiama and Hualou Long
Received: 1 December 2020
Accepted: 25 January 2021
Published: 29 January 2021

Publisher's Note: MDPI stays neutral with regard to jurisdictional claims in published maps and institutional affiliations.

1. Introduction

Farmland fragmentation has generally been considered as negative for agricultural production and food security and equivalent to the increase in production costs leading to farm inefficiency [1–11]. Consequently, most contemporary agricultural land policies aim to reduce fragmentation through land consolidation as a panacea to this quandary [12–18]. Besides the classical land consolidations programs, other instruments such as land banking [19–21], voluntary parcel exchange, land restrictions, cooperative farming, and land use consolidation (LUC) in Rwanda and Malawi [11,22–27] have been applied in some specific areas and situations. The success of each strategy depends on local conditions of a country and specific management and governance factors, since the strategy which works well in one country might not succeed in another [11]. Such idiosyncrasies necessitate each time a careful and substantive assessment of how and where farmland fragmentation patterns (forms, causes and both problematic and beneficial impacts) are similar or different. This assessment similarly applies to the success requirements and operational conditions of farmland fragmentation management strategies (as specific farmland management instruments) and their anticipated impacts prior to their transfer between countries [11,12,16,28]. The documented experience shows that the disregard of local conditions when designing and implementing land consolidation programs in sub Saharan Africa (Kenya, Malawi, Rwanda, Tanzania) and India led to failures and unintended harmful consequences in some areas [11,12,24,29–35].

On the other hand, there are counter arguments which consider farmland fragmentation as a demand-driven farmer's choice and strategy for risk management, exploitation of multiple ecological zones, labor bottlenecks management and self-sufficiency or independency in food production in subsistence communities through crops diversification for household food security [6,31,33,36–42]. These advocates argue that not all land fragmentation forms are equally problematic or defective. There might indeed be situations where the benefits of fragmentation outweigh the costs of consolidation, especially when it comes to areas which are overpopulated by communities relying on self-subsistence (agroecological) agriculture and/or mountainous characterized by diverse crop-growing conditions, socio-ecological heterogeneities and small farm sizes [11]. In such cases, fragmentation would be more favorable to support the management and mitigation of food production and market risks for the motives of local food security through its components of quality, accessibility, quantity and sustainability [4,11,30,31,43–45] and food sovereignty [46]. For these research scholars, a strategy which favors diverse and multi-cropping systems (polyculture) under varying crop-growing conditions manages better the risks of total crops failure and production loss resulting from the consequences of the ever-increasing climate change scenarios (manifested in changes in rainfall patterns and temperatures) leading to environmental hazards (droughts, floods, winds, etc), diseases outbreak, and food price fluctuations, than a mono-cropping systems-based one (monoculture) [11,33]. Furthermore, being one of the key agroecological principles and elements, the spatial and temporal crops diversification at both field/plot, farm and landscape levels increases the resilience of local farmers against various climate change, prices fluctuations and other global risks thereby acting as the sustainable strategy for achieving food diversity, self-sufficiency in the production of culturally acceptable food diets and food sovereignty as the local approach of achieving food security [46–48]. Hence, farmland fragmentation in this situation is rather viewed as a rational choice which adapts to the environmental variations and generates local food security than a drawback [11]. Moreover, in the line of disheartening land consolidation initiatives, various studies over time disclosed negative correlations between farm sizes and crop yields when labor market management conditions are unfavorable [11,31,49,50]. According to Ntihinyurwa and de Vries [11], this stance explains why farmland fragmentation persists, and the choice dilemma of farmers about fragmentation conservation for its positive sides and/or its prevention and banishment for its negative sides, in spite of various consolidation initiatives to combat it. In this respect, de Vries and Chigbu [51] and Ntihinyurwa and de Vries [11] posit that both land fragmentation and land consolidation are equally responsible land management instruments, given the circumstances in which they are carried forward.

As part of the same debate, there exist various contrasting social, economic, and ecological theoretical constructs and models, which favor, explain and support the claims of each side, i.e., deriving and proving the benefits of land fragmentation or of land consolidation. On land consolidation side, Ntihinyurwa and de Vries [11] highlight for example the economies of scale theory which states that farm size and crops yield or output are positively related [17,52], the Gestalt theory stipulating that the whole is greater than the sum of its parts [53], along with the Malthusian theory which stipulates the existence of an inverse relationship between the population growth and food supply [54–56]. On the other hand, the economies of scope theory asserting that the volume of production is the result of many heterogeneous factors [30,31,57], the complexity theory which argues for adaptation to emerging unpredictable complex phenomena [58–60], the ecological resilience theory which highlights the role and relevance of biodiversity conservation as an adaption to nature shocks [61,62], the agroecology stipulating the crops diversification, resilience to natural shocks, and responsible governance of land and natural resources [47,48], and the Boserup's theory which stipulates the existence of a proportional relationship between the population growth and agriculture intensification [55,56,63], support land fragmentation position. This polarized duality poses a crucial dilemma to policy makers and research scholars about whether they should devise and advise policies in favor of defragmentation (consolidation) or fragmentation conservation [11]. As stated by Ntihinyurwa and de Vries [11], this dilemma sometimes leads to the design of irrelevant farmland fragmentation control strategies which overlook the idiosyncrasies of specific fragmentation scenarios and its both contextual problems and benefits, and as a consequence derive disputed results leading to the failures.

Despite the subjectivity and the contradictions of various studies in literature, none of them has previously attempted to reconcile the above polarized views about farmland fragmentation and consolidation, and devise an explicit comprehensive relationship between these two concepts and food security as an end result instead of the existing focus on agriculture production and food quantity, since food security goes beyond the quantity. Chigbu et al. [32], Maxwell and Smith [64], Pinstrup-Andersen [65], Manjunatha et al. [66], Ntihinyurwa et al. [33], and Ntihinyurwa and de Vries [11] argue that although the popular logic is that land consolidation (especially due to increased farm size and reduced distances) has direct positive effects on increasing food security by boosting food production from conventional agriculture, this makes sense when food security is viewed from the lens of quantity. However, food security is much beyond the quantity of food production. It has the quality, accessibility, utilization, acceptability, sustainability, and sovereignty perspectives which can be achievable even under land fragmentation scenarios [11,32,33,46,67]. It is about more than growing enough food, since it implies the demand for it as well as the supply, the quality as well as quantity, an adequate diet (culturally acceptable quality and quantity meeting the local food preferences and needs) today and assurance of one tomorrow [11,32,33,46,54,65]. Following Sen's food entitlement theory [68], food security is achieved when everyone has access to regular, safe, nutritious and enough food [11,32,65,69]. For the advocates of food sovereignty, food security is achieved when local peasants have self-sufficiency in the production of their own food based on their cultural food preferences through local and sustainable agroecological approaches [46,70]. Furthermore, only few sporadic studies such as Bentley [30], Blarel et al. [31], Abubakari et al. [12], Kadigi et al. [50], Ntihinyurwa et al. [33], and Ntihinyurwa and de Vries [11] explicitly show when, where, how, and why one should keep fragmentation or opt for consolidation approaches, thereby calling for more comprehensive and holistic studies about this subject. In light of these arguments, there is a need to identify and compare categories and attributes of farmland fragmentation scenarios. For each of such scenarios one can describe which causes their constitution have, their impacts (positive and negative) on external variables like food security, and which control strategies would be most appropriate to them.

To address this specific research lapse and respond to these research calls, this study aims to:

❖ critically review (by exploring and synthesizing) the existing documented conceptual relationships between farmland fragmentation and its control interventions (including land consolidation), and food security;

❖ identify the knowledge gaps and openings for further research;

❖ reconceptualize the relationships between farmland fragmentation, its control strategies, and food security;

❖ propose a new theoretical model of farmland fragmentation management which may better help policy makers than current subjective and disaggregated ones, and guide and inform future solutions-oriented and evidence-based studies about appropriate and suitable alternatives for dealing with farmland fragmentation.

It explicitly results in a substantive explanation of different farmland fragmentation scenarios, the conditions under which they become defective or beneficial, and proposes the suitable potential strategies for their sustainable management under various specific circumstances. Moreover, the development and comparison of farmland fragmentation scenarios and food security extends the existing debate about farmland fragmentation and consolidation, and multiple UN sustainable development goals, namely SGDs 1, 2, 12, 13 & 15, versus the global trends towards market-oriented agriculture. Specifically, SDGs 1.4 and 2.3 address land rights and how farmers own, access, secure and control land resources among all the heirs; SDG 12.2 refers to the sustainable management and efficient use of natural resources (including land); the diversification of crops in different fragmented and scattered areas with diverse growing conditions as an adaptive strategy (climate smart, agroecological or resilient agriculture) to the ongoing new global challenging realities of climate change and the core of SDGs 2.4,5; 13.1 to end hunger and malnutrition resulting from food insecurity, is addressed by SDG 2.1, 2, 3, 4 & 5); and the agrobiodiversity and ecosystems conservation through the protection of their natural habitats on land comes as focus of SDG 15.3,4,5 & 9), in the framework of the Agenda 2030 [11,71]. The decisions about farmland use either in fragmented or consolidated forms can be most directly linked with these five SDGs whose specific targets capture the sustainable land management (ownership and use) and climate change adaption and mitigation, as key factors of sustainable agriculture production and food security to end hunger, malnutrition and poverty, even though land management as a scientific discipline may be connected with all the SDGs [11]. Since the terms of land fragmentation, land consolidation, and food security are variously conceptualized in different contexts, scientific disciplines and levels of analysis, in this article, only their meaning in the context of agriculture production at all levels is followed. The focus is given to the concept of food security from the lens of agriculture-based food stuffs, with little attention on the animal-based ones for the purpose of nutritional balance and food quality. Irrespective of the spatial and temporal limitations, only the literature about this topic in English language is considered.

The article is shaped in the following structure: The first section introduces the concepts of farmland fragmentation, farmland consolidation, and food security. The second section addresses the methodology of the literature identification, review, analysis, synthesis and reconceptualization. The subsequent third section categorizes and discusses farmland fragmentation scenarios and how these relate to their existing generic control (management) strategies (instruments) and interventions. Thereafter in fourth section, the concept of food security is discussed, and its relationships with farmland fragmentation and farmland consolidation approaches are assessed and synthesized. This section subsequently derives the new model of farmland fragmentation management and the reconceptualized relationships which are discussed and motivated in Section 5. Finally, the conclusions and implications of the study for further research and policies are drawn.

2. Methodology

2.1. Research Approach and Boundaries

As an integrative review article (relying only on secondary data), this research opts for an integrative concept-centric qualitative approach which draws on the deductive logical reasoning following the rationalist theory through the exploratory research design, to create new scientific knowledge from the existing general facts in literature and inform future research and policies [11,72–75]. This approach is considered by Ntihinyurwa and de Vries [11] and Ntihinyurwa and de Vries [75] as the most suitable research epistemology for this kind of study, since it deductively uses the researchers' own reasoning (abstract way of reasoning) without sensory experiences or empirical data to create novel scientific knowledge. The researchers use their own knowledge about the topic to critically analyze and synthesize the existing knowledge about different concepts, theories and principles, and deduct their own new and particular conceptualizations (models or frameworks) from the reviewed general facts [11,72–76]. Webster and Watson [74] and Torraco [73] argue that this approach fosters the critical review, analysis and synthesis of existing knowledge about the topic under research, with the objective of devising possible relationships among various research variables, identify knowledge gaps and contradictions, and seek opportunities for future research. The main aim of this approach is the re-conceptualization of the topic in a more understandable way for the guidance of future perspectives and expansion of the existing theories or creation of new knowledge in a particular scientific domain [11,72–76]. Hence, given the scope of the study of understanding various farmland fragmentation scenarios and proposing their suitable management strategies and interventions to achieve food security, only the literature about the forms, causes and impacts of agricultural land fragmentation and its alternative control measures across contexts and disciplines at all spatial levels was considered as a contextual boundary of the study, since the required information can mostly be derived from the relationships among these research variables. The use of multiple spatial levels of analysis is explained by the fact that farmland fragmentation itself is a multi-level phenomenon, whose causes, impacts as well as control strategies can be identified from the local (individual, household, family, village) to regional and national levels [11]. Spatial and temporal limitations (boundaries) were not considered throughout the review process for internal data validity purposes. This led to the review of both old and new geographically unlimited available literature materials on the topic, as a suitable method for this case of research approach which requires a comprehensive and broad literature. This review technique adopts a synthetic strategy of sense making which suggests the use of multiple cases and broad selection criteria to create a more comprehensive knowledge [11,74,75,77]. Nevertheless, for the purpose of preventing various conceptual divergences, misuse, and linguistic bias, both empirical (primary) and review (secondary) literature only in English language in which a large body of extensive literature on this topic exist [11,75], was considered for this review. This approach was recently used in quite similar studies and contexts by Asiama et al. [29], Asiama et al. [28], Ntihinyurwa and de Vries [11], and Ntihinyurwa and de Vries [75]. The following subsection explains the processes and methods for literature identification (search, selection criteria and its sources, scientific repositories or databases), review, analysis and synthesis techniques, and the reconceptualization or modelling methods and procedures as summarized in the research design (see Figure 1).

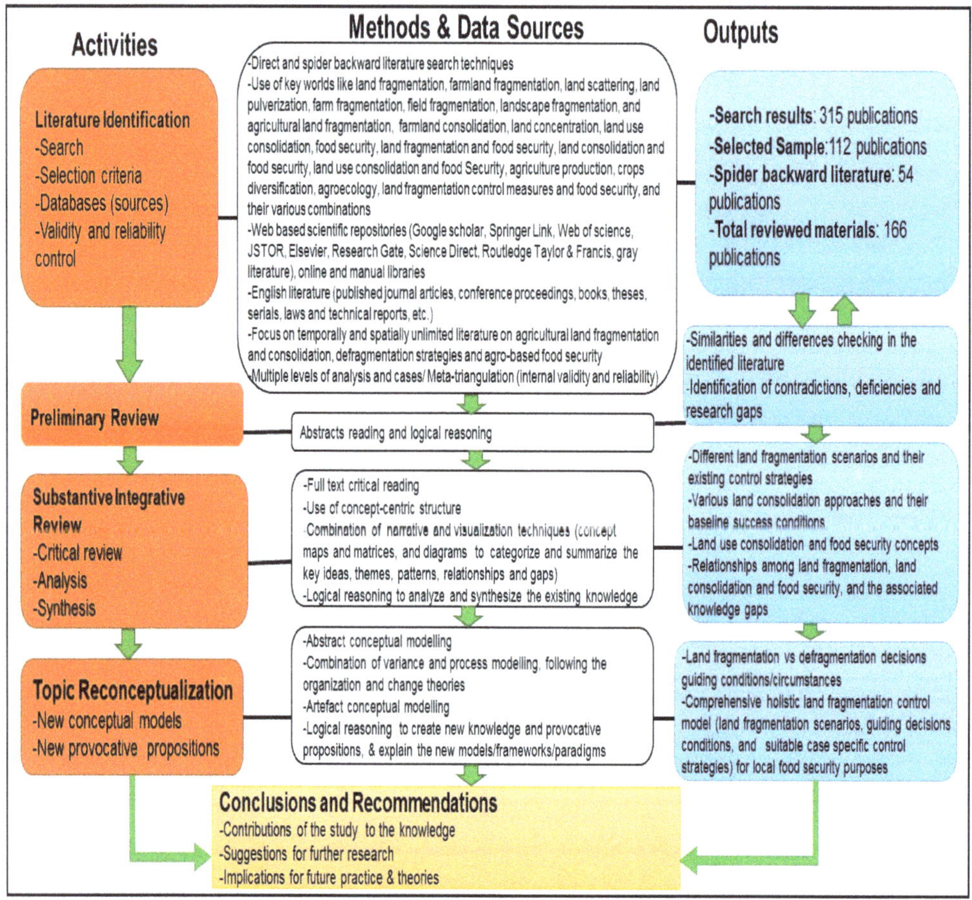

Figure 1. Overview of the research process and design. **Source**: Adapted from Ntihinyurwa and de Vries [11] and Ntihinyurwa and de Vries [75].

2.2. Data Sources and Research Methods

Once the boundaries of the literature were set up, we proceeded with the literature identification (search and selection). The following key words combinations were the basis for the search strategy: farmland fragmentation, farm fragmentation, land fragmentation, landscape fragmentation, field fragmentation, land pulverization, agricultural land fragmentation, land scattering, land fragmentation control measures, land consolidation, farmland consolidation, land concentration, land use consolidation (LUC), farm land use consolidation, crop consolidation, food security, farmland fragmentation and food security, farmland consolidation and food security, land use consolidation and food security, land banking and food security, agriculture production and food security, crops diversification and food security, land fragmentation control measures and food security, agroecology and food security, and agriculture intensification and food security. These key word combinations were chosen based on their closeness to the topic and the likelihood of generating the desired information. Individual instances of these key words and their diverse combinations were the systematic search strategies across different well-known web based scientific repositories (for soft documents) and the online and physical library visits (for hard documents) in English language (see Section 2.1 and Figure 1 for detailed search and selection criteria). These web-based scientific repositories include among others: Web of Science, Google Scholar, Springer Link, Research Gate, Routledge (Taylor & Francis), JSTOR, and Journals websites. Additionally, throughout the literature search process, relevant grey literature (published and unpublished non-commercial literature materials) from various governmental and non-governmental multilateral and bilateral organizations and institutions (such as FAO, GLTN, USAID, IFPRI, World Bank Group, UN and UN-Habitat amongst others) was taken into account. According to Webster and Watson [74], Torraco [73], Ntihinyurwa and de Vries [11], and Ntihinyurwa and de Vries [75], the use of multiple synonymous and diverse key words across many different data sources in literature identification provides a benefit of offering a large variety of documents about the topic for the motives of data and findings validity and authenticity. The search was nearly complete when no new concepts were found in the records set [11,73–75]. The literature identification process resulted in the retrieval of 315 written records including 292 soft documents varying from published peer reviewed journal articles and review papers, magazine articles, books, book sections, conference proceedings, laws and acts, technical reports, theses to press releases and serials, and 23 hard documents from visited libraries. The screening and preliminary review of the search results was done through critically scanning all titles and abstracts of the retrieved literature materials taking into account the above-mentioned review boundaries (see Section 2.1 and Figure 1) and elimination of duplicates. This process resulted in the selection of 112 relevant materials eligible for a full text review.

In the light of the aim and the scope of the study, a concept-centric (thematic) approach was adopted as the most suitable organization strategy for integrative literature review, analysis and synthesis [11,72–75]. Following this approach, all the articles with similar claims and views were grouped together and categorized through the combination of textual (narrative) and visual representations [11,73–75]. In this regard, throughout the reading session, a concept matrix [78,79] was developed to categorize different ideas and themes across various research variables encapsulating the concepts of farmland fragmentation, farmland consolidation, and food security in a more understandable, precise and narrow way. The content review consisted of both the analysis and synthesis of key and critical aspects of the research variables, and a listing and display of new relationships and research gaps [11,75]. During the review process, new seminal articles and frequently cited relevant references were identified and traced backward from their original materials using a spider backward literature search technique for further consideration in the review [11,75]. This technique resulted in the selection of 54 additional eligible documents for full text review, which therefrom generated the total number of 166 reviewed materials. In order to identify the strengths, weaknesses, deficiencies, contradictions, problematical situations and research gaps which need to be closed by the new knowledge [11,75], various farmland fragmentation scenarios and their existing control strategies were thoroughly reviewed, and their spotlighted relationships with food security critically analyzed using our existing knowledge about the topic through the logical reasoning following the rationalist theory [11,73–75]. This approach of conceptual reasoning is suitable for integrative theoretical studies which seek to analyze insights from past experiences and views for the preparation of future perspectives and guidance [11,72–75], and has been previously used by many research scholars in quite similar context with this one including McPherson [80], Bentley [30], Asiama et al. [29], Asiama et al. [28], Ntihinyurwa and de Vries [11], and Ntihinyurwa and de Vries [75]. The identified theoretical relationships and research gaps from the critical analysis were exhaustively summed-up in different diagrams and alternative models, or weaved together in a unique synthesis for a better presentation of the situation and basis for a reconceptualization of farmland fragmentation, farmland consolidation and food security nexuses. This also helped to inform the design of a new comprehensive and holistic conceptual thinking about farmland fragmentation management to support the achievement of the sustainable development goals (SDGs 1, 2,12,13, and 15) within the existing climate change realities. An abstract conceptual modelling combining both variance and process graphical models in artefact format [77,81] and textual models following the organization theory, theory of change, complexity theory and the soft systems methodology of thinking (SSM) [59,60,82,83], and the dynamic systems theory [84–86], was used to develop a new model. This model shows different farmland fragmentation scenarios, their proposed specific managerial decisions and strategies under different conditions, and their hypothetical impacts on food security and other aspects of livelihoods. The logic conceptual reasoning approach [74,87] combined with the reviewed theoretical foundations and documented empirical findings of the reviewed materials were used to justify various combinations and propositions of the model [75]. Finally, the implications of the new model to the existing knowledge and decision and policy makers were explained, and suggestions for future research to fill the newly identified gaps and empirically test the new relationships were derived [11,72–74,76].

3. Farmland Fragmentation Scenarios and its Management Strategies: Land Consolidation as a Controversial Multi-layered and Progressive Panacea to a Multidimensional Quandary

The concept of farmland fragmentation may at first glance seem very complex, fluid and multidisciplinary, as it refers to both a spatial structure and a management strategy [75]. As a multidimensional concept, it has been variously and subjectively defined in the existing literature [30,88]. Some research scholars commonly define it as the situation where a single farm consists of numerous spatially separated (non-contiguous) small parcels often scattered over a wide area [2,4,30,37,75,88–93]. Igbozurike [94] provides a more holistic and objective conceptualization by defining it as the process by which a contiguous block of land is split into two or more parts [75]. It has been simultaneously described as a natural and socio-economic phenomenon that occurs at different spatial levels (parcel, farm, land block and landscape). Thus, its conceptualization and derived forms should draw from the existing relationship between land parcel (object) and people (subject) in land management paradigms [33,75]. Following this approach, any fragmentation in the physical characteristics of a land parcel as an object (size, use, shape, type, location) dictate the existence of different physical fragmentation forms [75]. Similarly, any fragmentation derived from the social relationships (rights, restrictions and responsibilities) between land parcel (object) and people (subject) implies the occurrence of various social or tenure fragmentation forms (both visible and hidden ownership and usership) at different spatial levels [75]. Moreover, the economic characteristics of land (value and market) often dictated by social and physical traits may also imply some fragmentation forms and scenarios. Therefore, referring to the study of Ntihinyurwa and de Vries [75], physical farmland fragmentation stands for any type of fragmentation in physical characteristics of land either internal or external at all spatial levels, while tenure fragmentation refers to any fragmentation form derived from the split in the social characteristics of land in terms of its relationships with people, irrespective of the exclusive internality and externality criteria. Hence, in this context, farm fragmentation (often referred to as internal or within farm fragmentation) denotes the situation when a single farm is physically split into many relatively small plots (parcels) either spatially dispersed or contiguous (physical fragmentation), or shared by many undocumented co-owners or co-users (hidden tenure fragmentation in terms of ownership or usership) [33,75]. On the other hand, according to Ntihinyurwa and de Vries [75], farmland fragmentation refers to the split of the farming structure in a relatively small land block or region into many small farms (visible and hidden tenure fragmentation in terms of ownership and usership), or into many small plots or parcels (physical fragmentation). When the split into many plots happens at the parcel level, this phenomenon is denoted as parcel or field fragmentation. The land value (social and economic) fragmentation and land market fragmentation exist when a land block is split into smaller subunits like land parcels and plots with different socio-physical peculiarities dictating the diversity in value and market of the land. Irrespective of various contradictory theories of Earth creation, it is obvious that the landscape is naturally fragmented (in soil type, size, location, shape, topography), which ontologically explains the existence of physical fragmentation at the landscape level as a natural phenomenon, independent of human activities and land-people relationship. This relationship is often defined at the parcel level [95,96], which dictates the existence of physical, social (tenure), and economic fragmentation forms. Despite the dynamic nature of this fragmentation concept, King and Burton [97] and Ntihinyurwa and de Vries [75] assert that the above fragmentation forms can coexist in the same area at different levels, and its extent is determined by the local conditions in specific countries and areas [75]. This entails the existence of different possible generic and specific fragmentation scenarios from various combinations of its indicators [75]. Notwithstanding various conceptualizations of this fragmentation phenomenon from the socio-economic and physical perspectives and different subjective levels of analysis, all the analyzed literature materials have a commonality of referring to agriculture land fragmentation (see Ntihinyurwa and de Vries [75] for more details on various farmland fragmentation forms and scenarios).

The causes and impacts of farmland fragmentation in the literature have always been subject to contradictory and multidisciplinary debate by considering it either beneficial (voluntary) to farmers (as risk management strategy for household food security) or defective (derived from external imposition which leads to the reduction of farm efficiency through the increase of production costs) (see Ntihinyurwa and de Vries [11] and Ntihinyurwa et al. [33] for more details). However, recent studies revealed that its problematical and beneficial scenarios are dictated by a combination of local specific external conditions, varying from economic, socio-cultural, political, ecological, technical to environmental ones, which therefore similarly implies the variation of their management strategies [11,12,17,18,30,33,90]. According to [75], with reference to the problems linked with farmland fragmentation, its problematic forms can be categorized into four distinct groups: (i) *farmland location or spatial fragmentation* (problems of long distance between plots and farmstead); (ii) *farmland size fragmentation* (problems of small plots and farm sizes); (iii) *farmland shape fragmentation* (problems of shape irregularity), and (iv) *farmland use fragmentation* (problems of multiple mixed uses or multiculture).

Whether problematic or beneficial, agricultural land fragmentation needs a certain level of management for sustaining the quality and quantity of agriculture production for food security purposes in a given area. In this regard, various strategies have been developed over time to control this complex phenomenon. Demetriou [98] grouped them into three main categories as follows:

- Legal provisions and restrictions relating to inheritance, minimum size of parcel subdivision, joint ownership, absentee landowners, prevention of transfer to non-farmers, leasing, and imposition of a maximum limit on the size of a holding to prevent the rational drivers of fragmentation phenomenon from worsening the situation.
- Land management approaches including land consolidation, land funds and land banking, voluntary parcel exchange, and cooperative farming to reverse and inhibit the harm of the existing fragmentation.
- Agricultural land protection policies which embrace the Purchase of Development Rights (PDR), the Transfer of Development Rights (TDR), and the Cluster Development Programs (CDP) in USA to prevent its use for other purposes or development activities like residential, commercial, etc. These are described below.

(a) *Land Consolidation*: Even though the concept of land consolidation has its roots in the medieval ages with the first initiative in the 1750's in Denmark as a social reform [4] and was implemented in different countries for millennia, there is no common definition for it, as it varies across contexts and by country with respect to the end goals and objectives. It is generally known as a process of arranging parcels together in order to make them more productive and reduce the adverse effects of fragmentation in agriculture [4,19,98,99]. In the German Land Consolidation Act (1976), it is considered as an instrument of improving production and working conditions in agriculture and forest lands as well as promoting the general use and development of land in rural areas and the living conditions of rural livelihood, through the re-arrangement of agricultural land by restructuring the shape, size, ownership and location of farmland parcels and forestry [10,16,97,100–102].

FAO [4] defines it as a land management activity that involves all the procedures for exchanging, rearranging, realigning, and expanding farm parcels in rural areas with the goal of increasing food productivity. In this context, the parcel boundaries, ownership, size and location of the land are restructured for its best use and management. Since the reallocation of new parcels as a core for land consolidation procedures is rather value than shape, use, size and location based, land valuation based on soil evaluation is considered as the basic activity which should be given a special attention and management to reduce the pace of resultant conflicts in this strategy [12,15–17,103–106]. Land consolidation started as a monofunctional concept with a single objective of improving agriculture production which is still kept in Scandinavian countries like Sweden (*fastighetsreglering*), Norway and Denmark. It gained its momentum in 1970s, and started to integrate other objectives of rural development like village renewal, landscape and natural resources management, and forests management afterwards [16,20,107]. It is currently implemented with success as a comprehensive rural development strategy in many western European countries like the Netherlands (*ruilverkaveling*), Germany (*Flurbereinigung*), France (*remembrement*), Luxembourg, Spain, Belgium, Switzerland and Austria, as well as Finland (*uusjako*), and in Asian countries like China, India, Nepal, South Korea and Japan, where it is embedded in large national and regional development programs [16,101,102,107]. Van der Molen et al. [20] argue that land consolidation also called concentration [39] as an ambivalent concept (instrument and principle) following different principles (parcel reallocation and improvement of physical conditions) has a common objective of making the parceling of one farm more compact (few contiguous parcels close to the homestead) and the farming structure in a given region denser (few farms per land block with higher average farm size), in order to create more operational and viable farm units. In this context, it is theoretically and commonly understood as a process of making the parceling of one farm or a farming structure of any region more compact with few parcels per farm or land block, and higher average parcel and farm sizes respectively [4,16,18,20,108,109]. FAO [4] advances that any modern land consolidation should follow the following principles:

- The objective should be to improve the rural livelihoods rather than only the primary production of agricultural products.
- The end result should be the whole community renewal through its sustainable economic and political development, and the protection and sustainable management of natural resources.
- The process should be participatory, democratic and community-driven not only in concept, but also in practice.
- The interventions should be to assist the community to define new uses for its resources and then reorganize the spatial components accordingly.
- The approaches should be comprehensive and cross-sectoral, integrating elements of rural and broader regional development including the rural-urban linkages.

Several other studies indicate that the local agricultural, economic, social, cultural, environmental, agroecological and political conditions of the area dictate the procedures, objectives and models of land consolidation in different countries, although the implementation principles remain the same everywhere [12,13,16,18,19,28,33,110]. They argue that the variations in local conditions make it necessary to allow the creation of different local versions or approaches of land consolidation, based on the available problematic land fragmentation forms and needs of rural local population. A successful consolidation approach of flat areas cannot necessarily apply in mountainous areas. This idea is guided by the FAO principles stipulating that a good land consolidation strategy must recognize the diversity of rural society and the non-problematic scenarios of land fragmentation [4,111]. This dictates the need for diverse local solutions, including keeping the beneficial fragmentation for crops diversification motives for food security and sovereignty, and risks and labor management under multiple agroecological zones [30]. In this vein, different land consolidation forms (approaches) have been developed over time. The most commonly known include the comprehensive, simplified, voluntary, individual, government-led, private company-dominated, and farmland use consolidation among others [4,16,23,99,108,112,113] as described below:

- ➢ *Comprehensive land consolidation*: It embraces the re-shaping and re-allocation of parcels together with a broad range of other measures and activities that support and promote the rural development [4,16,114]. Examples of such activities include extension services for rural communities, the village renewal, the construction of rural roads and water infrastructure, the co-construction and support to community based alternative agro-processing techniques, the erosion control measures, the construction and rehabilitation of irrigation and drainage systems, the creation of social infrastructure including sports grounds and other public facilities, along with the environmental protection and improvement measures including the designation of nature reserves [4,16,114]. This model prevailing in Germany and the Netherlands presents the drawback of taking too long in implementation, due to the complexity of involved activities and large coverage. It is mostly *government-led*, and somehow involves a certain level of compulsion in participation [4,16,114]. It is more effective when it is combined with land banking programs to counter the challenges of unwilling participants in order to enlarge the parcels and landholdings [4,16,114]. Although its implementation procedures vary from country to country, they generally involve the following phases: initiation or the design of the project (feasibility study); inventory of existing 3Rs (rights, restrictions and responsibilities like ownership, tenancies, easements, usufructs, mortgages and conflicts) and values over land (land valuation); elaboration of the detailed consolidation plan showing the new parcels layout and their reallocation which shall be presented to the public for claims consideration and accepted by all land owners before the final plan; implementation of the final plan and appeal proceedings; and finally a concluding phase in which the final records are produced [4,16]. Drawing from the recent study of Veršinskas et al. [114], the *mandatory* and *majority-based* (the decisions to compulsorily consolidate are based on the votes of the majority) *land consolidation* types fall in this category. The same study groups the consolidation process in this model in three phases of the feasibility phase, the re-allotment phase, and the registration and implementation phase. Notwithstanding its multifunctionality, when flexible and participatory, the comprehensive land consolidation can be subject to different changes and take different approaches to adapt it to the local collective needs and objectives, contrary to the government-centered one [115].
- ➢ *Simplified land consolidation*: To overcome the challenges of long duration due to the complexity of activities in comprehensive consolidation models, the simplified land consolidation has been created to optimize the conditions in agricultural sector through the exchange or re-allocation of parcels, and the provision of additional lands from land banks [4,16]. These simplified projects are often combined with minor public works like the rehabilitation of infrastructure and sometimes the provision of minor facilities with the primary objective of improving the working conditions in agriculture. They are mostly implemented on a small coverage and follow similar but simplified procedures as comprehensive land consolidation [4,16]. This is the case of German special land consolidation proceedings and Swedish forest re-allotment projects [4,16].
- ➢ *Voluntary group consolidation*: It is based on the mutual agreement among close land owners to consolidate their adjacent plots with no element of compulsion in some countries [4]. Since the consolidation is entirely voluntary, during the process, all participants must fully agree with the proposed project [4,16,114,116]. In the light of this, such voluntary projects tend to be small, usually with less than ten participants and best suited to address small and localized fragmentation problems with less harm to the environment [4,16,28]. In Denmark, this option is most common and almost all land consolidation projects are carried out in a completely voluntary process, and typically involve the negotiations with up to 50 land owners, even though some

few projects may involve about 100 participants [4]. Countries like Lithuania and others are currently following this approach [4,16,28,116].

➤ *Individual consolidation*: In this form, the consolidation of holdings takes place on an informal and sporadic basis without a direct involvement of the state and the provision of public facilities [4,117,118]. Nevertheless, the state can play a significant role in encouraging consolidations that improve agriculture by promoting instruments such as joint land use agreements like cooperative farming, scattered parcels exchanges among farmers to create compact farms, farmland use or crop consolidation, and leasing and retirement schemes [4,16,117].

➤ *Land Use Consolidation (LUC) or Consolidation of crops*: LUC program also known as Farm Land Use Consolidation in USAID reports, and land consolidation in the Ministerial order on land consolidation models in Rwanda (2010), refers to the consolidation of the use of farmlands where all farmers with close parcels grow one same crop in a synchronized way up to the minimum size of 5ha from the 8 priority food crops (maize, beans, wheat, rice, Irish potatoes, banana, cassava and soybeans) chosen by the government at the national level based on the Agro Ecological Zones (AEZ) of the country [23,26,32,33,117,119]. Contrary to other land consolidation programs, the individual land rights in LUC remain intact [32,33,117]. It is a national program implemented in the whole country as one of the pillars of the Crop Intensification Program (CIP) with the objectives of increasing agricultural production, improving the living conditions in rural areas, and meeting food security [23,26,28,32,33,117]. Huggins [24] and Pritchard [34] call it "Crop Consolidation". Similar programs have been previously documented in Malawi, and in Europe in case of viniculture consolidation programs [16,26,120].

Although land use consolidation (LUC) is conceptually considered as a special form or approach of land consolidation, from the practical perspective in Rwanda and Western Europe, the two terms do not have much in common in terms of activities involved. While in land consolidation the sizes, shapes, boundaries, locations and ownership of land parcels are rearranged with no control on the use, and the parcel values kept intact, only the use of farmlands for priority crops is consolidated in the case of land use consolidation in Rwanda, with all the other attributes remaining unchanged. Nevertheless, the two strategies share the same objective of improving agriculture production and the rural livelihoods, even though LUC has been criticized to only lead to the monoculture (mono-cropping) system resulting in food insecurity at the household level in case of climate change, natural shocks, and market imperfection scenarios [24,32–34,119,121]. Furthermore, one could wonder whether it is the most suitable strategy to the problematic land fragmentation scenarios in Rwanda, considering the heterogeneous local social, economic, physical and ecological conditions of the country. In support to this doubt, recent findings of Isaacs et al. [122] revealed the benefits of improved intercropping system to outperform the ones from the government-led mono-cropping through LUC in terms of household food security and risks management insurance. Niyonzima [123] found that the national farming programs including LUC encouraging the monoculture and environmental policies have failed to address the local farmers needs in the Eastern Province of Rwanda mainly due to the market imperfections, thereby recommending the support to mixed farming systems as a promising solution for agricultural production and household food security concerns. Therefore, contrary to the studies of Laepple [120], Vitikainen [16], Musahara et al. [26], and Asiama et al. [28], we claim that there is no rationale for considering *land use consolidation* as part of conventional *land consolidation approaches*, rather a particular type of agricultural land use management, and a tool for farmland management like land consolidation as well.

With regard to the emergence of new issues in the implementation of *government-led land consolidation* projects in China, a new approach of *company-dominated pattern of land consolidation* programs [113] has been developed as an efficient strategy for both physical and tenure fragmentation problems. In this approach, the private companies act as land bank institutions and acquire large lands through the negotiations-based expropriation programs from small farmers to create big land funds which could later be farmed as single consolidated viable operational units, or leased to big farmers [113]. The commonality of these consolidation models is that most of them are regulated and facilitated by land professionals [114].

Whereas the success conditions of different land consolidation approaches vary from country to country, the common key feature is that the relative economic value and ownership of land should be kept constant before and after consolidation following the surrogate principle of land valuation, with the benefits from such consolidation exceeding the costs of its establishment [16,30]. Similarly, Van Dijk [18], Hartvigsen [19], Asiama et al. [28], and Asiama et al. [17] argue in their respective studies on Central and Eastern Europe and Ghana, that the economic value of land should exceed its social value (perfect land market leading to high land mobility) as a key precondition for success of land consolidation programs. In this regard, various researchers have highlighted and documented the general baseline conditions which need to be considered before the development of any specific land consolidation approach in an area [4,12,18,19,28,30,107,108,124–127]. These include:

-*Land tenure system*: It dictates the decisions about the parcels reallocation process as a core for land consolidation. The customary or community land tenure system has been considered to be an obstacle to this activity, since farmers only have use rights over their lands, with no allocation rights without the consent of the chiefs who hold the custodian rights (allodial title) to control and allocate the use of land on behalf of their community [12,17,28]. Furthermore, in customary tenure systems, land is considered as a sacred property of the family which should be preserved for future generations (ibid). This increases the social attachment to land and social land value, which in turn reduces land mobility as an obstacle to land reallocation and land market [18,19]. Likewise, the absentee owners in case of usership fragmentation obstruct the reallocation process since they do not find any direct interests from consolidation. On the other hand, the users (tenants) do not have the ownership and allocation rights, which decreases their willingness to invest in long term projects like land consolidation [11,18]. The failure of previous land consolidation attempts in customary lands has been attributed to the focus on technical and economic aspects, thereby ignoring this important social benefit [28]. Asiama et al. [17] found that the exchange of parcels in the Ghanaian customary lands is only possible among family members within the same community, with very limited mobility among communities. For this, the statutory individual private tenure system with individual ownership rights has been pinpointed as a suitable success condition for modern land consolidation projects by facilitating the decision making about reallocation with consent from one or few owners [12,18,19,28,104].

-*Economic status and land market*: They dictate the approach of land consolidation to be adopted, and the reallocation process. A perfect land market increases the economic value of land (land as an economic commodity), which in turn reduces the social attachment to land (social value), thereby easing land mobility and the reallocation of land during land consolidation. This is explained by the theory of land mobility stipulating that when the economic value of land is higher than its social value, the mobility of land through any kind of transfer increases [18,28,109]. Furthermore, the macro economic conditions have been found to facilitate the adoption of modern comprehensive land consolidation approach, which needs considerable financial capacities from both farmers and the state, while the subsistence economies favor more simplified and cheap approaches [16,28].

-*Willingness of farmers to participate*: This is crucial for the success of any land consolidation project and the type of approach to follow. It is dictated by land psychology (i.e., sense of social attachment to land), economic status, land availability, and land market. From a rational perspective, farmers accept to participate when the economic benefits from the project outweigh the costs and its social ones. Participation also relates to the degree to which project managers have an affinity with the area [128]. FAO [4] suggests that land consolidation process should be demand-driven by farmers, and the government should intervene to assist them in choosing the suitable approaches to their land use needs.

In case a big number of farmers accept the participation, the reluctance of few farmers is overcome by a certain level of compulsion in some cases and the expropriation processes through land banking. The subsistence farmers in risks prone areas with scarce land and absentee owners often resist land consolidation programs which rearrange the ownership structure, sizes, locations and boundaries, due to the fear of losing their original rights over land [4,12,17,30,97,129].

-*Availability of land banks*: Although not a sine qua non condition for all land consolidation projects, it is very important during the reallocation process, as it provides additional lands from the governmental land funds to bridge the lapse of unwilling participants. Land banks provide an opportunity for expansion, shaping of farmlands and creation of adjoining infrastructure; facilitates the increase in land mobility; and creates the room for a flexible land consolidation design and reallocation process [4,12,21,28,109,125,127].

-*Existence of a legal framework*: It determines the success of land consolidation projects by regulating the whole process from the initiation to the concluding stage. Since land consolidation projects involve the exchange and reallocation of land rights, there is a need for a strong legal basis to regulate the interferences among different private property rights by the state, for the sake of transparent protection of the rights of landowners and users, and prevention of the prevalence of any conflicts from the process. It also provides the procedures for resolution of any conflict resulting from the sensitive land valuation and reallocation processes, and regulates the modalities of participation in the projects [4,12,28,106]. According to Bullard [99], the legislation is not only meant to address land fragmentation, but also to prevent its reoccurrence in future. For this, the absence of the legal frame is considered as a major obstacle to the success of any land consolidation project.

-*Level of political structure*: It determines the political will of the state to support land consolidation projects, which in turn dictates the type of approach to follow, the duration of the project, its implementation procedures and success. When there is a high level of political will, the government takes a primary initiative to finance land consolidation programs at large scale, which in turn stimulates the willingness of voluntary participation of farmers and reduces the duration and costs of implementation. In contrast, the lack of political will slows down the process, and induces farmers to adopt cheap approaches on voluntary basis with no direct influence of the state [16,28].

-*Existence of problematic land fragmentation*: Since land consolidation is designed to solve the existing problems of land fragmentation, there is a need to know the available forms of land fragmentation in a specific local area, and their problematic circumstances to inform the suitable land consolidation and other approaches, since not all land fragmentation problems need land consolidation control strategies, neither are all fragmentation forms problematic [4,11,20,21,30,31,33,75,130]. The review of existing documents has revealed that the modern land consolidation is only suitable for physical (internal) fragmentation problems of big farms. Expanding the stipulations of Abubakari et al. [12] and Asiama et al. [28] for the availability of a certain type of land fragmentation as a precondition for an introduction of land consolidation projects in a given area, we argue that there should be a problematic land fragmentation suitable for land consolidation strategies since some fragmentation forms like tenure fragmentation might need different other strategies for their control. It has further been found that the adoption of non-suitable land consolidation strategies to the existing local land fragmentation problems has led to their failure in many countries like Kenya, Malawi, Japan, and others (ibid).

-*Biophysical/geographical/agroecological/environmental conditions*: Variations in topography (slope distribution), soil quality and water distribution, and the microclimatic conditions determine the forms of land fragmentation and which control strategies are suitable in a given area with respect to the benefits and costs associated with the valuation and reallocation activities. Contrary to hilly and mountainous topographies characterized by high diversity or heterogeneous microclimatic conditions and soil qualities, flat terrains with quite homogeneous conditions make it easy to exchange parcels with similar characteristics and values [12,28,30]. Furthermore, the hilly and mountainous areas with sharp variations in surface characteristics hinder the creation of regular shapes and infrastructures as land boundaries may naturally follow the physical characteristics of the terrain like hill tops or valleys [12,13,28,40,107]. King and Burton [97], Bentley [30], and Janus et al. [108] argue that due to the sharp variations in soil quality, and agroecological conditions in mountainous and hilly lands, the costs of consolidation may exceed its benefits, which dictates the development of different consolidation approaches rather than focusing on economic profitability, or keeping fragmentation in such areas. Prior to the development of any land consolidation approach, one needs to consider its anticipated effects on the environment, since previous experience has shown that large comprehensive consolidation projects have led to the loss of biodiversity. There should be measures to conserve the environment within the projects, or the development of environmental friendly approaches like simplified or voluntary or small localized land consolidation projects involving few people and activities [16,28,131–133].

-Technical aspects (existence of land information system and consolidation experts): Since the consolidation of parcels involves the restructuring and rearrangement of socio-spatial land characteristics like ownership, use, size, shape, location, value and boundaries, there is a need to have a well-functioning and updated land information system (LIS) to provide such information for a successful reallocation [4]. However, it is not a prerequisite prior to the establishment of consolidation projects, since the experience has shown that this database can be created later within the project [4]. Furthermore, since the implementation procedures of land consolidation vary from country to country with specific success conditions, the creation or adoption of new specific approaches adapted to the local societal needs requires some technical capacity and infrastructure, which can be provided from experts' technical knowledge [28,134]. Therefore, a team of experts made of land use planners, land surveyors, estate valuation surveyors, land administrators, land managers, agricultural engineers and agronomists, lawyers, socio-economists, agroecologists and environmentalists need to be in place to assist the farmers in the preparation and execution of the suitable land consolidation projects tailored to the local needs [12,16,28,30].

(**b**) *Land Banking:* It is explained as the process of transferring and acquiring the ownership of small parcels from small farmers to big farmers to enlarge their holdings through sales, and/or to the government or private investors through expropriation procedures in order to use them as land funds (land banks) for the development of infrastructure and land buffer during land consolidation projects, with an overall objective of creating more operational and viable farm units [18,98,125,127]. Land banks provide an opportunity for expansion and shaping of farmlands, and the creation of adjoining infrastructure [18,125,127]. It follows the principle of ownership exchange, and targets to eliminate the size related land fragmentation problems and reduce the number of boundaries and its related conflicts [20,21]. It has been implemented in Western Central European countries like Germany and the Netherlands, often integrated in large land consolidation projects, although recent studies have also found it suitable to the Eastern and Central European land fragmentation problems [4,19,21,98,125,127]. It can be voluntary by old farmers to young active farmers willing to enlarge their farms, or compulsory through governmental agencies for agriculture and infrastructure developments projects (ibid).

(**c**) *Voluntary parcel exchange:* It involves the exchange of distant non-contiguous parcels from the farmsteads among two or more landowners, resulting in more compact farms from adjacent parcels of each landowner with more efficient spatial layout [30,98]. The main target is to reduce the distance related costs, irregular shapes, and the number of boundaries by decreasing the number of scattered plots per farm under the circumstances of subsistence economies and scarce land. This strategy has been used with success in smaller land consolidation areas with a limited number of farmers in Germany (§ 103a FlurbG-) and the Netherlands (through a notarial agreement) where the primary benefit is in agriculture, and can be combined with land banking activities for its effectiveness [17,21].

More recently, it has also been considered for areas with other land uses than agriculture, most notably to suit the preservation of nature and merging ecological areas.

(d) *Restrictions of the minimum parcel size subdivision, Joint ownership and Cooperative farming*: For the purpose of reducing the negative effects of small farm sizes, different countries have established the legal provisions restricting the subdivision of parcels into small non-economically viable units and partible inheritance, thereby encouraging their joint ownership by many co-owners or heirs, and their cooperative farming. In *joint ownership*, a big piece of land is owned by many co-owners but operated by one or few farmers, where in many cases one of the co-owners (heirs) later buys the shares of other heirs, or the co-owners lease their shares to one big farmer or cooperative (tenants) under specific use rights, restrictions and responsibilities [30,97,135,136]. Farmers may prefer the subdivision of the title over a piece of land in terms of shares without affecting its physical characteristics (ibid). With regard to the *cooperative farming*, a group of farmers jointly operate one big co-owned or leased farm together or agree to cultivate one type of crop on their small plots in a given area in order to create big and more economically operational and viable farms. In both cases, farmers retain their rights over land, whereas in the latter case, the boundaries of their parcels are kept intact, which has been found as a barrier to agriculture mechanization since it is difficult to move the machinery on small separately owned plots with many boundaries [3,97,98]. This is the case of Rwanda, where the article 30 of the law governing land forbids the subdivision of agricultural and livestock land which would result into small pieces of less than 1ha, thereby encouraging the joint ownership of such parcels and their cooperative farming through *land use consolidation (LUC) program* or cultivation of the same priority crop [26,33,117] as explained above in Section 3 (a). However, although these strategies have been used with success in many countries (i.e., joint ownership in Taiwan) with subsistence economies and growing population under land scarcity conditions to tackle and reverse the problems of land fragmentation [136], different studies report their failure in countries like India, Nepal and Rwanda, as a result of the reluctance of farmers against them [8,24,30,32–34,97,98,123,137,138]. These studies decry these strategies to obstruct/deprive the full enjoyment of use rights over land for independent purposes, thereby inducing many ownership and use related conflicts viewed as a result of spatial injustices [139] leading to tenure and food insecurity in cases of compulsory participation and compliance to them. Moreover, the minimum parcel size subdivision restrictions have been criticized to lead to hidden ownership fragmentation thereby increasing farmland use fragmentation and the ownership and use related conflicts over land [32,33].

(e) *Land realignment*: It refers to the adjustment of land boundaries between two or more land parcels with the aim of remedy to the existing encroachment problems and or land management problems. It only implies minor changes in boundaries structure of adjacent plots thereby affecting the changes in sizes and shapes of parcels. It has been applied with success under the circumstances of internal fragmentation with contiguous parcels under the same operatorship to eliminate the problematic boundaries for the purpose of merging small plots into larger economically operational units [20,28].

Besides the above commonly known strategies to control the problematic land fragmentation, various studies have documented several other socio-economic and agronomic strategies to optimize the benefits from fragmented holdings by minimizing their defective effects on agriculture production without their elimination. These include different agriculture intensification programs (intensive use of labor and inputs in small heterogeneous farms and parcels); risks management strategies (agricultural insurance, agroecological approaches, food storage, pests control measures, credits, high yielding and resistant varieties) [11,30]; on-field harvest sales and off-farm employment [30,130]; the relocation of very distant farmsteads to close the best farms [97]; and many different case specific strategies parallel to the rational farmland fragmentation conservation under different circumstances like the cases of consolidation of one agricultural use type or crop (land use consolidation in Rwanda and viniculture in Europe) [16,24,27,120] as explained above in previous paragraphs.

Despite the variety of these documented land fragmentation control strategies, land consolidation has been broadly and commonly used as a panacea to this quandary, regardless of its different forms and specific cases [21,33]. Although both land consolidation and land fragmentation are theoretically considered as land management instruments, the dominant discourse in the literature and practice presents a common weakness of tending to show the traditional land consolidation as the appropriate tool and solution to land fragmentation problems thereby ignoring the possible benefits of the later [30,33]. However, some studies revealed that land consolidation alone does not solve all land fragmentation problems. Whereas it is suitable to address land fragmentation problems of Western European and Scandinavian countries in areas characterized by big farms with many irregularly shaped and spatially dispersed parcels (internal fragmentation), it does not suit the Central and Eastern European countries which have many small farms (small size problems as an indicator of tenure fragmentation), and failed to be adapted to some African and Asian countries with complex traditional land tenure systems [20,21,28,33,88,109]. Furthermore, empirical evidence has critically proven that it tends to favor (benefit) big farmers with many scattered plots by increasing their income from agriculture at the expense of small farmers with small plots as an important pitfall, probably due to the diseconomies of scale [27,30,119,121]. In this vein, Nilsson [27] and Muyombano and Espling [121] found land use consolidation (LUC) not relevant to the fragmentation problems of small farms in Rwanda. Besides that, it has been largely criticized by many researchers for resulting in the loss of employment in case of its introduction in densely populated countries with subsistence economies thereby leading to the increase in rural urban migration, the loss of agrobiodiversity, and food insecurity through monoculture [30,31,37,38,124,133,140–142]. For this, land banking, voluntary parcel exchange, land realignment, joint ownership and cooperative farming have been proposed as suitable strategies for other land fragmentation problems than internal fragmentation of big farms [12,16,17,20,21,26,98,127]. Apart from that, Bentley [30] and Blarel et al. [31] argue that the problems of land fragmentation should be eliminated by focusing on fighting its root causes through curbing the population growth, creation of off-farm employment, and increasing agriculture technology.

However, in spite of the large body of literature about land fragmentation control strategies, only few studies explicitly address how, when, where and why different land fragmentation forms and specific control strategies can be inter-related and mutually conducive. The hesitation to study these interlinkages are connected to the inherent complexity and country-specificity of land fragmentation problems. These studies argue that land fragmentation issues are complex and vary from country to country and case to case with strong dependency on local social, economic, political, cultural, agricultural, agroecological and environmental conditions. Hence, there is no standard objective strategy or measure to control this phenomenon, nor is there a successful transfer of specific strategies in different areas with different characteristics [12,19–21,28,127]. This makes it difficult to objectively compare and assess the effectiveness of these strategies [108]. The empirical evidence revealed that the failure to consider the local conditions prior to the transplantation of land consolidation programs has previously led to their failure in some African (Kenya, Malawi, Tanzania) and Asian countries dominated by customary and communal land tenure systems [12,18,21,29]. Therefore, there is a need to take into account specific local land fragmentation forms, their causes and impacts (problematical and beneficial) under specific conditions, and analyze the similarities and differences prior to any attempt to transfer any fragmentation control strategy among different areas, and/or create new progressive tools and responsible approaches suitable (adapted and updated) to the existing dynamic local conditions [12,18,28,143]. The idea behind is that a successful strategy in one area might not succeed in another due to the differences in operational conditions. One needs to know the factors of its success prior to its broad transplantation elsewhere. Abubakari et al. [12] in their study on land consolidation in the Ghanaian customary lands strongly argue that the success of any land consolidation program depends on the suitability of local conditions with its baseline conditions, with respect to land characteristics like its tenure, use, value, location, size and shape. For them, the information about the convergence or divergence of these conditions needs a careful feasibility study in specific areas under consideration. Bentley [30], Van Dijk [18], and Van Dijk [21] note that land fragmentation is minimized or reduced, when the number of owners, users or farmers (tenants) and farms in a given area (tenure fragmentation) declines, the number of irregularly shaped parcels per farm and the overall distance between them and the farmstead (physical fragmentation) drops, the number of uses/crops per farm (use fragmentation) declines, and the number of farmers who are operating/using their own lands (discrepancy between usership and ownership or tenure fragmentation) increases.

To this end, our review of the defragmentation strategies shows that land consolidation instruments are suitable to control internal (location and shape) land fragmentation of big farms through the creation of compact farms with one or few close regularly shaped big parcels, and tenure fragmentation (ownership and usership) through the creation of compact farming structure in a given region by reducing the number of owners and increasing farm sizes with regular shapes). Voluntary parcel exchange suits for internal fragmentation of small farms through the reduction of distances between parcels and homestead. Land banking is suitable for size or tenure fragmentation to reduce the number of farms/owners and increase the farm sizes in a given area, while cooperative farming is suitable for internal fragmentation in case of boundaries and shapes realignment through the joint ownership, and tenure fragmentation in case of consolidation of one use type or crop as it happens in land use consolidation in Rwanda and viniculture in Europe. The risks management strategies (insurance, resistant varieties, etc) and on field sales can be used to minimize internal fragmentation. agriculture intensification programs (inputs and labor use intensity) and off-farm employment can be suitable for reduction of land tenure fragmentation problems, while agricultural land protection policies can be suitable to prevent and reduce land tenure and size fragmentation problems. Finally, the restrictions about partible inheritance, minimum size of parcel subdivision and absentee landowners, the prevention of transfer to non-farmers and leasing suit for dealing with land tenure fragmentation, whilst the imposition of a maximum limit on the holding size suits for preventing internal physical fragmentation [4,16,18–20,30,98,120,127,144,145]. These strategies can be categorized into two groups of preventive (legal provisions and protection policies for agricultural land) to spot the root causes of fragmentation), and mitigation (land management approaches, socio-economic and agronomic measures) to manage the impacts of an already occurred fragmentation.

Recognizing both the potential benefits and problematic situations of land fragmentation, Bentley [30] and Asiama et al. [17] suggested a specific model of land consolidation in blocks or localized land consolidation where only spatially dispersed parcels within the same microzone with homogenous soil and agroecological conditions are consolidated. This helps to conserve and give farmers access to all types of parcels in different sites with diverse microclimates and growing conditions, for both increasing the agricultural production efficiency as well as crops diversification for risks and labor management and food security purposes through food sovereignty at the local level. In this case, land consolidation does not necessarily result in a single parcel, rather in few parcels located in different sites to keep the topographical advantages of fragmentation. Likewise, Cholo et al. [41] proposed a consolidation of small parcels into larger heterogeneous plot clusters to enhance food security by exploiting synergies between agroecological adaptation practices and land fragmentation. Adversely, Ntihinyurwa et al. [33] proposed a consolidation approach which provides farmers with single contiguous farmlands or parcels that can be cultivated with multiple crops to answer the desire to meet food diversification, risks management, labor bottlenecks management as land fragmentation claims, and agriculture production efficiency by minimizing the time and travel costs as land consolidation claims. For this, there is a need for a strong objective land capability and suitability classification prior to the development of any local specific land consolidation approach. Figure 2 summarizes our findings on various documented instruments, strategies and policies to control different problematic land fragmentation scenarios.

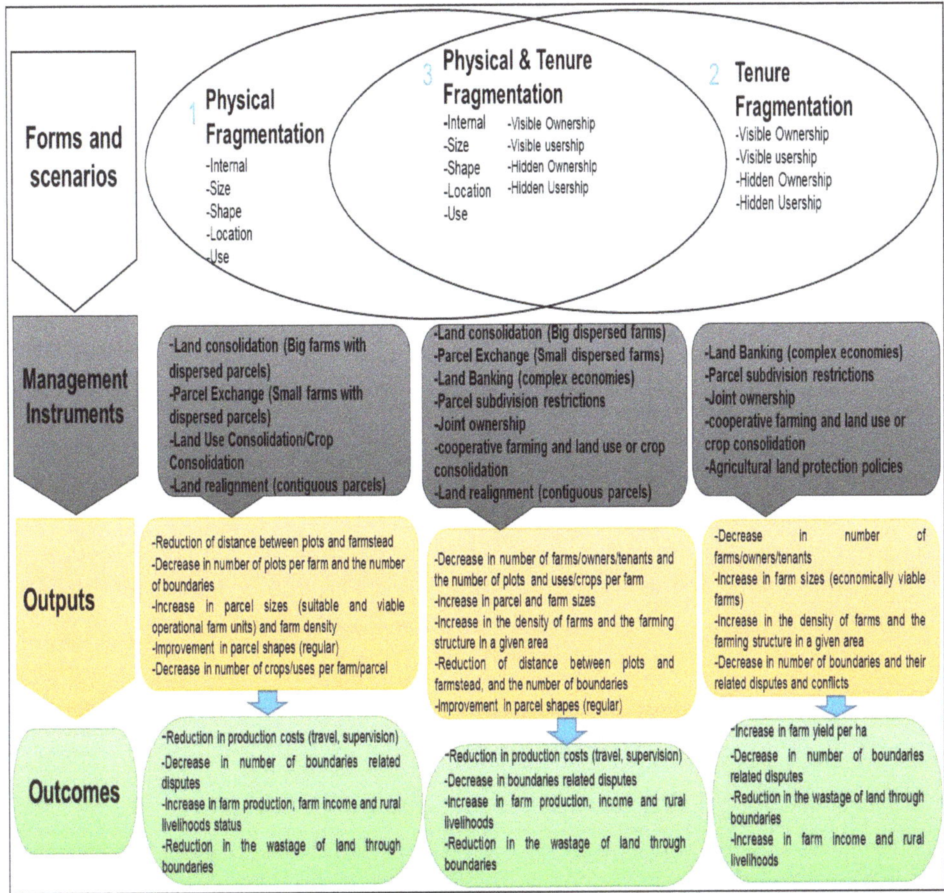

Figure 2. Synthesis of the problematic land fragmentation control strategies (instruments). **Source**: Developed from the reviewed literature

In summary, both the problematic and beneficial or rational land fragmentation scenarios need a certain level of management in order to optimize the income from agriculture. However, the complexity of this phenomenon makes it difficult to choose the suitable instruments (strategies) for specific circumstances, which calls for trade-offs among different alternatives and their right combinations under various local conditions. For this, the analyzed literature has on one hand revealed that the problems related to *physical land fragmentation* (internal, location or distance, shape, use, plot or parcel size and boundaries) can be minimized by land consolidation for large heterogeneous farms (under complex strong economies), voluntary parcel exchange and on-field harvest sales for small heterogeneous farms (under moderate and subsistence economies), land realignment for homogenous farms (contiguous plots) to eliminate and reorganize boundaries, and land use consolidation or crop consolidation for multiple uses on small plots and farms. *Land tenure fragmentation* problems (ownership, usership, small farm sizes, and boundaries) are reduced by land banking for small farms under complex strong economies, parcel sizes subdivision and absentee owners' restrictions, joint ownership, cooperative farming, land use (crop) consolidation, and agricultural land protection policies for small farms under moderate and subsistence economies. Furthermore, the introduction of insurance systems and mechanization in agriculture, and the market perfection (for food and labor) have been used as strategies to eliminate the fundamental reasons for internal fragmentation in complex strong economies with market-oriented agriculture (use of multiple zones for production diversification, risks and labor management strategies). On the other hand, different agriculture intensification programs such as the combined use of soil mineral and organic fertilizers and amendments, pests control measures, labor use efficiency and intensity, and high yielding and resistant crop varieties have been documented as suitable strategies to maximize the income from agriculture on fragmented plots under the beneficial or rational fragmentation, subsistence and moderate economic conditions (market imperfections) for risks management, labor schedule, production diversification and control of ownership and use related conflicts over land. In some special cases like mountainous areas, the costs of alleviating land fragmentation may far exceed its benefits [11,30]. In such cases, keeping fragmentation is more beneficial than its alleviation (ibid). Therefore, Bentley [30], Van Dijk [18], Van Dijk [21], Asiama et al. [28], Asiama et al. [17], and Ntihinyurwa and de Vries [11] suggest that any attempt to control land fragmentation should consider different local and case specific social, cultural, economic, political, environmental, and agroecological conditions of an area, and the benefit cost analysis to guide the decisions about the suitable strategies for the sake of their success. In this regard, there is a need to develop strategies which simultaneously minimize the defective effects of fragmentation thereby optimizing or without jeopardizing its potential benefits [33,75] for food security purposes. Local agroecological approaches tailored to the needs of local peasants (farmers) should be given a key place in the management of local farmland fragmentation scenarios of subsistence communities for sustainable agriculture production, farm resilience, self-sufficiency in culturally acceptable (desirable) food production, increase in food sovereignty and the household food security motives. The next section discusses this food security concept.

4. Food Security as a Multidimensional, Multilevel and Multisource Concept

The concept of food security has been variously defined over time across different disciplines for particular interests and goals at different spatial levels and social scales. Chigbu et al. [32] and Dam Lam et al. [146] found that by the end of 2015, there were more than 200 different definitions of food security. However, despite the subjective and sometimes contradictory conceptualizations of the term, most of these definitions are oriented towards the supply of sufficient (enough) food availability (quantity of calories) at all times (stability) to meet the needs (demands) of the growing population from domestic and wild production, stocks, food imports or purchase from the markets, and food aids [32,54,147–151]. The majority of them were following the 1789 Malthusian food availability theory, stipulating the balance between the population growth and food availability (food growth rate should not be below the population growth rate) at the macro and meso spatial levels (community, regional, national, global). This tendency persisted till the introduction of Sen's theory of food entitlement in 1981 stipulating the notions of access, affordability, allocation (distribution) and utilization of food at the micro spatial levels (household and individual) [54,64,65,68,150]. With an attempt to reconcile different conceptualizations of this term, FAO [69] from the World Food Summit (WFS) in November 1996 developed a more comprehensive widely accepted definition of food security as a status/situation: "when all people, at all times, have physical, social and economic access to sufficient, safe and nutritious food which meets their dietary needs and food preferences for an active and healthy life", and vice-versa for food insecurity. The same definition was extended later in 2009 in the world summit on food security, where the four pillars (dimensions) of food availability, food accessibility, food usage (utilization) and food stability (sustainability) through which food security can be measured at both national, regional, community, household and individual levels were linked to this concept, while the nutritional dimension was added to it as an integral part [33,54,65,146–148,150–152]. With respect to the nutritional dimension, the concept implies that food and nutrition security is achieved when adequate food (in terms of quantity, quality, safety, and socio-cultural acceptability as components of food security) is available and accessible, and satisfactorily utilized by all individuals at all times to live a healthy and happy life [54,67,150,151]. This stipulates the consideration of the aspects of availability, accessibility and stability of food of acceptable quantity, quality, safety and diversity, based on the social and cultural preferences of any society or an individual at all levels [67]. Whenever one of these aspects is not met, people may suffer from hidden and visible hunger [33] and malnutrition, which negatively affect the health and livelihoods of the population. Since the aspect of availability in this definition stands for the supply of enough food of acceptable quality and quantity to broadly meet the demands of the population, it is mostly used to measure food security at the meso and macro levels (community, global and national); Whilst the accessibility and utilization entailing the capacity of individuals or households to meet their preferred food needs for an active healthy life, stand for the micro levels (household and individual). The same concept of food security highlights the chronic and the transitory food insecurity at all these levels/scales, as a result of instability of all the other aspects/pillars. The household food security is the application of this concept to the family level, where individuals within the households are the hub of concern [33,54,65,69,148,151,153].

The achievement of food security at all levels following the FAO definition is function of different factors including the economic status of the household, socio-cultural norms and values, demographic characteristics, agricultural system, education level, and environmental and agroecological characteristics of the area, to cite only few [32,33,65,69,154]. Surprisingly, the popular logic of achieving food security has over time focused on reducing the population growth through different family planning policies, and boosting agriculture production to keep the balance between the food demands of the growing population and food availability (supply) at the macro and meso levels (national, regional and community), thereby ignoring its entitlement and sovereignty at the local (household and individual) levels [46,47,54,65,68,149,155]. However, since food security is a very complex, multidimensional and multilevel concept, difficult to achieve in silos, this can only be possible if other external economic, socio-cultural, political, agroecological, and environmental factors are overlooked. Food security entails more than growing enough food, since it implies the demand for it, as well as the supply, the quality as well as quantity, diversity as well as accessibility, an adequate diet (culturally acceptable quality and quantity) today and assurance of one tomorrow [11,32,33,54,65,67–69,148,150,153]. It has the aspects of quality, access/affordability, acceptability, utilization/usage and stability/sustainability which can only be achieved when everyone in the household has access to regular, safe, nutritious and enough acceptable food to meet his/her food preferences [11,32,33,65,68,69,150]. Therefore, in the existing critical context of the ever growing fluctuations in climate and food prices which directly affect the household's food acquisition (domestic and wild agriculture and animal production, purchase, aids, and imports) and allocation (distribution and usage), and food safety concerns, the achievement of food security at the micro levels requires the change of food production paradigms. This needs the shift from the mass food production systems through conventional agriculture and monoculture, and consumption patterns prioritizing the quantity and availability, towards more diversified and locally produced food stuffs through sustainable, climate or natural risks resilient and smart agriculture systems, following various agroecological approaches including the polyculture (growing wide diversity of food crops in space and time) [11,32,33,37,41,45–48,71,156,157]. This can help to sustainably meet the cultural dietary needs and food preferences of acceptable quantity, quality and safety for all local people, as the suitable method of achieving food sovereignty, an adaptation strategy to the existing climate change realities for ending hunger and malnutrition, and local approach of meeting food security stipulated by many policy initiatives and goals like the SDGs 1.4; 2.1,2,3,4,5; 12.2; 13 & 15.3,4,5,9 [11,32,37,38,41,46–48,71,141,155,157–165].

Being the main factor of food production in many countries, the agriculture production of enough staple food crops as the basic component of food systems for food security (food supply side) requires the focus on agriculture intensification of small scale farms or agroecological strategies on fragmented land, and agriculture expansion of large scale farms on consolidated land, to meet the local needs and food preferences of the growing population. Recent studies and social movements advocate for the achievement of local food security by focusing on the concept of food sovereignty, which stipulates the self-sufficiency and autonomy in food production by local small scale farmers through various agroecological methods and agricultural systems tailored to their needs, knowledge, cultural values and traditions, and other particular circumstances [46–48,155,165]. Nonetheless, this does not alone guarantee the complete solution to the problem of food insecurity, since other aspects like food utilization and food market entail more than that [32,65,150]. The evidence has shown that food insecurity may exist in cases of high availability and accessibility of food in sufficient quality and quantity, mainly due to the lack of knowledge about the right preparation and combination of balanced nutritional diets, and the basic health and hygienic services like clean water (ibid). To this end, one needs to focus on a holistic and careful assessment of food security status, by considering all its underpinning factors at all levels. Figure 3 summarizes these various factors of food security

As Figure 3 summarizes, food security as a multidimensional and multilevel concept cannot only be achieved by a single instrument. It requires a holistic approach which considers the contributions of different factors at different levels of analysis to create food systems that offer the possibilities to meet the availability of qualitatively and quantitatively acceptable food in a given area, accessible (affordable) to all people, with the best and balanced combinations (utilization) to meet the nutritional diets/needs and food preferences of the ever growing population at the regular basis (sustainability/stability) with scarce resources or production factors (land and capital) for an active and healthier life. Considering the growing challenges of climate change and other natural shocks from food production side, an attention should be focused on the trade-offs between the role of some agroecological principles like crops diversity on food stuffs diversification as a source of qualitative, sustainable, acceptable and resilient food systems on one hand, and the quantity of agriculture production to meet the food needs and demands of the growing population irrespective of its quality on the other hand, on either fragmented or consolidated land parcels at the local levels (community, village, household and individual) [11]. The growing tendency is that poor people are choosing to compromise to food quality and quantity aspects for the benefit of food stability in case of shortages of food availability and accessibility as a result of climate change and price fluctuations realities, by creating more sustainable resilient farms through local agroecological farming systems [46,47,156]. The next section establishes the relationship among farmland fragmentation, farmland consolidation and food security concepts.

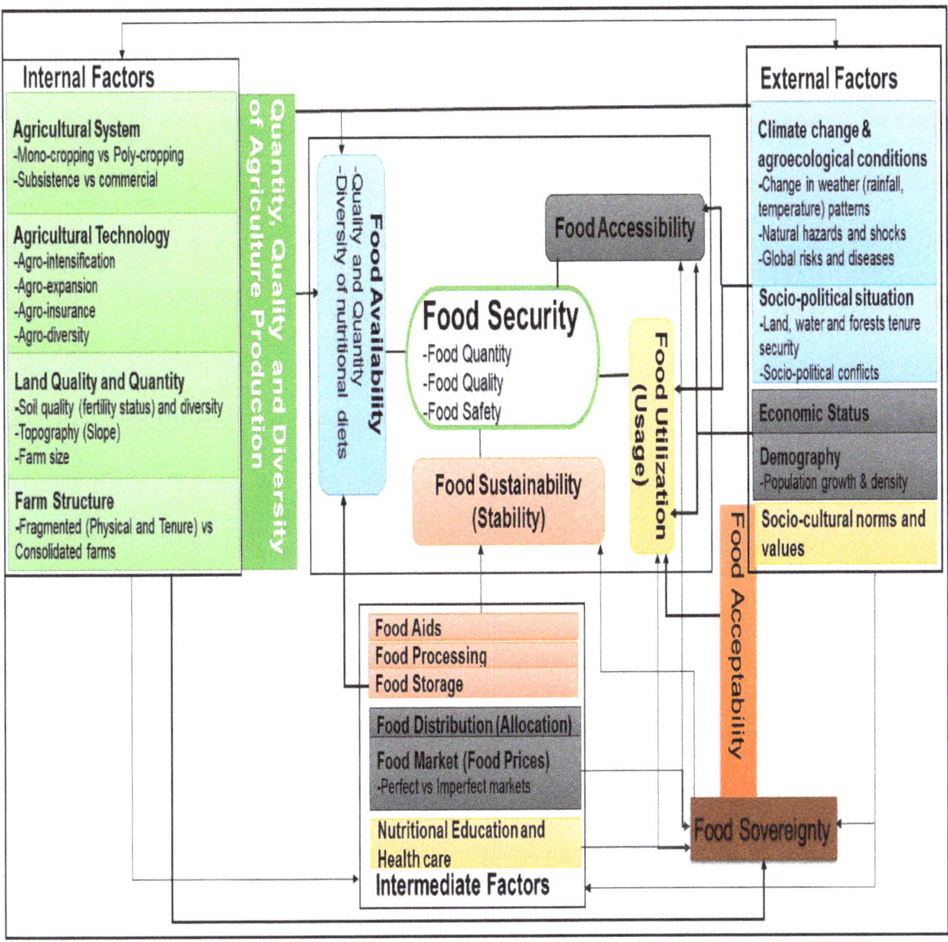

Figure 3. The multidimensionality of food security as influenced by internal and external factors. **Source:** Developed from the reviewed literature

5. Farmland Fragmentation, Farmland Consolidation and Food Security Nexuses: Relationships, Overlaps, Research Gaps and Future Perspectives

5.1. Relationships, Overlaps and Research Gaps

Farmland fragmentation and farmland consolidation are two interlinked concepts theoretically considered as instruments of agricultural land management for food security purposes [11,33,51]. In this vein, regardless of the fragmentation or consolidation statuses and scenarios, the farmland remains a fundamental asset for food security [33]. However, while there is enough empirical evidence and substantial literature about the relationship amid land fragmentation, land consolidation, land productivity, agricultural production, and farm profitability and efficiency, only few disaggregated studies address the linkages between farmland fragmentation, farmland consolidation and food security. Furthermore, from our critical review, there is a lack of comprehensive studies that have documented the linkages between the two concepts and food security as a multidimensional, multilevel, and multisource concept. Therefore, this section builds on the existing disaggregated studies about these three concepts and adopts the conceptual reasoning approach which follows the rationalist theory to fill this litterature and knowledge gap through the analysis of different related theoretical connotations.

Foremost, a large category of studies shows the physical land fragmentation (internal, location, shape, use) as a defective phenomenon, and a major threat to agriculture production, farm productivity and profitability, by hampering farm efficiency and the economies of scale, thereby positing land consolidation approaches as a panacea to this problem [2,3,6–8,10,12,14,29,52,66,89,90,105,107,118,166–174]. These studies broadly argue that farming on small scattered and irregularly shaped plots increases the travel and supervision costs due to long distances between parcels and the household, which reduces the yield per hectare, farm profitability, and abandonment of farming activities on very distant parcels in some cases. Furthermore, regardless of the distance and adjacency of parcels, land tenure fragmentation leads to small farms with small non-economically viable land units, which in turn hinders the economies of scale, since the mechanization and expansion of market-oriented agriculture is not tenable on such units [4]. This reduces the quantity of agricultural production and food availability or supply in a given area as a pillar of food security. Therefore, the majority of these studies propose land consolidation as the appropriate solution to these issues. They back this position by showing how consolidated and compact big farms with larger farm units and parcels or plots decrease the agriculture production costs and increase the yields per hectare and farm profits as the key characteristics of the economies of scale, which in consequence positively impacts on the supply (availability) and sustainability of enough quantity of food to meet the food demands of the population for food security in a given area. For Lerman and Cimpoieş [118], big consolidated farms offer higher agriculture production and econimic performance than small fragmented ones, which incresases the food security status and the well-being of the rural population. From the proponents of this view, the quantity of agriculture production of food crops matters most, in order to satisfy the food demands of the growing population following the Malthusian theory of population growth and food availability or supply of 1798 [54] and fill the gap stipulated by the reverse relationship between the population growth and the limited food productive capacity of land resources in this theory. Likewise, this claim is shared by the advocates of other alternative strategies like land banking, cooperative farming, and joint ownership against the tenure or size fragmentation of farmland. They tend to believe that food security of the growing population can be met by producing enough quantity of food crops through agriculture mechanization and the economies of scale, which can only be achieved on consolidated or big farms [21,106,109,125,127,136,175].

Nevertheless, different studies have criticized some farmland consolidation approaches to lead to the establishment of monoculture systems which result in production of single or few types of food stuffs, thereby negatively affecting food diversity and balanced nutritional diets and inducing food insecurity [30,31,37,38,176,177]. This is the case of Land Use Consolidation program in Rwanda, criticised of worsening food insecurity issues by promoting the monoculture system at the expense of multicultural one and its irreplaceable adaptive benefits, through the reduction of agriculture production diversification as a source of food diversity aspect at the household level, despite its major outcomes in terms of boosting the national production of 8 priority crops grown in this program [24,27,32–34,119,121–123,137,178]. These studies posit that the availability of enough quantity of food of some priority crops at the national level through LUC does not necessarily mean that the needs and food preferences of households members are met, while the practical evidence has revealed the increase in households vulnerability to food insecurity since the introduction of this program in 2008 (ibid). Combined with the consequences of climate change (droughts and floods from changes in rainfall patterns) and imperfect food market, this LUC program has been pointed out to worsen the problem of food insecurity at the household and individual levels, by reducing its quality, accessibility, acceptability, and sustainability aspects in some parts of the country [32,33].

Despite these findings, the debates over what to do with farmland fragmentation, farmland consolidation and food security have often been disassociated from those related to climate change and agroecology, and only linked with agriculture production, due to the presumed negative impacts of farmland fragmentation by policy makers. Following the multidimensional nature of food security, the prevalent justification for land consolidation is that it increases the farm size and reduces the producion costs associated with the distance, and thus contributes to food security given higher quantities in food production from food crops. Nonetheless, this logic makes sense when food security is viewed from the lens of quantity, since food security is much more than the quantity of food. It includes the aspects of diversity, quality, access, sustainability, acceptability and utilization of food [64,65,69] along with food sovereignty [46,47,155], which are also achievable under the conditions of land fragmentation scenarios [11,32,33,66].

Therefore, a different category of studies witnesses the evidence of positive relationships between physical farmland fragmentation and food security. These studies argue that physical farmland fragmentation contributes to the improvement of the aspects of food quality through the diversity of nutritional diets, and the regular (sustainable) availability and accessibility of food at the household and individual levels under the conditions of subsistence economy, climate and food prices fluctuations, and vice versa for farmland consolidation [11,30–34,37,38,41,42,50,57,94,122,141,157,176,177,179,180]. These advocates of this standpoint commonly argue that farming on spatially and topographically fragmented and dispersed parcels with irregular shapes offers farmers the possibilities to grow a wide range of diverse crops in areas with different crops suitabilities and growing conditions for the purposes of food stuffs diversity production, farm resiliency, and the management of risks of climate change and food prices fluctuations. This in turn increases food diversity, quality, accessibility, acceptabiliy, and sustainability of subsistence households, thereby inducing the likelihood of meeting food sovereignty and food security at the local levels [11,46], following the economies of scope, resilience, agroecology, and complexity theories. In the same vein, Blarel et al. [31], Alexandri et al. [179], Ciaian et al. [37], Cholo et al. [41], Knippenberg et al. [45], Ntihinyurwa et al. [33], and Ntihinyurwa and de Vries [11] advance that farmland fragmentation leads to the cultivation of diversified food crops and the production of a diversity of food basket for self-sufficiency of subsistence farmers in order to meet their nutritional demands and food security at cheap prices, as the cheapest strategy to achieve the household food security under the circumstances of climate change, land scarcity and food market imperfections. This claim coincides with the advocates of agroecology and food sovereignty, which posit the achievement of food security at the local levels through self-sufficiency and autonomy in food production tailored to the needs of cultural and traditional diets of local subsistence farmers using various local agroecological methods (temporal and spatial crops diversification through polycultures, and the knowledge of local peasants) on more resilient small scale farms [46–48,70,155,165]. The collective of these studies stipulates that, the more the differences and high diversity or heterogeneity in land and soil qualities; the higher the variety of soil-crop suitability classes and production potentials; the higher the crops diversification (agrodiversity), farm resiliency and food stuffs diversity; the higher the self-sufficiency in food production, the higher the nutritional balance; the higher the food quality and sustainability; the higher the food acceptability and sovereignty; the higher the food security [33]. Furthermore, contrary to the principles of the economies of scale theory, the proponents of this view counter argue that land tenure or size fragmentation (small farm sizes) backs the diseconomies of scale theory stipulating the inverse farm size and agriculture production relationships, following the Boserup's theory of population growth and agriculture intensification of 1965, probably due to imperfections in labor market in subsistence economies, and the growth of technology in agriculture [31,49,50,63,181,182]. This implies that the intensification of agriculture leads to better outputs in terms of agricultural production on small farms than on bigger ones, which directly impacts on food availability (quantity and quality). TWN and SOCLA [155] argue that small farms are more productive than large farms, if the total output is considered rather than yields from a single crop. Nevertheless, this Boserup's theory stipulating the proportional relationship between the population growth and agriculture intensification [11,63] has shown its limitation at a certain critical threshold of very high population density, thereby giving a reason to the Malthusian theory in such circumstances [55,56].

However, in case the consolidation practices offer to farmers the options of growing multiple crops on consolidated plots (voluntary land consolidation models), and the provision of agricultural insurance services and resistant crop varieties, there are no more reasons for keeping fragmentation. Such consolidated parcels lead to high agriculture production of diverse crops, which in turn results in the regular and adequate availability and accessibility of food of acceptable quality and quantity, thereby contributing to the improvements in household food security [4,16,30,111,157]. Moreover, comprehensive land consolidation models may integrate some specific programs of food processing, food storage and nutritional education to contribute to the improvement in food quality through more balanced nutritional diets and food accessibility and stability aspects as a support to food availability, to meet the household food security in its multidimensional conceptualization [4,16].

Besides the effects of farmland fragmentation and consolidation strategies on food security, the status of the latter may also determine the kind of decision about the fragmentation management approaches. Since the primary objective of consolidation approaches is to increase the food security status by sustaining food availability (supply) to meet the food demands of the growing population through agriculture production, these approaches may not be necessary in case of the lack of food insecurity problems in a given area (when food security already exist under fragmentation scenarios of big farms) [4,18,21,30,31,37,38,41,50]. This is the case of countries with abundant land and low population densities like the USA, Russia, Canada and many others.

Finally, the critical review has drawn the reciprocal relationship between farmland fragmentation and farmland consolidation concepts. Farmland fragmentation is documented as a precondition and milestone for an establishment of any farmland consolidation program in a given area [4,12,18,19,98], and exists in an area which was previously consolidated according to the Gestalt theory of a whole [53]. Notwithstanding their reverse theoretical meanings, the two concepts share the same practical measurement indicators (see Ntihinyurwa and de Vries [75] for more details). Figure 4 summarizes the theoretical relationships between farmland fragmentation, farmland consolidation and food security. These relational linkages in literature lay a foundation for a theoretical model which could be adapted or used by scholars for building frameworks on the subject (Figure 4).

Simply put, Figure 4 shows that any farmland fragmentation and consolidation scenario which engages the multi-cropping system (agroecological approach) and agricultural intensification, food processing and storage, and nutritional education programs may lead to the achievement of food security at the household level, except in cases of lack of those intensification programs on small non-resilient farms. On the other hand, the consolidation programs implying the mono-cropping systems are susceptible to lead to food insecurity status through malnutrition (under and/or over nutrition), especially when combined with external factors like market imperfections, climate change, natural shocks, and the absence of the above-mentioned supporting programs. In this respect, the concepts of farmland fragmentation, farmland consolidation and food security are interlinked. The type of this interlinkage is determined by external factors like climate change, socio-economic status, agrobiodiversity (agroecology), demographic aspects, and land characteristics. Therefore, for the purpose of achieving the sustainable development goals (SDGs 1,2,12,13 and 15) stipulating the attainment of the multidimensional food security through sustainable (climate resilient) land management strategies (see Paragraph 6 of Section 1 for specific targets), any attempt to achieve food security through agricultural production should consider the importance of all the above-mentioned factors at the local levels for its success.

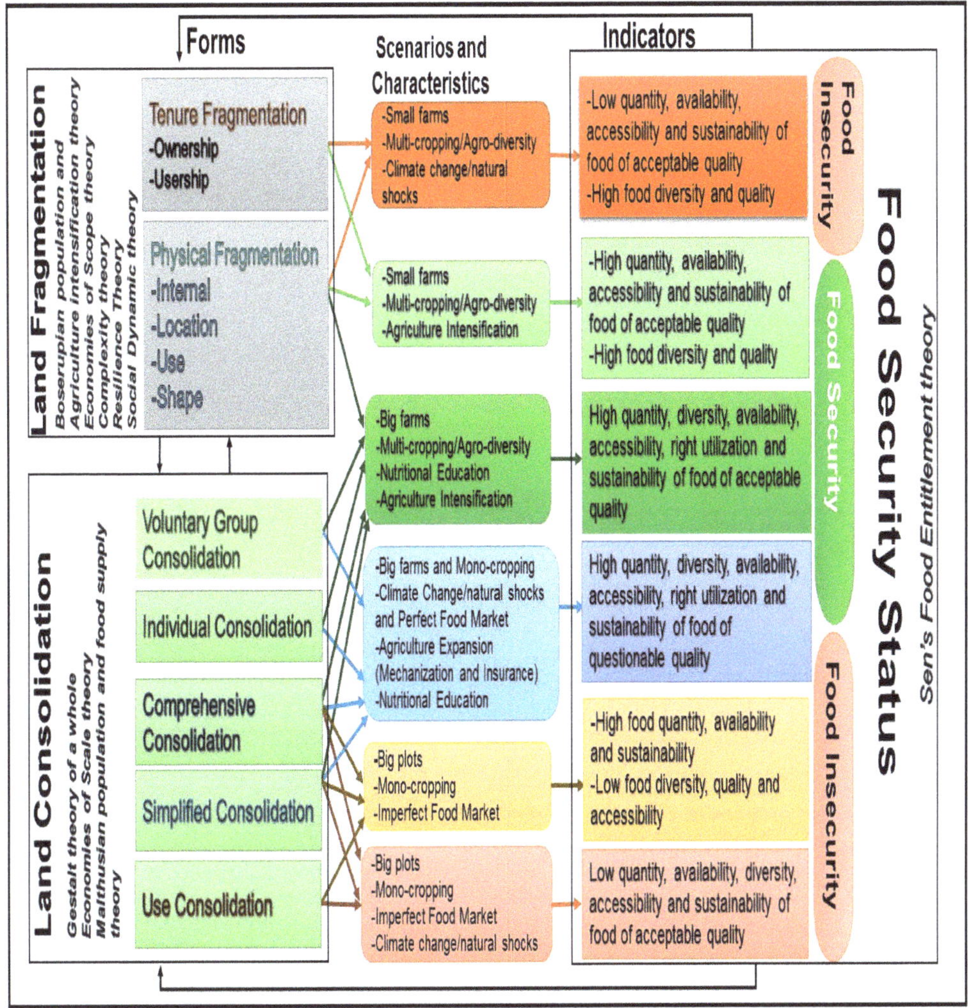

Figure 4. Theoretical relationships between farmland fragmentation, farmland consolidation and food security. **Source:** Developed from the reviewed literature

5.2. Reconceptualization of Farmland Fragmentation Management for Food Security

Recognizing the complexity, polarity and multidimensional nature of farmland fragmentation and food security concepts, there is a need to develop the local context specific and progressive farmland fragmentation management strategies, which consider both its defective and beneficial sides following the dynamic systems theory [84–86] and agroecological approaches (elements, principles, and methods) [47,48,155,165], rather than focusing on the blind subjective and irrelevant decisions of either defragmentation through different consolidation programs for food quantity and availability, or fragmentation conservation for food quality, accessibility, sustainability, acceptability and sovereignty purposes. Since our critical review of the literature has shown that both the defective and beneficial fragmentation forms may coexist in the same area, the identification of those forms, their causes and impacts, and assessment of the local social, cultural, economic, political, biophysical, agroecological and environmental conditions in a given area along with the benefits-costs analysis prior to any decision, would give an insight on the suitable combinations of strategies. This would further serve as an important guidance to policy makers and research scholars, and the best approach for the optimum management of this phenomenon. This position is theoretically and empirically supported by previous studies of Bentley [30], Van Dijk [18], Van Dijk [21], Asiama et al. [28], Asiama et al. [17], Ntihinyurwa and de Vries [11], and Ntihinyurwa and de Vries [75] in Sub-Saharan Africa and Europe. Being progressive, flexible and fit for specific situations and scenarios, this approach can accommodate different emerging solutions to new problematical situations, under the dynamic climate change realities and changes in local conditions in a given area. Building from this approach, we propose the following conceptual model for farmland fragmentation management in Figure 5 to refresh the existing sporadic and outdated conceptualizations about the management of this phenomenon, considering the major global threats of climate change, natural shocks, population growth, and urbanization. The model implicitly shows when, where, how and why one could opt for defragmentation or fragmentation conservation, through the hypothetical relationship between various farmland fragmentation scenarios [11,75], the proposed suitable management strategies or solutions, and food security and the general livelihoods of the rural (local) farming population. It results from a combination of variance and process models through abstract modelling techniques and the deductive logic conceptual reasoning approach [72–74,77,87]. After being empirically tested in different local areas, the outcomes from this model will be translated into suggestions for farmland fragmentation management strategic options under different specific local conditions.

For the sake of optimizing the income from agriculture and meeting food security, both the problematic and rational farmland fragmentation scenarios need a certain level of management. In this regard, as Figure 5 shows, the problems related to *Physical Farmland Fragmentation* (internal, location or distance, shape, use, small parcel and plot sizes, and boundaries) can be minimized by farmland consolidation in case of large heterogeneous farms (under complex strong or market-oriented economies); voluntary parcel exchange and on-field harvest sales in case of small heterogeneous and homogenous farms (under moderate and subsistence economies); land realignment in case of homogenous farms (contiguous plots) to eliminate and reorganize the boundaries; and farmland use consolidation or crop consolidation in case of multiple agricultural uses on small plots and farms. On the other hand, *Farmland Tenure Fragmentation* problems (ownership, usership, small farm sizes, and boundaries) can be reduced by land banking and off-farm employment in case of small farms under complex strong or market-oriented economies; restrictions about the parcel sizes subdivision and absentee owners, joint ownership, cooperative farming, farmland use (crop) consolidation, and agricultural land protection policies in case of small farms under moderate and subsistence economies. Furthermore, to prevent the worsening of the tenure fragmentation situation, where possible, the combination of these strategies with strong family planning measures that curb the population growth following the Malthusian theory of population and food supply, could generate good results [54–56].

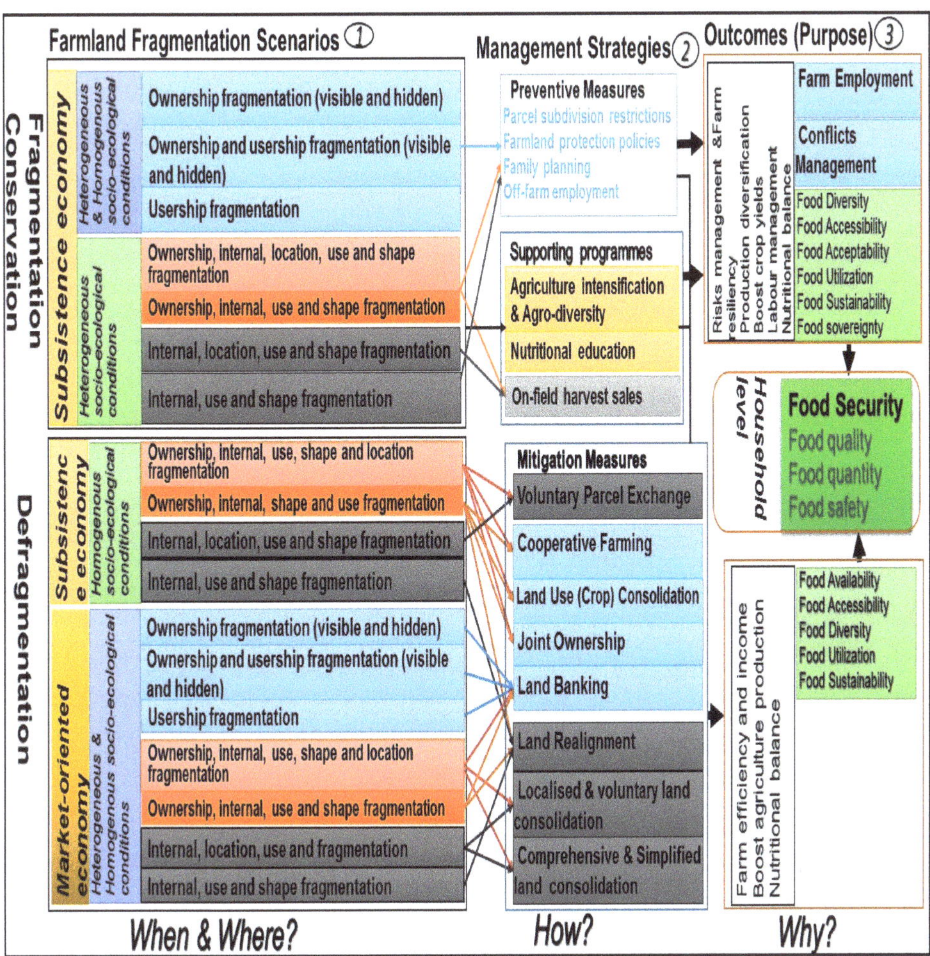

Figure 5. Farmland Fragmentation Management Conceptual Model. **Source:** Developed from the reviewed literature and the logic conceptual reasoning and modelling [11,73–75,77,81,87].

Considering the coexistence of both rational and defective fragmentation in the same area, there is a need to develop strategies that simultaneously minimize the negative effects of fragmentation thereby optimizing or without jeopardizing its potential benefits. To this end, the following specific consolidation models suggested by different researchers would apply in different specific cases after a careful benefits-costs analysis:

❖ *Land consolidation in blocks or localized land consolidation*: This model is suggested by Bentley [30] and Asiama et al. [17] in areas where only spatially dispersed parcels within the same micro-zone characterized by homogenous soil and ecological conditions are consolidated. This helps to conserve and give farmers access to all types of parcels in different sites with diverse microclimates and growing conditions, for both increase in agricultural production efficiency as well as crops diversification for risks and labor management, and food security purposes. In this case, land consolidation does not necessarily result in a single big parcel, rather in few big and medium size parcels located in different sites to keep the topographical advantages of fragmentation. This would apply to cases of physical (internal) fragmentation under subsistence and developing (middle-income) economies characterized by high heterogeneity of agroecological conditions. This fits with the consolidation model of small topographically dispersed parcels into larger heterogeneous plot clusters proposed by Cholo et al. [41] to enhance food security through the exploitation of synergies between adaptation practices and farmland fragmentation in Ethiopia. The voluntary group, simplified and individual land consolidation models, and voluntary parcel exchange strategies would also apply to this case.

❖ *Multicropping-based land consolidation approach*: Suggested by Bentley [30] and Ntihinyurwa et al. [33], this model provides farmers with single contiguous farmland parcels which can be cultivated with multiple crops to answer the desire for food diversification, production risks and labor bottlenecks management as farmland fragmentation claims on one hand, and agriculture production efficiency by minimizing the time and travel costs as farmland consolidation claims on the other hand. This applies to the cases of small parcels spatially scattered in the same topography with quite homogenous agroecological conditions, in both subsistence and strong economies for both food quantity and diversity (quality) purposes. The comprehensive, simplified and voluntary land consolidation models, along with land banking programs would also fit for this case, if farmers fully enjoy the use rights over their lands.

The success of these specific models requires a strong objective land capability and suitability classification prior to their development, based on a functional soil information system (SIS), which lacks in many developing and underdeveloped countries. Moreover, drawing from the study of Chigbu et al. [32] on tenure and food security responsive land use consolidation in Rwanda, the consolidation of land for agriculture expansion through market-oriented and monoculture based systems in more homogenous areas with less variability in agro-ecological, physical (soil, slope, water, etc), socio-economic, cultural, and climatic conditions for food quantity and availability; and the conservation of multiculture based systems on either consolidated or fragmented land in more heterogeneous conditions through various agroecological approaches as a risk management strategy, climate change resilience and adaptation strategy, and food crops diversification for food diversity and quality, accessibility and sustainability, cultural acceptability and sovereignty, could offer optimal solutions to farmland fragmentation and food insecurity problems.

In developed countries characterized by complex strong economies with market-oriented agriculture and perfect food, land and labor market, the fundamental reasons for internal fragmentation conservation (use of multiple zones for production diversification, production risks management, and labor management strategies) are always removed and compensated by the introduction of insurance systems and mechanization programs in agriculture [18,30,98]. In this case, keeping fragmentation would be useless. In contrast, different agriculture intensification and agroecological programs including the combined use of soil mineral and organic fertilizers and amendments, pests control measures, labor use efficiency and intensity, crops diversity, and high yielding and resistant crop varieties could be the most suitable strategies to maximize the income from agriculture on fragmented plots under the circumstances of beneficial fragmentation in subsistence and moderate economies characterized by high population densities and market imperfections. This can offer the benefits of risks management, labor schedule, agriculture production diversification, and control of land ownership and use related conflicts, thereby by increasing food sovereignty and the local (household and individual) food security, following the Boserup's theory of 1965 on population growth and agriculture intensification [63] below a certain critical threshold. This has empirically been evidenced by various studies in different countries [30,31,33,37,41,50,55,56,94,155]. Furthermore, in some particular cases like mountainous areas under subsistence economies where the costs of alleviating farmland fragmentation outweigh its benefits, keeping fragmentation would be more beneficial than its alleviation [11,30].

6. Conclusions

In the context of contrasting advocacies for farmland fragmentation, farmland consolidation, farmland use consolidation and food security nexuses, this study extends the discourse by explicitly and comprehensively displaying different conditions under which and how one could choose between farmland fragmentation conservation and defragmentation (consolidation) policies or both, as responsible farmland management tools to achieve food security. With logical reasoning, the study critically analyzed different documented farmland fragmentation scenarios, their problematical (defective) and rational (beneficial) situations under different circumstances, and proposed their suitable specific management models to achieve the multidimensional food security at the micro levels (household and individual) as a new contribution to the knowledge in the field of farmland management.

In contrast to the dominant standpoint of the current literature, this study reveals that both farmland fragmentation and farmland consolidation impact on food security in different ways at different levels. Therefore, for the purposes of achieving food availability, accessibility and sustainability for food security at all levels, the defragmentation process to minimize the problems related to physical farmland fragmentation (internal, location or distance, shape, use, small plot sizes and boundaries) can take the form of farmland consolidation for large heterogeneous farms under complex strong or market-oriented economies with high land availability; voluntary parcel exchange and on-field harvest sales for small heterogeneous farms under moderate and subsistence economies with land scarcity; land realignment for homogenous farms with contiguous plots to eliminate and reorganize the boundaries; and farmland use consolidation or crop consolidation for multiple uses on small plots and farms. Similarly, farmland tenure fragmentation problems (ownership and usership, small farm sizes and boundaries) can be prevented and minimized by land banking for small farms under complex strong economies; restrictions about the parcel sizes subdivision and absentee owners, joint ownership, cooperative farming, farmland use (crop) consolidation, agricultural land protection policies, and family planning measures (to curb the population growth) in the case of small farms under moderate and subsistence economies. This hypothetical stance is backed by the Malthusian theory of population and food supply, economies of scale and Gestalt theories, which commonly advocate in favor of agriculture expansion on bigger consolidated farms than on fragmented ones. On the other hand, for the purposes of food diversity, quality, accessibility, independency, acceptability, sovereignty, and sustainability for food security, different agriculture intensification and agroecological programs, and other land saving technologies could be the most suitable strategies to maximize the income from agriculture on fragmented plots under the circumstances of beneficial fragmentation in subsistence and moderate economies characterized by high population densities, market imperfections and land scarcity. These include the combined use of soil mineral and organic fertilizers and amendments, crops diversification, pests control measures, labor use efficiency and intensity, and resistant and high yielding crop varieties. This position is supported by the Boserup's theory on population growth and agriculture intensification below a certain critical threshold. In case of the coexistence of both rational and defective fragmentation scenarios in an area, various specific strategies which could simultaneously minimize the defective effects of fragmentation thereby optimizing or without jeopardizing its potential benefits can give better and more balanced or optimal solutions. These include land consolidation in blocks or localized land consolidation models for internally fragmented subsistence farms with plots spatially scattered in different heterogeneous topographies, and multicropping-based land consolidation approaches for fragmented farms with parcels spatially dispersed in homogenous topography, in combination with or without agriculture intensification programs.

In order to empirically test and evaluate how farmland fragmentation can be best managed for food security motives, prior to the design of any policy and strategy in favor of either farmland fragmentation conservation or defragmentation (consolidation) or both as land management tools, this study recommends the identification of all the possible farmland fragmentation scenarios (forms, causes, and their positive and negative impacts) in a given area, and the conditions dictating their problematic and beneficial status quos. In this line, for the sake of assessing the problematic and rational farmland fragmentation forms and scenarios under distinctive local circumstances, a rigorous feasibility study should be conducted, before the development of their suitable, flexible (dynamic), desirable, climate resilient, sustainable, feasible and multidimensional food security responsive coping strategies, policies and interventions at the household and individual levels. Instead of the existing focus on food productivity at the community, regional and national levels, the efforts should be oriented towards the improvement of food security status at the household and individual levels and the consideration of agroecological approaches in local food production on either fragmented or consolidated land. Therefore, further research should focus on the scrutiny and the development of more detailed and comprehensive indicators which can facilitate the trade-offs between farmland fragmentation conservation and defragmentation policies and interventions for food security motives under various particular local contexts.

The novel insights of this study can inform and guide policy makers, research scholars and the general scientific community for the devise of the suitable policies, interventions, tools and strategies for the best management of local farmland fragmentation scenarios. Moreover, contrary to the existing popular and global logic favoring the market-oriented agriculture often combined with agriculture expansion on big consolidated farms to achieve food security, this novel knowledge about the necessity of the variety of farmland management instruments to address particular farmland fragmentation scenarios contributes to the achievement of the sustainable development goals (SDGs 1.4; 2.1,2,3,4 and 5; 12.2; 13.1; and 15.3,4,5 and 9) of ending hunger, malnutrition, and poverty in the framework of the agenda 2030. As stipulated by these SDGs specific targets, this can be possible through the diversification of crops in diverse fragmented and scattered areas with various crop-growing conditions; equal distribution, ownership, access and control, sustainable management and efficient use of land resources; and agrobiodiversity and ecosystems conservation on land as an adaptive strategy to the global climate change realities and challenges (through climate smart or resilient agriculture), often combined with agriculture intensification programs to increase the agriculture production of small farms (see Paragraph 6 of Section 1 for these specific SDGs targets), for food sovereignty and the household food security motives.

Author Contributions: Conceptualization, P.D.N. and W.T.d.V.; methodology, P.D.N.; software, P.D.N.; validation, P.D.N. and W.T.d.V.; formal analysis, P.D.N.; investigation, P.D.N.; resources, P.D.N.; data curation, P.D.N.; writing—original draft preparation, P.D.N.; writing—review and editing, W.T.d.V.; visualization, P.D.N.; supervision, W.T.d.V.; project administration, P.D.N.; funding acquisition, P.D.N. All authors have read and agreed to the published version of the manuscript.

Funding: This research received no special external funding. The Article Processing Charge (APC) was funded by the Technische Universitaet Muenchen (TUM) Open Access Publishing Fund.

Acknowledgments: The authors would like to acknowledge the Editors, anonymous reviewers and colleague research scholars who contributed to shaping and improving the quality of this review paper through their insightful comments. We are also grateful to the Katholischer Akademischer Auslaender Dienst (KAAD) and the Government of Rwanda through the Ministry of Education for their various supports towards the accomplishment of this research.

Conflicts of Interest: The authors declare no conflict of interest. The funders had no role in the design of the study; in the collection, analyses, or interpretation of data; in the writing of the manuscript, or in the decision to publish the results.

References

1. Akkaya, A.S.T.; Kirmikil, M.; Gündoğdu, K.S.; Arici, I. Reallocation model for land consolidation based on landowners' requests. *Land Use Policy* **2018**, *70*, 463–470. [CrossRef]
2. Alemu, G.T.; Berhanie Ayele, Z.; Abelieneh Berhanu, A. Effects of Land Fragmentation on Productivity in Northwestern Ethiopia. *Adv. Agric.* **2017**. [CrossRef]
3. Bizimana, C.; Nieuwoudt, W.L.; Ferrer, S.R. Farm size, land fragmentation and economic efficiency in southern Rwanda. *Agrekon* **2004**, *43*, 244–262. [CrossRef]
4. FAO. *The Design of Land Consolidation Pilot Projects in Central and Eastern Europe*; FAO: Rome, Italy, 2003; Volume 6, ISBN 9251050015. Available online: http://www.fao.org/3/a-Y4954E.pdf (accessed on 20 April 2018).
5. Jürgenson, E. Land reform, land fragmentation and perspectives for future land consolidation in Estonia. *Land Use Policy* **2016**, *57*, 34–43. [CrossRef]
6. Kawasaki, K. The costs and benefits of land fragmentation of rice farms in Japan. *Aust. J. Agric. Resour. Econ.* **2010**, *54*, 509–526. [CrossRef]
7. Latruffe, L.; Piet, L. Does land fragmentation affect farm performance? A case study from Brittany, France. *Agric. Syst.* **2014**, *129*, 68–80. [CrossRef]
8. Niroula, G.S.; Thapa, G.B. Impacts and causes of land fragmentation, and lessons learned from land consolidation in South Asia. *Land Use Policy* **2005**, *22*, 358–372. [CrossRef]
9. Tan, S.; Heerink, N.; Qu, F. Land fragmentation and its driving forces in China. *Land Use Policy* **2006**, *23*, 272–285. [CrossRef]
10. Hiironen, J.; Riekkinen, K. Agricultural impacts and profitability of land consolidations. *Land Use Policy* **2016**, *55*, 309–317. [CrossRef]
11. Ntihinyurwa, P.D.; de Vries, W.T. Farmland fragmentation and defragmentation nexus: Scoping the causes, impacts, and the conditions determining its management decisions. *Ecol. Indic.* **2020**, *119*, 106828. [CrossRef]
12. Abubakari, Z.; Van der Molen, P.; Bennett, R.; Kuusaana, E.D. Land consolidation, customary lands, and Ghana's Northern Savannah Ecological Zone: An evaluation of the possibilities and pitfalls. *Land Use Policy* **2016**, *54*, 386–398. [CrossRef]
13. Demetriou, D.; Stillwell, J.; See, L. Land consolidation in Cyprus: Why is an Integrated Planning and Decision Support System required? *Land Use Policy* **2012**, *29*, 131–142. [CrossRef]
14. Keeler, M.E.; Skuras, D.G. Land fragmentation and consolidation policies in Greek agriculture. *Geogr. Assoc.* **1990**, *75*, 73–76. Available online: https://www.jstor.org/stable/40571937 (accessed on 15 April 2018).
15. Louwsma, M.; Lemmen, C.; Hartvigsen, M.; Hiironen, J.; Du Plessis, J.; Chen, M.; Laarakker, P. Land Consolidation and Land Readjustment for Sustainable Development:the Issues to be Addressed. In Proceedings of the FIG Working Week, Surveying the World of Tomorrow—From Digitalisation to Augmented Reality, Helsinki, Finland, 29 May–2 June 2017; FIG: Lausane, Switzerland, 2017; p. 19. Available online: http://www.fig.net/resources/proceedings/fig_proceedings/fig2017/papers/p06g/P06G_louwsma_lemmen_et_al_8973.pdf (accessed on 23 May 2019).
16. Vitikainen, A. An overview of land consolidation in Europe. *Nord. J. Surv. Real Estate Res.* **2004**, *1*, 124–136. Available online: https://journal.fi/njs/article/view/41504 (accessed on 14 July 2019).
17. Asiama, K.; Bennett, R.; Zevenbergen, J.; Mano, A.D.S. Responsible consolidation of customary lands: A framework for land reallocation. *Land Use Policy* **2019**, *83*, 412–423. [CrossRef]
18. Van Dijk, T. Scenarios of Central European land fragmentation. *Land Use Policy* **2003**, *20*, 149–158. [CrossRef]
19. Hartvigsen, M.B. Land Reform and Land Consolidation in Central and Eastern Europe after 1989: Experiences and Perspectives. Ph.D.-Serien for Det Teknisk-Naturvidenskabelige Fakultet, Aalborg Universitet. Ph.D. Thesis, Aalborg Universitetsforlag, Aalborg, Denmark, 2015. [CrossRef]
20. Van der Molen, P.; Lemmen, C.; Van Dijk, T.; Uimonen, M. Introducing the subject of modern land consolidation and symposium report. In *Modern Land Consolidation: Symposium in Volvic (Clermont-Ferrand), Proceedings of the Symposium on Modern Land Consolidation Conference Amenagement Foncier Moderne, Clermont-Ferrand France, 10–11 September 2004*; FIG Commission 7; International Federation of Surveyors: Frederiksberg, Denmark, 2004; pp. 4–18. Available online: https://research.utwente.nl/en/publications/introducing-the-subject-of-modern-land-consolidation-and-symposiu (accessed on 11 June 2019).
21. Van Dijk, T. Land consolidation as Central Europe's panacea reassessed. In *Modern Land Consolidation: Symposium in Volvic (Clermont-Ferrand), Proceedings of the Symposium on Modern Land Consolidation Conference Amenagement Foncier Moderne, Clermont-Ferrand France, 10–11 September 2004*; International Federation of Surveyors: Frederiksberg, Denmark, 2004; pp. 31–50. Available online: http://www.fig.net/commission7/france_2004/papers_symp/ts_01_vandijk.pdf (accessed on 11 June 2019).
22. Bizoza, A.R.; Havugimana, J.M. Land use consolidation in Rwanda: A case study of Nyanza district, Southern province. *Int. J. Sustain. Land Use Urban. Plan.* **2013**, *1*, 64–75. Available online: https://www.sciencetarget.com/Journal/index.php/IJSLUP/index (accessed on 15 April 2018).
23. Kathiresan, A. *Farm Land Use Consolidation in Rwanda*; Assessment from the Perspectives of Agriculture Sector, Republic of Rwanda, Eds.; Ministry of Agriculture and Animal Resources, Republic of Rwanda: Kigali, Rwanda, 2012. Available online: https://www.minagri.gov.rw/fileadmin/user_upload/documents/agridocs/Farm_Land_Use_Consolidation_in_Rwanda.pdf (accessed on 15 April 2018).
24. Huggins, C. *Consolidating Land, Consolidating Control: State-Facilitated 'Agricultural Investment' through the 'Green Revolution' in Rwanda*; Working paper 16; Land Deal Politics Initiative (LDPI): The Hague, The Netherlands, 2013; pp. 1–26. Available online: https://assets.publishing.service.gov.uk/media/57a08a28e5274a31e0000464/LDPI_WP_16.pdf (accessed on 15 April 2018).

25. Mbonigaba, J.; Dusengemungu, L. *Farm Land Use Consolidation-a Home Grown Solution for Food Security in Rwanda*; Ministry of Agriculture and Animal Husbandry, Ed.; Rwanda Agricultural Board, Ministry of Agriculture and Animal Husbandry: Kigali, Rwanda, 2013; p. 8. Available online: https://www.academia.edu/download/55849911/Farm_Land_Use_Consolidation-_a_Home_Grown_Solution_for_Food_Security_in_Rwanda.pdf (accessed on 18 April 2018).
26. Musahara, H.; Nyamulinda, B.; Bizimana, C.; Niyonzima, T. Land use consolidation and poverty reduction in Rwanda. In Proceedings of the 2014 Annual World Bank Conference on Land and Poverty: Integrating Land Governance into the Post-2015 Agenda: Harnassing Synergies for Implementation and Monitoring Impact, Washington, DC, USA, 24–27 March 2014; The World Bank: Washington, DC, USA, 2019. Available online: https://www.oicrf.org/documents/40950/43224/Land+use+consolidation+and+poverty+reduction+in+Rwanda.pdf/d89771cb-b975-6f89-ac94-fb6a19fdc093 (accessed on 7 August 2019).
27. Nilsson, P. The Role of Land Use Consolidation in Improving Crop Yields among Farm Households in Rwanda. *J. Dev. Stud.* **2019**, *55*, 1726–1740. [CrossRef]
28. Asiama, K.; Bennett, R.; Zevenbergen, J. Land consolidation on Ghana's rural customary lands: Drawing from the Dutch, Lithuanian and Rwandan experiences. *J. Rural Stud.* **2017**, *56*, 87–99. [CrossRef]
29. Asiama, K.; Bennett, R.; Zevenbergen, J. Land Consolidation for Sub-Saharan Africa's Customary Lands: The Need for Responsible Approaches. *Rural Dev.* **2017**, *5*, 39–45. [CrossRef]
30. Bentley, J.W. Economic and Ecological Approaches to Land Fragmentation: In Defense of a Much-Maligned Phenomenon. *Annu. Rev. Anthropol.* **1987**, *16*, 31–67. [CrossRef]
31. Blarel, B.; Hazell, P.; Place, F.; Quiggin, J. The Economics of Farm Fragmentation: Evidence from Ghana and Rwanda. *World Bank Econ. Rev.* **1992**, *6*, 233–254. [CrossRef]
32. Chigbu, U.E.; Ntihinyurwa, P.D.; de Vries, W.T.; Ngenzi, E.I. Why Tenure Responsive Land-Use Planning Matters: Insights for Land Use Consolidation for Food Security in Rwanda. *Int. J. Environ. Res. Public Health* **2019**, *16*, 1354. [CrossRef]
33. Ntihinyurwa, P.D.; de Vries, W.T.; Chigbu, U.E., Dukwiyimpuhwe, P.A. The positive impacts of farm land fragmentation in Rwanda. *Land Use Policy* **2019**, *81*, 565–581. [CrossRef]
34. Pritchard, M.F. Land, power and peace: Tenure formalization, agricultural reform, and livelihood insecurity in rural Rwanda. *Land Use Policy* **2013**, *30*, 186–196. [CrossRef]
35. King, R.; Burton, S. Structural change in agriculture: The geography of land consolidation. *Prog. Geogr.* **1983**, *7*, 471–501. [CrossRef]
36. Ali, D.A.; Deininger, K.; Ronchi, L. *Costs and Benefits of Land Fragmentation: Evidence from Rwanda*; Policy Research Working Papers of the World Bank Group; World Bank Group: Washington, DC, USA, 2015; p. 7290. [CrossRef]
37. Ciaian, P.; Guri, F.; Rajcaniova, M.; Drabik, D.; Gomez y Paloma, S. Land fragmentation and production diversification: A case study from rural Albania. *Land Use Policy* **2018**, *76*, 589–599. [CrossRef]
38. Di Falco, S.; Penov, I.; Aleksiev, A.; van Rensburg, T.M. Agrobiodiversity, farm profits and land fragmentation: Evidence from Bulgaria. *Land Use Policy* **2010**, *27*, 763–771. [CrossRef]
39. Molle, F.; Srijantr, T. Between concentration and fragmentation: The resilience of the land system in the Chao Phraya Delta. In *Perspectives on Social Agricultural Change in the Chao Phraya Delta*; White Lotus Press: Bangkok, Thailand, 2003; Volume 12, pp. 77–107. ISBN 9744800259. Available online: http://horizon.documentation.ird.fr/exl-doc/pleins_textes/divers15-08/010033857.pdf (accessed on 27 July 2019).
40. Sklenicka, P.; Salek, M. Ownership and soil quality as sources of agricultural land fragmentation in highly fragmented ownership patterns. *Landsc. Ecol.* **2008**, *23*, 299–311. [CrossRef]
41. Cholo, T.C.; Fleskens, L.; Sietz, D.; Peerlings, J. Land fragmentation, climate change adaptation, and food security in the Gamo Highlands of Ethiopia. *Agric. Econ.* **2019**, *50*, 39–49. [CrossRef]
42. Knippenberg, E.; Jolliffe, D.; Hoddinott, J. Land Fragmentation and Food Insecurity in Ethiopia. *Policy Res. Work. Pap. World Bank Group* **2018**, 8559. [CrossRef]
43. Ciaian, P.; Guri, F.; Rajcaniova, M.; Drabik, D.; Paloma, S. Land Fragmentation, Production Diversification, and Food Security: A Case Study from Rural Albania. In Proceedings of the Land Economics/Use Conference, Milan, Italy, 9–14 August 2015; International Association of Agricultural Economists (IAAE): Toronto, ON, Canada, 2015; p. 28. [CrossRef]
44. Ciaian, P.; Miroslava, R.; Guri, F.; Zhllima, E.; Shahu, E. The impact of crop rotation and land fragmentation on farm productivity in Albania. *Stud. Agric. Econ.* **2018**, *120*, 116–125. [CrossRef]
45. Knippenberg, E.; Jolliffe, D.; Hoddinott, J. Land Fragmentation and Food Insecurity in Ethiopia. *Am. J. Agric. Econ.* **2020**, *102*, 1557–1577. [CrossRef]
46. Leventon, J.; Laudan, J. Local food sovereignty for global food security? Highlighting interplay challenges. *Geoforum* **2017**, *85*, 23–26. [CrossRef]
47. Wezel, A.; Herren, B.G.; Kerr, R.B.; Barrios, E.; Gonçalves, A.L.R.; Sinclair, F. Agroecological principles and elements and their implications for transitioning to sustainable food systems. A review. *Agron. Sustain. Dev.* **2020**, *40*, 1–13. [CrossRef]
48. Thomas, V.G.; Kevan, P.G. Basic principles of agroecology and sustainable agriculture. *J. Agric. Environ. Ethics* **1993**, *6*, 1–19. [CrossRef]
49. Ali, D.A.; Deininger, K. *Is There a Farm-Size Productivity Relationship in African Agriculture? Evidence from Rwanda*; Policy Research Working Papers of the World Bank Group; World Bank Group: Washington, DC, USA, 2014. [CrossRef]

50. Kadigi, R.; Kashaigili, J.; Sirima, A.; Kamau, F.; Sikira, A.; Mbungu, W. Land fragmentation, agricultural productivity and implications for agricultural investments in the Southern Agricultural Growth Corridor of Tanzania (SAGCOT) region, Tanzania. *J. Dev. Agric. Econ.* **2017**, *9*, 26–36. [CrossRef]

51. De Vries, W.T.; Chigbu, U.E. Responsible land management-Concept and application in a territorial rural context. *fub. Flächenmanag. Bodenordn.* **2017**, *2*, 65–73. Available online: https://mediatum.ub.tum.de/1360258 (accessed on 23 July 2019).

52. Stigler, G.J. The economies of scale. *J. Law Econ.* **1958**, *1*, 54–71. [CrossRef]

53. Wertheimer, M. *Gestalt Theory*; Kegan Paul, Trench, Trubner & Company: London, UK, 1938. [CrossRef]

54. Burchi, F.; De Muro, P. From food availability to nutritional capabilities: Advancing food security analysis. *Food Policy* **2016**, *60*, 10–19. [CrossRef]

55. Demont, M.; Jouve, P.; Stessens, J.; Tollens, E. Boserup versus Malthus revisited: Evolution of farming systems in northern Côte d'Ivoire. *Agric. Syst.* **2007**, *93*, 215–228. [CrossRef]

56. Desiere, S.; D'Haese, M. Boserup versus Malthus: Does population pressure drive agricultural intensification? Evidence from Burundi. In Proceedings of the 89th Annual Conference on Agricultural Economics Society, University of Warwick, Coventry, UK, 13–15 April 2015; Agricultural Economics Society: Banbury, UK, 2015. [CrossRef]

57. Teece, D.J. Economies of scope and the scope of the enterprise. *J. Econ. Behav. Organ.* **1980**, *1*, 223–247. [CrossRef]

58. Salvati, L.; Tombolini, I.; Gemmiti, R.; Carlucci, M.; Bajocco, S.; Perini, L.; Ferrara, A.; Colantoni, A. Complexity in action: Untangling latent relationships between land quality, economic structures and socio-spatial patterns in Italy. *PLoS ONE* **2017**, *12*, e0177853. [CrossRef]

59. Wim, T.; Ónega, L.F.; María, T.J.; Crecente, M.R. A Complexity perspective on institutional change: Dealing with land fragmentation in Galicia. *Soc. Evol. Hist.* **2015**, *14*, 77–107. Available online: https://www.sociostudies.org/journal/files/seh/2015_2/077-107.pdf (accessed on 23 September 2019).

60. Norberg, J.; Cumming, G. *Complexity Theory for a Sustainable Future*; Columbia University Press: New York Chichester, NY, USA, 2008; ISBN 9780231508865.

61. Gunderson, L.H. Ecological resilience in theory and application. *Annu. Rev. Ecol. Syst.* **2000**, *31*, 425–439. [CrossRef]

62. Lengnick, L. *Resilient Agriculture: Cultivating Food Systems for a Changing Climate*; New Society Publishers: Gabriola Island, BC, Canada, 2015; p. 288. ISBN 9781550925784. Available online: https://books.google.de/books?id=n-D0AgAAQBAJ (accessed on 29 September 2019).

63. Boserup, E. *The Conditions of Agricultural Growth: The Economics of Agrarian Change under Population Pressure*; Transaction Publishers, Rutgers: New Brunswick, NJ, USA; London, UK, 2011; ISBN 9780202307930.

64. Maxwell, S.; Smith, M. *Household Food Security: Concepts, Indicators, Measurements: A Conceptual Review*; International Fund for Agricultural Development: Rome, Italy, 1992; pp. 1–72. ISBN 9280620215.

65. Pinstrup-Andersen, P. Food security: Definition and measurement. *Food Secur.* **2009**, *1*, 5–7. [CrossRef]

66. Manjunatha, A.; Anik, A.R.; Speelman, S.; Nuppenau, E. Impact of land fragmentation, farm size, land ownership and crop diversity on profit and efficiency of irrigated farms in India. *Land Use Policy* **2013**, *31*, 397–405. [CrossRef]

67. Leroy, J.L.; Ruel, M.; Frongillo, E.A.; Harris, J.; Ballard, T.J. Measuring the Food Access Dimension of Food Security: A Critical Review and Mapping of Indicators. *Food Nutr. Bull.* **2015**, *36*, 167–195. [CrossRef] [PubMed]

68. Sen, A. *Poverty and Famines: An Essay on Entitlement and Deprivation*; Oxford University Press Inc.: New York, NY, USA, 1981; ISBN 0198284632.

69. FAO. *World Food Summit: Rome Declaration on World Food Security and World Food Summit Plan of Action*; FAO: Rome, Italy, 1996; Available online: http://refhub.elsevier.com/S0264-8377(18)30887-1/sbref0095 (accessed on 15 April 2018).

70. Komatsuzaki, M. Agro-ecological Approach for Developing a Sustainable Farming and Food System. *J. Dev. Sustain. Agric.* **2011**, *6*, 54–63. [CrossRef]

71. UN. *Transforming Our World: The 2030 Agenda for Sustainable Development*; UN General Assembly: New York, NY, USA, 2015; Available online: www.un.org/en/development/desa/population/migration/generalassembly/docs/globalcompact/A_RES_70_1_E.pdf (accessed on 29 September 2019).

72. Torraco, R.J. Writing integrative literature reviews: Guidelines and examples. *Hum. Resour. Dev. Rev.* **2005**, *4*, 356–367. [CrossRef]

73. Torraco, R.J. Writing integrative literature reviews: Using the past and present to explore the future. *Hum. Resour. Dev. Rev.* **2016**, *15*, 404–428. [CrossRef]

74. Webster, J.; Watson, R.T. Analyzing the past to prepare for the future: Writing a literature review. *Mis Q.* **2002**, *26*, 13–23.

75. Ntihinyurwa, P.D.; de Vries, W.T. Farmland fragmentation concourse: Analysis of scenarios and research gaps. *Land Use Policy* **2021**, *100*, 104936. [CrossRef]

76. Snyder, H. Literature review as a research methodology: An overview and guidelines. *J. Bus. Res.* **2019**, *104*, 333–339. [CrossRef]

77. Langley, A. Strategies for Theorizing from Process Data. *Acad. Manag. Rev.* **1999**, *24*, 691–710. [CrossRef]

78. Klopper, R.; Lubbe, S.; Rugbeer, H. The matrix method of literature review. *Alternation* **2007**, *14*, 262–276. Available online: http://alternation.ukzn.ac.za/Files/docs/14.1/12%20Klopper%20.pdf (accessed on 3 June 2019).

79. Salipante, P.; Notz, W.; Bigelow, J. A matrix approach to literature reviews. In *Research in Organizational Behavior: An Annual Series of Analytical Essays and Critical Reviews*; Elsevier: Amsterdam, The Netherlands, 1982; Volume 4, pp. 321–348.

80. McPherson, M.F. Land fragmentation: A selected literature review. In *Development Discussion Paper*; Harvard Institute for International Development, Harvard University: Cambridge, MA, USA, 1982; Volume 141.

81. Iwasaki, Y. Reasoning with multiple abstraction models. In *Recent Advances in Qualitative Physics*; Faltings, B., Struss, P., Eds.; The MIT Press: London, UK, 1992; pp. 67–82. ISBN 0262061422. Available online: https://books.google.rw/books?hl=en&lr=&id=xmoSnz47ouUC&oi=fnd&pg=PA67&ots=PT-3zwUmiB&sig=iCRwC4Zr3E4ZllWzZ5A76-G9Gi0&redir_esc=y#v=onepage&q&f=false (accessed on 27 October 2019).

82. Dhillon, L.; Vaca, S. Refining theories of change. *J. Multidiscip. Eval.* **2018**, *14*. Available online: https://pdfs.semanticscholar.org/6874/1e95cb3e6405fa114c00886f2d11cea2faff.pdf (accessed on 27 October 2019).

83. Lewis, M.W.; Grimes, A.J. Metatriangulation: Building theory from multiple paradigms. *Acad. Manag. Rev.* **1999**, *24*, 672–690. [CrossRef]

84. Boulding, K.E. General Systems Theory—The Skeleton of Science. *Inf. Pubsonline* **1956**, *2*, 197–208. [CrossRef]

85. Smith, L.B.; Thelen, E. *A Dynamic Systems Approach to Development: Applications*; Hardcover; The MIT Press: Cambridge, MA, USA, 1993; Volume 414, p. xviii. ISBN 0262193337.

86. Thelen, E. Dynamic Systems Theory and the Complexity of Change. *Psychoanal. Dialogues* **2005**, *15*, 255–283. [CrossRef]

87. Robinson, S.; Arbez, G.; Birta, L.G.; Tolk, A.; Wagner, G. Conceptual modeling: Definition, purpose and benefits. In Proceedings of the 2015 Winter Simulation Conference, Huntington Beach, CA, USA, 6–9 December 2015; IEEE Press: Piscataway, NJ, USA, 2015; pp. 2812–2826. [CrossRef]

88. Sabates-Wheeler, R. Consolidation initiatives after land reform: Responses to multiple dimensions of land fragmentation in Eastern European agriculture. *J. Int. Dev.* **2002**, *14*, 1005–1018. [CrossRef]

89. Binns, S.B.O. *The Consolidation of Fragmented Agricultural Holdings*; Food and Agriculture Organization of the United Nations(FAO): Rome, Italy, 1950; Volume 11.

90. McPherson, M.F. Land fragmentation in agriculture: Adverse? Beneficial? And for whom? In *Development Discussion Paper*; Harvard Institute for International Development, Harvard University: Cambridge, MA, USA, 1983; Volume 145, p. 81.

91. Simons, S. Land fragmentation in developing countries: The optimal choice and policy implications. In Proceedings of the 19th International Conference of Agricultural Economists, Malaga, Spain, 26 August–4 September 1985; International Association of Agricultural Economists: Milwaukee, WI, USA, 1985; pp. 703–716. Available online: https://ageconsearch.umn.edu/record/183032/files/IAAE-CONF-228.pdf (accessed on 7 May 2018).

92. Dhakal, B.; Khanal, N. Causes and Consequences of Fragmentation of Agricultural Land: A Case of Nawalparasi District, Nepal. *Geogr. J. Nepal* **2018**, *11*, 95–112. [CrossRef]

93. Sundqvist, P.; Andersson, L. A Study of the Impacts of Land Fragmentation on Agricultural Productivity in Northern Vietnam. Bachelor's Thesis, Uppsala University, Uppsala, Sweden, 2006. Available online: http://www.diva-portal.org/smash/get/diva2:131275/FULLTEXT01.pdf (accessed on 13 July 2019).

94. Igbozurike, M.U. Fragmentation in tropical agriculture: An overrated phenomenon. *Prof. Geogr.* **1970**, *22*, 321–325. [CrossRef]

95. Brown, P.M.; Moyer, D.D. *Multipurpose Land Information Systems: The Guidebook*; Federal Geodetic Control Committee (FGCC): Rockville, NY, USA, 1994; Volume 2, Available online: https://books.google.rw/books?id=VG-avwEACAAJ (accessed on 14 July 2019).

96. Henssen, J. *Land Registration and Cadastre Systems: Principles and Related Issues*; Technische Universität München: Munich, Germany, 2010.

97. King, R.; Burton, S. Land fragmentation: Notes on a fundamental rural spatial problem. *Prog. Geogr.* **1982**, *6*, 475–494. [CrossRef]

98. Demetriou, D. *The Development of an Integrated Planning and Decision Support System [IPDSS] for Land Consolidation. Ph.D. Thesis, University of Leeds: Leeds, UK*; Springer International Publishing: Cham, Switzerland, 2014; Available online: http://www.springer.com/series/8790 (accessed on 14 May 2018).

99. Bullard, R. *Land Consolidation and Rural Development*; Home, R., Ed.; Papers in Land Management: No. 10; Anglia Ruskin University: Cambridge, UK; Chelmsford, UK, 2007; p. 148. Available online: http://citeseerx.ist.psu.edu/viewdoc/download?doi=10.1.1.600.2529&rep=rep1&type=pdf (accessed on 17 June 2019).

100. Stroessner, G. Land consolidation in Bavaria: Support given to rural areas. *J. Irrig. Eng. Rural Plan.* **1986**, *1986*, 53–59. [CrossRef]

101. Meuser, F.J. *European Expert Meeting on Land Consolidation, 1988. Analysis of the Results*; Lehrstuhl fuer Bodenordnung und Landentwicklung der TU Muenchen: Muenchen, Germany, 1992; p. 195. Available online: http://agris.fao.org/agris-search/search.do?recordID=DE94R0107 (accessed on 23 February 2020).

102. Magel, H. The change of paradigms in European rural development and land consolidation. *Z. Fur Kult. Landentwickl.* **2001**, *42*, 4–9. Available online: https://www.researchgate.net/publication/242280883_The_change_of_paradigms_in_European_rural_development_and_land_consolidation (accessed on 23 February 2020).

103. Asiama, K.O.; Bennett, R.; Zevenbergen, J.; Asiama, S.O. Land valuation in support of responsible land consolidation on Ghana's rural customary lands. *Surv. Rev.* **2018**, *50*, 288–300. [CrossRef]

104. Lemmen, C.; Jansen, L.J.; Rosman, F. Informational and computational approaches to Land Consolidation. In Proceedings of the FIG Working Week: Knowing to Manage the Territory, Protect the Environment, Evaluate the Cultural Heritage, Rome, Italy, 6–10 May 2012; FIG: Lausane, Switzerland, 2012; p. 16. Available online: http://www.fig.net/resources/proceedings/fig_proceedings/fig2012/papers/ts02e/TS02E_lemmen_jansen_et_al_6049.pdf (accessed on 23 February 2020).

105. Vasiljevic, D.; Radulovic, B.; Babovic, M.; Todorović, S. *Land Consolidation as Unused Potential: The Effects of Implementation, Barriers and Potential Relevance of Agricultural Land Consolidation in Serbia*; Nacionalna Alijansa za Lokalni Ekonomski Razvoj: Belgrade, Serbia, 2018; ISBN 9788680128047.

106. Zhang, X.; Ye, Y.; Wang, M.; Yu, Z.; Luo, J. The micro administrative mechanism of land reallocation in land consolidation: A perspective from collective action. *Land Use Policy* **2018**, *70*, 547–558. [CrossRef]

107. Sonnenberg, J. Fundamentals of land consolidation as an instrument to abolish fragmentation of agricultural holdings. In Proceedings of the FIG XXII International Congress, Washington, DC, USA, 19–26 April 2002; FIG: Lausane, Switzerland, 2002; pp. 1–12. Available online: http://citeseerx.ist.psu.edu/viewdoc/download;jsessionid=B38B18AD2AABDA7FE7DC2DE424B79FA0?doi=10.1.1.489.4061&rep=rep1&type=pdf (accessed on 30 January 2020).

108. Janus, J.; Łopacka, M.; John, E. Land consolidation in mountain areas. Case study from southern Poland. *Geod. Cartogr.* **2017**, *66*, 241–251. [CrossRef]

109. Hartvigsen, M. Land reform and land fragmentation in Central and Eastern Europe. *Land Use Policy* **2014**, *36*, 330–341. [CrossRef]

110. Cay, T.; Ayten, T.; Iscan, F. Effects of different land reallocation models on the success of land consolidation projects: Social and economic approaches. *Land Use Policy* **2010**, *27*, 262–269. [CrossRef]

111. FAO. *Voluntary Guidelines on the Responsible Governance of Tenure of Land, Fisheries and Forests in the Context of National Food Security*; FAO: Rome, Italy, 2012; p. 47. ISBN 9789251072776. Available online: http://www.fao.org/3/a-i2801e.pdf (accessed on 30 January 2020).

112. Thapa, G.B.; Niroula, G.S. Alternative options of land consolidation in the mountains of Nepal: An analysis based on stakeholders' opinions. *Land Use Policy* **2008**, *25*, 338–350. [CrossRef]

113. Zhang, B.; Niu, W.; Ma, L.; Zuo, X.; Kong, X.; Chen, H.; Zhang, Y.; Chen, W.; Zhao, M.; Xia, X. A company-dominated pattern of land consolidation to solve land fragmentation problem and its effectiveness evaluation: A case study in a hilly region of Guangxi Autonomous Region, Southwest China. *Land Use Policy* **2019**, *88*, 104115. [CrossRef]

114. Veršinskas, T.; Vidar, M.; Hartvigsen, M.; Mitic Arsova, K.; van Holst, F.; Gorgan, M. *Legal Guide on Land Consolidation: Based on Regulatory Practices in Europe*; FAO: Rome, Italy, 2020; Volume 3, p. 193. ISBN 9789251328583. [CrossRef]

115. Johansen, P.; Ejrnæs, R.; Kronvang, B.; Olsen, J.; Præstholm, S.; Schou, J. Pursuing collective impact: A novel indicator-based approach to assessment of shared measurements when planning for multifunctional land consolidation. *Land Use Policy* **2018**, *73*, 102–114. [CrossRef]

116. Haldrup, N.O. Agreement based land consolidation—In perspective of new modes of governance. *Land Use Policy* **2015**, *46*, 163–177. [CrossRef]

117. Ntihinyurwa, P.D.; Masum, F. Participatory Land Use Consolidation in Rwanda: From Principles to Practice. In Proceedings of the FIG Working Week: Surveying the World of Tomorrow—From Digitalisation to Augmented Reality, Helsinki, Finland, 29 May–2 June 2017; FIG: Lausane, Switzerland, 2017. Available online: https://www.fig.net/resources/proceedings/fig_proceedings/fig2017/papers/ts04h/TS04H_ntihinyurwa_masum_9008.pdf (accessed on 25 June 2017).

118. Lerman, Z.; Cimpoieş, D. Land consolidation as a factor for rural development in Moldova. *Eur. Asia Stud.* **2006**, *58*, 439–455. [CrossRef]

119. USAID. *Assessment of the Economic, Social, and Environmental Impacts of the Land Use Consolidation Component of the Crop Intensification Program in Rwanda*; U.S. Agency for International Development(USAID/Rwanda), Democracy and Governance Office: Kigali, Rwanda, 2014; Available online: https://pdfs.semanticscholar.org/38dd/9d6c9d35e7b77ab59ca9431c581c91ceb3e9.pdf (accessed on 15 April 2018).

120. Laepple, E. *Land Consolidation in Europe*; 3-7843-2525-4; Schriftenreihe des Bundesministers fuer Ernaehrung, Landwirtschaft und Forsten, Reihe B: Flurbereinigung: Münster-Hiltrup, Germany, 1992; pp. 5–12.

121. Muyombano, E.; Espling, M. Land use consolidation in Rwanda: The experiences of small-scale farmers in Musanze District, Northern Province. *Land Use Policy* **2020**, *99*, 105060. [CrossRef]

122. Isaacs, K.B.; Snapp, S.S.; Chung, K.; Waldman, K.B. Assessing the value of diverse cropping systems under a new agricultural policy environment in Rwanda. *Food Secur.* **2016**, *8*, 491–506. [CrossRef]

123. Niyonzima, T. Land Use Dynamics in the Face of Population Increase: A Study in the Districts of Gatsibo and Nyagatare, Eastern Province, Rwanda. Ph.D. Thesis, School of Business, Economics and Law, Göteborg University, Göteborg, Sweden, 2009. Available online: https://www.dissertations.se/dissertation/50a47cacc6/ (accessed on 5 July 2020).

124. Crecente, R.; Alvarez, C.; Fra, U. Economic, social and environmental impact of land consolidation in Galicia. *Land Use Policy* **2002**, *19*, 135–147. [CrossRef]

125. Damen, J. Land banking in The Netherlands in the Context of Land Consolidation. In Proceedings of the International Workshop: Land Banking/Land Funds as an Instrument for Improved Land Management for CEEC and CIS, Tonder, Denmark, 17–20 March 2004; FAO: Rome, Italy, 2004; pp. 1–5. Available online: http://www.fao.org/fileadmin/user_upload/reu/europe/documents/LANDNET/2004/Netherlands_paper.pdf (accessed on 26 June 2019).

126. Miranda, D.; Crecente, R. Suitability model for Land Consolidation projects: A case study in Galicia, Spain. In Proceedings of the Symposium on Modern Land Consolidation, Volvic (Clermont-Ferrand), France, 10–11 September 2004; FIG Commission 7. International Federation of Surveyors (FIG): Frederiksberg, Denmark, 2004; pp. 173–188. Available online: http://www.fig.net/resources/proceedings/2004/france_2004_comm7/papers_symp/ts_04_miranda.pdf (accessed on 12 September 2019).

127. Van den Berg, R.; Revilla, E.; Menken, M.; Verbeek, I. *Land Banking Principle: A Reconnaissance for Conditions and Practical Constrains for Application of the Land Banking Principle in the Netherlands*; University of Wageningen: Wageningen, The Netherlands, 2005.

128. De Vries, W.T.; Wouters, R.; Konttinen, K. A comparative analysis of senior expert experiences with land consolidation projects and programs in Europe. In Proceedings of the FIG Working Week 2019, Geospatial Information for a Smarter Life and Environmental

Resilience, Hanoi, Vietnam, 22–26 April 2019; FIG: Lausane, Switzerland, 2019. Available online: http://www.fig.net/resources/proceedings/fig_proceedings/fig2019/papers/ts05h/TS05H_de_vries_wouters_et_al_9781.pdf (accessed on 15 May 2020).

129. Bentley, J.W. Wouldn't you like to have all of your land in one place? Land fragmentation in Northwest Portugal. *Hum. Ecol.* **1990**, *18*, 51–79. [CrossRef]

130. Farmer, B.H. On Not Controlling Subdivision in Paddy-Lands. *Trans. Pap.* **1960**, *28*, 225–235. [CrossRef]

131. Lisec, A.; Pintar, M. Conservation of natural ecosystems by land consolidation in the rural landscape. *Acta Agric. Slov.* **2005**, *85*, 73–82. Available online: http://citeseerx.ist.psu.edu/viewdoc/download?doi=10.1.1.476.23&rep=rep1&type=pdf (accessed on 11 March 2020).

132. Muchová, Z.; Petrovič, F. Changes in the landscape due to land consolidations. *Ekológia* **2010**, *29*, 140–157. [CrossRef]

133. Zhang, Q.-q.; Luo, H.-b.; Yan, J.-m. Integrating biodiversity conservation into land consolidation in hilly areas—A case study in southwest China. *Acta Ecol. Sin.* **2012**, *32*, 274–278. [CrossRef]

134. Van Dijk, T. Complications for traditional land consolidation in Central Europe. *Geoforum* **2007**, *38*, 505–511. [CrossRef]

135. Downing, T.E. Partible inheritance and land fragmentation in a Oaxaca village. *Hum. Organ.* **1977**, *36*, 235–243. [CrossRef]

136. Vander Meer, P. Land Consolidation through Land Fragmentation: Case Studies from Taiwan. *Land Econ.* **1975**, *51*, 275–283. [CrossRef]

137. Lengoiboni, M.; Groenendijk, E.; Mukahigiro, A. Land tenure regularization in Rwanda: Registration of land rights for women and its impacts on food security. *J. Land Adm. East. Afr.* **2015**, *3*, 399–414. Available online: https://research.utwente.nl/en/publications/land-tenure-regularization-in-rwanda--registration-of-land-rights-for-women-and-its-impacts-on-food-security(424ca5fd-a967-4cdb-afa8-218c985a3317).html (accessed on 15 April 2019).

138. Ntihinyurwa, P. An Evaluation of the Role of Public Participation in Land Use Consolidation (LUC) Practices in Rwanda and Its Improvement. Master's Thesis, Technical University of Munich, München, Germany, 2016.

139. Uwayezu, E.; de Vries, W.T. Scoping land tenure security for the poor and low-income urban dwellers from a spatial justice lens. *Habitat Int.* **2019**, *91*, 102016. [CrossRef]

140. Jia, L.; Petrick, M. How does land fragmentation affect off-farm labor supply: Panel data evidence from China. *Agric. Econ.* **2014**, *45*, 369–380. [CrossRef]

141. Lazíková, J.; Bandlerová, A.; Rumanovská, L.; Takáč, I.; Lazíková, Z. Crop Diversity and Common Agricultural Policy—The Case of Slovakia. *Sustainability* **2019**, *11*, 1416. [CrossRef]

142. Van der Molen, P. Food security, land use and land surveyors. *Surv. Rev.* **2017**, *49*, 147–152. [CrossRef]

143. Pijanowski, J.M. Land consolidation development—Disscusion of a new approach recommended for Poland. *Geomat. Landmanag. Landsc.* **2014**, *2*, 54–65. Available online: http://yadda.icm.edu.pl/yadda/element/bwmeta1.element.baztech-243d0e8b-74c1-409d-b6f7-c85c6846fc7b/c/GLL-2-2014f.pdf (accessed on 25 January 2020).

144. Burton, S.; King, R. Land fragmentation and consolidation in Cyprus: A descriptive evaluation. *Agric. Adm.* **1982**, *11*, 183–200. [CrossRef]

145. Brabec, E.; Smith, C. Agricultural land fragmentation: The spatial effects of three land protection strategies in the eastern United States. *Landsc. Urban. Plan.* **2002**, *58*, 255–268. [CrossRef]

146. Dam Lam, R.; Boafo, Y.A.; Degefa, S.; Gasparatos, A.; Saito, O. Assessing the food security outcomes of industrial crop expansion in smallholder settings: Insights from cotton production in Northern Ghana and sugarcane production in Central Ethiopia. *Sustain. Sci.* **2017**, *12*, 677–693. [CrossRef]

147. Ecker, O.; Breisinger, C. *The Food Security System: A New Conceptual Framework*; 01166; International Food Policy Research Institute (IFPRI), Development Strategy and Governance Division: Washington, DC, USA, 2012; p. 24. Available online: http://cdm15738.contentdm.oclc.org/utils/getfile/collection/p15738coll2/id/126837/filename/127048.pdf (accessed on 13 April 2019).

148. FAO. *Global Strategic Framework for Food Security and Nutrition (GSF): "Making a difference in Food Security and Nutrition"*; Committee on World Food Security (CFS): Rome, Italy, 2014; Volume 41, Available online: http://www.fao.org/3/MR173EN/mr173en.pdf (accessed on 20 April 2020).

149. Maxwell, S. Food security: A post-modern perspective. *Food Policy* **1996**, *21*, 155–170. [CrossRef]

150. Pangaribowo, E.H.; Gerber, N.; Torero, M. Food and nutrition security indicators: A review. In *ZEF Working Paper Series, No. 108*; Econstor: Bonn, Germany, 2013; Available online: http://hdl.handle.net/10419/88378 (accessed on 24 February 2020).

151. Weingärtner, L. The concept of food and nutrition security. In *Achieving Food and Nutrition Security*; 3rd ed.; Klennert, K., Ed.; InWEnt—Internationale Weiterbildung gGmbH: Feldafing, Germany, 2009; Volume 3, pp. 21–52. ISBN 9783939394570. Available online: http://www.fao.org/docs/eims/upload/219148/food_reader_engl.pdf (accessed on 24 February 2020).

152. FAO. *The State of Food Insecurity in the World*; FAO & World Food Program of the United Nations: Rome, Italy, 2009; p. 61. ISBN 9789251062883. Available online: http://www.fao.org/docrep/pdf/012/i0876e/i0876e.pdf (accessed on 24 February 2020).

153. Campbell, C.C. Food Insecurity: A Nutritional Outcome or a Predictor Variable? *J. Nutr.* **1991**, *121*, 408–415. [CrossRef]

154. Holden, S.T.; Ghebru, H. Land tenure reforms, tenure security and food security in poor agrarian economies: Causal linkages and research gaps. *Glob. Food Secur.* **2016**, *10*, 21–28. [CrossRef]

155. TWN; SOCLA. *Agroecology: Key Concepts, Principles and Practices*; Third World Network (TWN) & Sociedad Científica Latinoamericana de Agroecología (SOCLA): Penang, Malaysia, 2015; p. 54. ISBN 9789670747118. Available online: http://foodfirst.org/wp-content/uploads/2015/11/Agroecology-training-manual-TWN-SOCLA.pdf (accessed on 29 December 2020).

156. Lake, I.; Abdelhamid, A.; Hooper, L.; Bentham, G.; Boxall, A.; Draper, A.; Fairweather-Tait, S.; Hulme, M.; Hunter, P.; Nichols, G. *Food and Climate Change: A Review of the Effects of Climate Change on Food within the Remit of the Food Standards Agency*; Food Standards Agency: London, UK, 2010. Available online: https://www.food.gov.uk/sites/default/files/media/document/575-1-1008_X02001_Climate_Change_and_Food_Report_28_Sept_2010.pdf (accessed on 25 June 2020).

157. Makate, C.; Wang, R.; Makate, M.; Mango, N. Crop diversification and livelihoods of smallholder farmers in Zimbabwe: Adaptive management for environmental change. *SpringerPlus* **2016**, *5*, 1135. [CrossRef]

158. Bailey, A. *Mainstreaming Agrobiodiversity in Sustainable Food Systems: Scientific Foundations for an Agrobiodiversity Index-Summary*; Bioversity International: Rome, Italy, 2016; ISBN 9789292550592. Available online: https://cgspace.cgiar.org/handle/10568/80360 (accessed on 25 June 2020).

159. Conceição, P.; Levine, S.; Lipton, M.; Warren-Rodríguez, A. Toward a food secure future: Ensuring food security for sustainable human development in Sub-Saharan Africa. *Food Policy* **2016**, *60*, 1–9. [CrossRef]

160. Fan, S.; Yosef, S.; Pandya-Lorch, R. Seizing the momentum to reshape agriculture for nutrition. In *Agriculture for Improved Nutrition: Seizing the Momentum*; eBook; CAB International: Wallingford, UK; International Food Policy and Research Institute (IFPRI): Washington, DC, USA, 2019; Volume 1, ISBN 9781786399328.

161. Lipper, L.; Thornton, P.; Campbell, B.M.; Baedeker, T.; Braimoh, A.; Bwalya, M.; Caron, P.; Cattaneo, A.; Garrity, D.; Henry, K. Climate-smart agriculture for food security. *Nat. Clim. Chang.* **2014**, *4*, 1068. [CrossRef]

162. Prasad, Y.; Maheswari, M.; Dixit, S.; Srinivasarao, C.; Sikka, A.; Venkateswarlu, B.; Sudhakar, N.; Prabhu Kumar, S.; Singh, A.; Gogoi, A. *Smart Practices and Technologies for Climate Resilient Agriculture*; Central Research Institute for Dryland Agriculture, Indian Council of Agricultural Research(ICAR): Hyderabad, India; New Delhi, India, 2014; p. 76. ISBN 9789380883304.

163. Rosegrant, M.W.; Cline, S.A. Global food security: Challenges and policies. *Science* **2003**, *302*, 1917–1919. [CrossRef] [PubMed]

164. Thornton, P.K.; Herrero, M. Adapting to climate change in the mixed crop and livestock farming systems in sub-Saharan Africa. *Nat. Clim. Chang.* **2015**, *5*, 830–836. [CrossRef]

165. FAO. *The 10 Elements of Agroecology: Guiding the Transition to Sustainable Food and Agricultural Systems*; FAO: Rome, Italy, 2018; p. 15. Available online: http://www.fao.org/3/i9037en/i9037en.pdf (accessed on 29 December 2020).

166. Dudzińska, M.; Kocur-Bera, K. Assessment of land fragmentation for the purpose of land consolidation works as exemplified by the Pasym commune. *Geomat. Land Manag. Landsc.* **2014**, 31–44. Available online: http://gll.ur.krakow.pl (accessed on 4 November 2019).

167. Edwards, C.J.W. The Effects of Changing farm size upon levels of farm Fragmentation: A Somerset Case Study. *J. Agric. Econ.* **1978**, *29*, 143–154. [CrossRef]

168. Papageorgiou, E. Fragmentation of land holdings and measures for consolidation in Greece. *Land Tenure* **1956**, 543–547.

169. Rahman, S.; Rahman, M. Impact of land fragmentation and resource ownership on productivity and efficiency: The case of rice producers in Bangladesh. *Land Use Policy* **2009**, *26*, 95–103. [CrossRef]

170. Sikor, T.; Müller, D.; Stahl, J. Land fragmentation and cropland abandonment in Albania: Implications for the roles of state and community in post-socialist land consolidation. *World Dev.* **2009**, *37*, 1411–1423. [CrossRef]

171. Tan, S.; Heerink, N.; Kruseman, G.; Futian, Q. Do fragmented landholdings have higher production costs? Evidence from rice farmers in Northeastern Jiangxi province, PR China. *China Econ. Rev.* **2008**, *19*, 347–358. [CrossRef]

172. Van Hung, P.; MacAulay, T.G.; Marsh, S.P. The economics of land fragmentation in the north of Vietnam. *Agric. Resour. Econ.* **2007**, *51*, 195–211. [CrossRef]

173. Sargent, F.O. Fragmentation of French land: Its nature, extent, and causes. *Land Econ.* **1952**, *28*, 218–229. [CrossRef]

174. Grammatikopoulou, I.; Myyrä, S.; Pouta, E. The proximity of a field plot and land-use choice: Implications for land consolidation. *J. Land Use Sci.* **2013**, *8*, 383–402. [CrossRef]

175. Grigg, D.B. *Population Growth and Agrarian Change: An Historical Perspective*; CUP Archive: Cambridge, UK, 1980; ISBN 0521296358.

176. Galt, A.H. Exploring the cultural ecology of field fragmentation and scattering on the island of Pantelleria, Italy. *J. Anthropol. Res.* **1979**, *35*, 93–108. [CrossRef]

177. Netting, R. Of men and meadows: Strategies of alpine land use. *Anthropol. Q.* **1972**, *45*, 132–144. [CrossRef]

178. Mutangiza, O. The Contribution of Land Use Consolidation Policy to Food and Nutrition Security of Women Farmers: A Case Study of the Farmers Involved in Potato Farming, Food Availability in Nyabihu District, Rwanda. Master's Thesis, Sveriges Lantbruksuniversitet (SLU), Swedish University of Agricultural Sciences, Uppsala, Sweden, 2019. Available online: http://urn.kb.se/resolve?urn=urn:nbn:se:slu:epsilon-s-10458 (accessed on 17 March 2020).

179. Alexandri, C.; Luca, L.; Kevorchian, C. Subsistence Economy and Food Security—The Case of Rural Households from Romania. *Procedia Econ. Financ.* **2015**, *22*, 672–680. [CrossRef]

180. Reddy, P.P. *Climate Resilient Agriculture for Ensuring Food Security*; Springer: New Delhi, India, 2015; Volume 373, ISBN 9788132221999. [CrossRef]

181. Paul, M.; wa Gĩthĩnji, M. Small farms, smaller plots: Land size, fragmentation, and productivity in Ethiopia. *J. Peasant Stud.* **2018**, *45*, 757–775. [CrossRef]

182. Wan, G.H.; Cheng, E. Effects of land fragmentation and returns to scale in the Chinese farming sector. *Appl. Econ.* **2001**, *33*, 183–194. [CrossRef]

 *land*

Review

Transparency of Land Administration and the Role of Blockchain Technology, a Four-Dimensional Framework Analysis from the Ghanaian Land Perspective

Prince Donkor Ameyaw * and Walter Timo de Vries

Land Management and Land Tenure, Technische Universität München (TUM), 80333 München, Germany; wt.de-vries@tum.de
* Correspondence: aprincedonkor.ameyaw@tum.de

Received: 6 November 2020; Accepted: 2 December 2020; Published: 3 December 2020

Abstract: Existing studies on blockchain within land administration have focused mainly on replacing or complementing the technology for land registration and titling. This study explores the potential of using blockchain technology to enhance the transparency of all land administration processes using an integrative review methodology coupled with a framework analysis. This study draws on the Ghanaian land administration perspective to make this insightful. It appears possible to apply a permissionless public blockchain across all land administration processes. This integrates all departments, processes, and stakeholders of land administration to enhance openness, improve availability and accessibility to information, and foster participation for transparency simultaneously. This can change the transparency variation in land administration to be more equal and homogenous regardless of land type. This, however, depends on the standardization of processes across the divisions, as well as negotiation and consensus amongst all stakeholders, especially with chiefs. Limitations include: limited storage and scalability, as well as huge electricity consumption for operation. This study's policy implications are a review of all paper-based land transactions, a comprehensive digitization of land administration processes, public–private partnership on blockchain-based land administration, and professionals and stakeholder education on the technology.

Keywords: land administration; blockchain technology; land tenure; land valuation; land use planning; land development; Ghana

1. Introduction

Land administration involves 'the process of determining, recording and disseminating information about the relationship between people and land' [1] (p. 2), [2]. UNECE defined it as involving the recording and dissemination of information about ownership, value and the use of land, as well as the associated resources, while implementing the land management policies [3]. This relationship between people and land, and the functions performed with regard to ownership, value and the use of land require transparency. The transparency of land administration depicts the situation where land transactions and services are carried out in openness, and with maximum participation by all the concerned stakeholders [4]. Transparency allows for land tenure security [4–8]. Land administration transparency enables landowners and prospective purchasers to know the exact status of their land rights and interests, as well as the relation that they have with other individuals concerning pieces of land. This enhances peoples' confidence to invest in land which improves economic conditions [9]. Transparency is noted as one of the key principles for good land governance [10]. A good governance in land administration is beneficial to societies in diverse ways as it ensures:

> *'Pro-poor support: rule of law is equal to all, and citizen has protected rights, Public confidence: greater public confidence, Economic growth: security of the land tenure and regulated transaction cost and taxation, Protection of state assets: legitimate use of state land for social and economic concessions, More effective and efficient public administration of land: formal market and reliable system, more revenue sharing for public services, Conflict prevention and resolution: equity, justice, and social stability'* [9] (p. 13)

Nevertheless, land administration across the world lacks transparency and is corrupt everywhere [10]. Land administration systems are considered to be among the most corrupt institutions in the world [11]. UN-HABITAT in 2007 observed that land offices in most countries are among the most corrupt institutions [12]. Corruption exists where there is lack of transparency [4]. This lack of transparency in land administration begets numerous land challenges which include; land tenure insecurity, high cost of land transactions due to informal payments, reduced private sector investment in land, less revenue for the state, increased land grabbing by officials, increased land conflicts, landlessness, and inequity in land distribution. These challenges promote social instability, exclusion and political instability through land conflicts, land poorness and landlessness. The situation leads to disregard for the ethics and standards of behavior as land titles, building permits and zoning regulations become no longer trusted by citizens [4,7,8]. These outcomes inhibit the overall development of societies. Most countries, particularly in Africa, face stunted development and impoverishment as land dominates the economy, and provides livelihoods to the majority of the continent's population [13]; thus, the focus of this study is from an African country's perspective. The need for the transparency of land administration has not received the needed attention in years past as [9] notes that the attention to the issue of transparency in land administration and land governance is recent.

In recent years, many studies on ways to enhance the transparency of land administration have shifted attention to the potentials of blockchain technology [11,14–17]. Blockchain is identified to enhance transparency in land administration processes and or functions though the integration of all land stakeholders, in a way that allows each stakeholder to be aware of, and to be involved in land transactions without intermediaries (land administration processes and functions have been used synonymously and interchangeably in this study). Blockchain helps to improve trust in the system and to enhance the confidence of citizens in the land administration system [17]. It is identified that despite the digitization of land records and diverse web applications, the system for land records management is weighed down by various kinds of errors and inconsistencies as well as a lack of transparency [18], the same problems which blockchain technology potentially eliminates [19]. Countries like Georgia, and Sweden, among others, have piloted blockchain technology to land administration and reported the successful outcomes of improved transparency and enhanced citizens' trust in the land institutions [20]. Several studies on the application of blockchain technology for land registration and land titling exist to show the potential of blockchain technology to improve transparency in these land administration processes [17,20–25]. These show the surging interest in blockchain technology in land administration. However, despite the rising interest in the potential of blockchain in land administration [19], there is to date no studies that have holistically assessed the transparency of all the processes of land administration, and how blockchain technology can help improve these. The existing studies mainly focus on land registration, and titling [14,17,20–22,25–29]. These, however, only fall under the land tenure processes and or functions [30]. Other land administration processes—including land value, land use planning, and land development—have not been sufficiently explored, if any, in order to see how blockchain technology can enhance the transparency of these processes and in a simultaneous way. This leaves a research gap. To focus only on land tenure processes and to conclude that blockchain technology enhances the transparency of land administration is to miss the vast land administration processes of land value, land use planning, and land development. This leads to missing the broad concept of land administration transparency. This also presents a challenge to aptly conceptualize blockchain technology and the transparency of land administration, and hence, this study aimed to fulfil this research gap. This study argues that understanding the transparency of land administration and the role of blockchain technology in this regard becomes incomplete if the processes of land administration are not holistically considered. To this end, this study was guided by these objectives:

1. To identify the essential elements and relations between blockchain technology and the transparency of land administration in the existing literature;
2. To assess the potential of blockchain technology to improve the transparency of land administration functions—based on the Ghanaian land administration context.

These objectives are particularly important as they fill a literature gap by looking at what transparency actually means in land administration discourses, and how the widely accrued technology of blockchain could potentially contribute to achieving this. Secondly, it helps in extending the literature on the potential of blockchain technology in the specific context of land administration in a more comprehensive approach. The paper starts by explaining the methodology applied to address both research questions. The subsequent section presents the elements and relations between blockchain technology and land administration processes and applies this from the Ghanaian perspective. The section that follows afterwards discusses the possible roles blockchain technology could play in enhancing or affecting the transparency of land administration processes. The final section reflects on the study's guiding framework and derives policy recommendations.

2. Materials and Methods

2.1. Research Approach and Boundaries

The novelty of blockchain's application in land administration opens it up for new discourses on its potentials to land administration. On this basis, more research studies are needed as conceptualizations and theoretical models in this regard are still preliminary. Methodologically, an integrative review is considered appropriate for such new topics as compared to a systematic, and semi-systematic review methods [31]. Given this, we apply a review methodology based on an integrative interpretation process of existing documentation and literature, with the aim of deriving an alternative conceptualization of transparency in land administration. Such an integrative literature review methodology is suitable when investigating the extent to which a new concept or technology fits in a new context. This approach has also been used in similar studies [32,33]. Furthermore, it is suitable for new and emerging topics that have not benefited from a large body of literature and conceptualizations [31]. Integrative reviews assess, critique, and synthesize existing literature on a topic in ways that evoke new theoretical frameworks, and perspectives [32,34]. Integrative reviews can follow rationalist theory as an appropriate epistemology and are based on an exploratory research design which deducts the scientific knowledge and new perspectives through the critical review, analysis and synthesis of existing literature [31,35,36]. According to [36], integrative literature review methodology, compared to systematic, and semi-systematic review methodologies offers a better opportunity to assess pending developments in a field and to identify factors that are shaping the future of ideas or issues in that field through critiquing, and analyzing relevant literature [31]. Doing this, however, requires prior understanding and knowledge on the topic to guide and facilitate the critique, analysis, and appraisal of existing relevant literature and concepts [35]. In this way, integrative review aids in identifying relationships, gaps, deficiencies, and opportunities for improvement on existing literature and concepts, thereby offering a possibility for rethinking the topic and improving scientific knowledge by extension (update) and or reconceptualization [31]. However, integrative review analysis is criticized for not being developed in accordance with any specific standard and is mostly not truly integrative but a mere summary of existing studies. This can lead to a lack of rigor as compared to systematic reviews [32,34]. The research underlying this paper overcame these potential critique points by employing and combining the method with the framework analysis (sometimes also referred to as qualitative content analysis). This provides a structured approach to analyze the main concepts and ideas which reveal relationships, divergences and gaps for critiquing, leading to a better synthesis of both the emerging perspectives and existing literature in a rigorous way [37].

Following the study objectives of identifying the conceptualization and relations in blockchain technology and land administration, and the potential of blockchain to enhance the transparency of land administration, a conceptual organizational structure of integrative literature review is used [31]. In this, the main concepts of the topic provide a framework around which the review is organized to help ensure coherence and clarity on what is being reviewed, and how the concepts of the topic enjoin into a unified idea [31]. In order to find contextual boundaries, the literature review focused on documents which specifically addressed ongoing research and practical advantages and disadvantages of blockchain in the context of land administration. The framework analysis creates a new structure for findings that help to summarize them in a way that supports answering the study questions [37]. This provides a clear stepwise approach to follow which produces a highly structured output of summarized findings and gives a holistic, descriptive overview that allows for easy critiquing and analysis [37]. The study's literature is not constrained by spatial and temporal boundaries. This allowed for the geographically unlimited literature review of all available and relevant data, from empirical, and review (secondary) literature in the English language in which a large volume of literature on the topic was found. This is not to conclude that literature did not exist in other languages, but the majority of returned literature was in English. Additionally, English is the language that the authors have mastery knowledge of. No linguistic biases were intended. The literature identification process and sources, review, analysis, synthesis, modelling/reconceptualization, and means of scientific knowledge extension on the topic follows in the next section and is summarized in Table 1 below.

Table 1. Research process and design overview.

	Steps	Activities	Output
1	Setting study boundaries	- Outlining spatial and geographical limits, language boundary, concepts under focus, and literature type and publication timeframe	- No geographical and spatial limitations. - Only data in English language were considered. - Focus centered on land administration processes and functions, and transparency, and blockchain's application in land administration.
2	Literature identification: - search strategy - selection - literature sources - validity and reliability	- Systematic literature search and spider backward search strategies. Direct typing into databases, the keywords and phrases like, land administration, land administration functions, land administration transparency, blockchain and land administration, blockchain for land registration, and blockchain and land transparency, and using their different combinations. - Focus on land administration functions or processes, transparency of land administration, and blockchain's application to land administration. - Scientific online databases including Google Scholar, Elsevier, Springer Link, Scopus, JSTOR, Research Gate, Web of Science, and Taylor & Francis. - Using different synonymous keywords and phrases, and their combinations across different scientific databases helped to check validity and reliability.	- Total search results = 195 online publications. - Selected publications = 81. - Spider backward publications = 17. - Total documents accepted and reviewed = 98 publications. - Final documents used for the study = 76 publications.
3	Initial literature review	Titles and abstract reading.	- Elimination of duplicated documents, and documents that did not meet the study focus, and boundaries.
4	Detailed integrative review: - critical review - analysis - synthesis	- Detailed and critical full text reading. - Use of concept structure and framework analysis. - Through textual narratives and visualizations techniques (framework diagrams, tables, procedural diagrams) the main ideas, themes, patterns, gaps, and relationships were categorized, and summarized. - Based on logical and deductive reasoning, the analysis and synthesis of emerging knowledge was made.	- Theoretical basis of blockchain technology. - Main potentials of blockchain technology. - Main land administration processes in Ghana and the challenges/gaps. - Potential relation between blockchain technology and transparency in the land administration processes.
5	Topic reconceptualization	- The use of rationalist theory of knowledge generation, logical, and deductive reasoning, and the authors' primary knowledge on the topic, new knowledge was created and implications explained.	- Ways of blockchain application to enhance transparency in all land administration processes of land tenure, land valuation, land use planning, and land development.
6	Conclusion and recommendations	- Study contribution to knowledge. - Areas of further research, and implications for practitioners, and policy makers.	- Extension of blockchain's potential to support land administration beyond just land tenure processes. - Revealing a new area for further research. - Explaining new knowledge's implication for practice and policy making.

2.2. Data Sources and Research Methods

The literature search was carried out systematically based on the main concepts and ideas in the topic using keywords, and phrases like, land administration, land administration functions, land administration transparency, blockchain and land administration, blockchain for land registration, blockchain and land transparency, land tenure, land valuation, land use planning, and land development. These keywords and phrases were searched for across different scientific databases including Google Scholar, Elsevier, Springer Link, Scopus, JSTOR, Research Gate, Web of Science, and Taylor & Francis. Searching with diverse synonymous keywords and phrases across the different scientific databases facilitates access to a large volume of documents on the topic and allows for a validity, and reliability check. The systematic literature search resulted in 102 documents on land administration. This number was based mainly on documents' titles and how they related to the land administration processes; land registration, land information, land valuation, land taxation, land use planning, and land development. This selection was based on sampling and is not considered to be representative of all land administration systems but for Ghana. This is because, although some general documents on land administration processes were considered, the main focus was on land administration processes from the Ghanaian perspective, and hence, more of the documents relating to this context were considered. Moreover, 26 documents on land transparency, 42 on blockchain's application in land administration and nine on the methodological approach, making a total of 179 retrieved documents. The initial critical reading of the documents' titles and abstracts in some instances, while being guided by the research boundaries, resulted in 81 documents for the detailed and critical full text reading and review. The full text reading helped to identify the extent to which the documents discussed the topic and revealed the missing gaps. A spider backward search strategy helped to find additional sources. Through the spider backward approach, new citations and references that come up in the full text reading of selected literature and have relevance to the study are traced back to their original documents for identification and review. This strategy resulted in 17 additional online documents making it a total of 195 documents in all. The spider backward retrieved documents were also subjected to review based on the study boundaries. In the end, 98 documents in total were critically reviewed and 76 accepted and used for this study.

Based on conceptual organizational structure, and the framework analysis approach, the main ideas were categorized under different broad themes of blockchain technology, and the transparency of land administration processes using text narratives and visual models, which are suitable for integrative literature analysis and synthesis [31]. This approach helped to compile the main ideas from the reviewed studies, and also evidence-based documented practical applications of blockchain, which were all used to summarize and synthesize the study findings with respect to the research objectives [38]. Abstractive textual and narrative modeling based on the rationalist theory, deductive reasoning, and the authors' knowledge on the topic were used to establish the potential relationship in blockchain technology and the transparency of the land administration functions. The rationalist approach guided the justification and explanation of this potential relationship. The implications of the new and extended scientific knowledge to existing literature, practitioners and policy makers is explained, the study limitation highlighted, and suggestions for future research directions made.

3. Results

3.1. Theoretical Basis of Blockchain Technology and Its Operation

Blockchain technology refers to a fully distributed crypto-graphical system that captures and stores a consistent, immutable and linear event log of the transactions between networked actors [39]. Blockchain technology allows for managing the records of transactions without a central server or authority [28]. Through this network, which is made of computers (for stakeholders) that operate on a blockchain system to execute transactions and are termed as *'Nodes'*, blockchain technology works, based on what is technically referred to as *'Blocks and Hashes'*. In blockchain's operation, transaction data are stored in digital containers called 'blocks'. The first block created is termed as the genesis block [40], and each block after it is created is linked to a parent block (the preceding block) through unique digital fingerprints termed as 'hashes' [14]. This is shown after Figure 1. The hashes are time-stamped in a header at the top of each block of information to give certainty on the order of transactions' creation. After creating a transaction, and before it is accepted onto the blockchain system, the majority of the nodes will have to verify and validate that it is accurate and authentic as exists in reality on the ground. This verification and validation process is done through a system termed as 'consensus mechanism' [1]. Once transactions are validated and accepted onto the blockchain system, the information in the blocks becomes immutable and resilient against tampering or falsification. In this way, not even the one that created it can manipulate the data; and, the transaction with its data can be accessed at any time by all stakeholders, which allows for transparency [16,20,41]. As compared to other land transaction management tools like modeling, database management, and workflow management, there are three main arguments for why the blockchain technology is considered a solution with great benefits and possibly no alternative. First, in blockchain land transaction, records, certificates, and digital IDs, cannot be manipulated. Second, there can be no double spending/sales of land since any purported attempt is automatically known to all stakeholders [42], and thirdly, land transaction rules and requirements can be embedded into the blockchain's 'smart contract' application which makes it difficult for anyone to manipulate the process, and it also reduces human error possibilities [16]. Smart contracts are blockchain applications which allow for a pre-programming of a contract by defining all the conditions and requirements, and when parties have met these conditions and requirements, the contract is executed automatically. The blockchain transaction steps are:

1. A node/stakeholder with an account signs digitally and initiates the transaction;
2. A timestamp is added to prove the time of transaction creation;
3. Transaction is broadcasted by decentralization to all other nodes on the network;
4. The transaction is mined ("which involves validation of a set of transactions (block) in the network by means of showing the computational proof of the work done") [24] (p. 20), by one of the nodes. After this, it is verified and validated as authentic, or declined if it is found otherwise by the majority of the nodes based on a consensus mechanism;
5. The validated transaction is then recorded in a new block and hashed to the previous block to form the chain of blocks as is shown after Figure 1.

Technically, three processes are identified in how a blockchain works. Digital time-stamping, distributed verification, and cryptographic hashing [43]. Figure 1, below, shows the blockchain transaction process, Figure 2 shows the blockchain structure, and Table 2 shows the inherent elements that make the blockchain beneficial to land administration transparency.

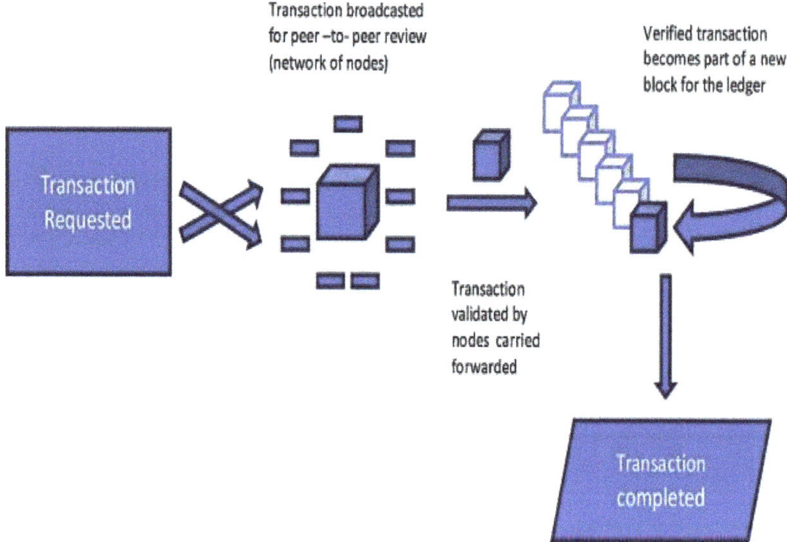

Figure 1. Blockchain transaction process. Source: [23].

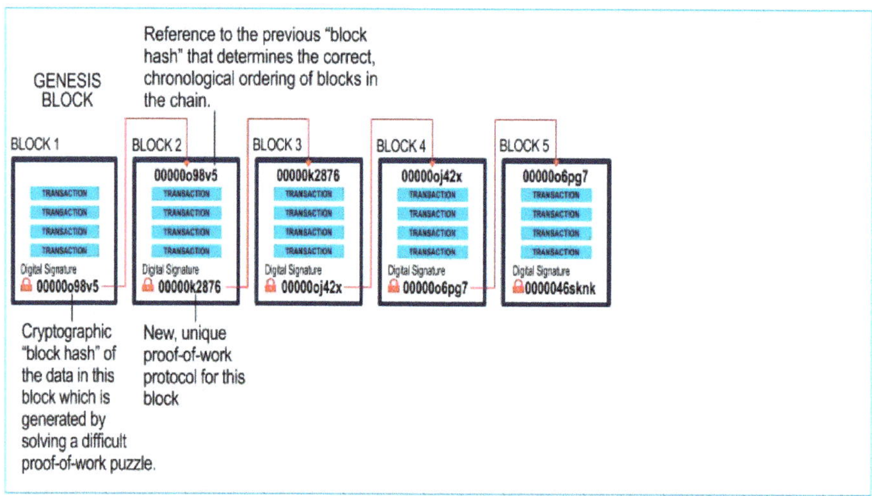

Figure 2. Blockchain structure. Source: [40].

Table 2. Blockchain potentials to land administration Source: authors' construct.

Potential Benefits of Blockchain	Elements that Make It Possible	References
Transparency	**Decentralization** of transactions across all nodes (stakeholders), possibility for evaluating the authenticity of the transaction by all stakeholders, and the access to all transactions and their historical records by all stakeholders.	[10,20,23,28,29]
Eliminating fraud and double sales	**Decentralization**—once a transaction has been completed, all stakeholders have copies and any further action on the transaction is known to the stakeholders. A second attempt to resell the land will be known to all stakeholders and therefore the transaction will not be validated since it has already been sold to another person.	[1,10,11,17]
Enhancing trust	**Immutability and consensus mechanism**—immutability of blockchain coupled with the consensus mechanism by majority stakeholders helps prevent manipulation of land data, as well as misrepresentation of land data in the system.	[1,20,23,26,44]
Establish clear ownership	**Hashing**—the existence and access to the historical facts on the transaction made possible by hashing allows ease in establishing the ownership status as well as all encumbrances.	[17]
Eliminating corruption	**Smart contracts**—land transactions based on and carried out using smart contracts helps to eliminate all forms of corrupt deals since all the procedures involved in the transaction are clear and can be carried out without human intervention thus leaving no room for corruption.	[20]
Eliminates manipulation	**Decentralization and immutability**—due to the distributed copies of transactions available to all nodes; and the difficulty to change blockchain data, any purported unauthorized changes or manipulation will be detected by all the stakeholders and will accordingly be declined or denied.	[10,23]
Easy information access	**Decentralization**—information stored is available to all the nodes at all times. This enhances the ease of access to land transaction information as there are no intermediaries.	[20,23,44]
Data quality, accuracy and integrity	**Consensus mechanism**—the verification and validation process inherent to the system ensures that the information accepted on the blockchain corresponds with reality on the ground. Any inconsistencies, and inaccuracies will lead to the rejection of the transaction.	[23]
Enhances high participation by all stakeholders	**Decentralization**—stakeholders become involved in the transaction at every stage due to the decentralized distribution across all the nodes. This allows all stakeholders to know about the transaction and to partake in it through the consensus mechanism.	[1,23,26,45]
Reduced human error possibilities	**Smart contracts** for land transactions help eliminate human involvement as all required actions necessary for carrying out transactions have been pre-programmed. Once a step is completed, the transaction moves to the next step without human actions until it is completed.	[10]
Security and resilience	**Decentralization and distribution**—due to the decentralized and distributed functionality, data are stored in multiple databases of different stakeholders which are tamper proof, immutable and encrypted. It is thus difficult to hack all the different databases at the same time.	[11,15,24]

The elements of the blockchain identified in Table 2 are reflected in the discussion section on how they help to achieve the transparency of the land administration.

It is important to point out that there are two main architectural categorizations of blockchain technology based on access and use possibilities. These are the public and private blockchain. These are further categorized into permissioned, and permissionless blockchains. The public and private categorizations determine who can access and read from the blockchain ledger, while the permissioned, and permissionless categorization determines who is able to introduce a transaction, and also participate in the consensus mechanism [16,46]. It is therefore important that the right blockchain architecture is selected depending on the purpose of application. Table 3 below shows the accessibility and use possibilities available in the different blockchain architectures.

Table 3. Blockchain architectural categorization Source: adopted from [16].

	Blockchain Architectural Categorizations			
	Public Blockchain		Private Blockchain	
	Permissionless	Permissioned	Permissionless	Permissioned
Participants	Anonymous	Identified	Identified	Identified
Data accessibility	Anyone	Anyone	Authorized participants	Authorized participants
Initiating transactions	Anyone	Authorized participants	Authorized participants	Network operator only
Participation in consensus mechanism	Anyone	Authorized participants	Authorized participants	Network operator only
Network types	Decentralized	Partly decentralized	Hybrid	Centralized

Some writers have advocated for the adoption of a private blockchain for land administration, specifically for land registration [44]. However, given the architecture categorizations in Table 3, this study considers a permissionless public blockchain more suitable for a land administration system. This is because, permissioned blockchains invade privacy/data protection policies with or without participants' consent since it allows participants to be automatically identified. Moreover, permissioned blockchains 'lose their decentralized, open nature, and become less transparent and more centralized', [47] (p. 152). These create difficulties in land data accessibility, lead to a lack of trust due to centrality and refute the transparency objective required in land administration. Public permissionless blockchain on the other hand helps to adhere to privacy/data protection policies. The anonymity of participants prevents the breach of privacy policies. In land administration, however, the question of who has what rights and to which land parcel is very critical, and therefore makes it important to be able to know participants' identity. To address this, a public permissionless blockchain has a way to allow participants' identity to be known where required. In [47], the authors noted that, in the public permissionless blockchain, although the users' identity is encrypted and hidden, there exists a possibility that in certain contexts, the identity of the participants can be inferred based on transaction patterns or other markers. This possibility helps to make inferences to participants and their actions whenever the need be, particularly where transactions or actions might appear suspicious. These functional possibilities of the public permissionless blockchain compared with the other architecture types make it more suitable for a public land administration system like the case in Ghana.

Notwithstanding these potentials and possibilities of blockchain enumerated, the technology, like any other technology, has its own flaws and or restrictive factors which must be taken into account before the decision to adopt and implement it. Generally, blockchain is criticized due to its limited storage capacity. Current public blockchains are unable to handle large volumes of land data such as deeds, titles, and maps [48]. This could cause problems in land administration since land transactions and data transactions occur daily. The authors in [17], however, recommend that an external storage for blockchain's smart contracts and documents can be created to support the system—see [17] for further details. Another challenge is scalability. Due to its nascent nature, and storage capacity limitation, there are challenges to scalability of the technology, particularly with increasing volumes of data and workload. This equally affects the speed of the system [47]. Moreover, blockchain technology consumes a huge amount of electricity, and this could be a potential challenge for some developing countries that do not have an equally huge electricity supply. Other adoption considerations of blockchain impede upon technological know-how. Blockchain in land administration is recent and immature [47]. Many land professionals are therefore not conversant with the use of the technology. It is important, therefore, to train professionals prior to blockchain adoption to be able to understand and use the technology. Finally, blockchain operation requires strong computational power and efficiency [49], coupled with strong and stable internet connectivity to be able to perform efficiently. These have to be considered in deciding on blockchain adoption.

3.2. Summary Overview of Land Administration Processes in Ghana

In assessing the extent to which transparency exists in a land administration system, it is important to know and recognize the differentiations and variations of the land administration processes. It is mainly assumed that the collective degree of transparency of each of the respective processes constitutes the variation of transparency of land administration as a whole. In [50], a land administration system is defined as a formal system that is used to locate and identify a real property, and to keep the records of past and current data regarding the ownership, value and use of that property. This definition is found to be suitable in this study's context as it highlights the different processes of land administration: land tenure, land valuation, land use planning, and land development. Few studies exist on the transparency of land administration in Ghana, and these have somewhat touched upon transparency issues in individual land administration processes of either land tenure, land valuation, land use planning, or land development [35,49,51–53]. No single study has concurrently assessed all processes of land administration, and the possibility of achieving a simultaneous transparency in these processes—which leaves a gap where data and research are missing. However, land administration, according to [54], must fulfil land title issuance, land taxation, land transaction registrations, changes in land use, resolving land disputes and handling complaints, and facilitate spatial and land use planning. These processes fall under the four broad land administration processes of land tenure, land value, land use planning, and land development [30]. This study thus argues that achieving a simultaneous transparency in all four main land administration processes has intrinsic and synergistic benefits that outweigh pursuing transparency in the individual processes separately. This is shown in Section 4 which comprehensively discusses the different land administration processes, and blockchain's potential to support and to achieve a simultaneous transparency across these processes.

3.2.1. Land Tenure Processes

Land tenure processes border on the land registration activities of securing and transferring rights in land and natural resources [55], and also on land information infrastructure. In these processes, [30] notes that land registration by means of land register establishment, creation of accessible land records, land transaction procedures, and the processing of information are the matters of interest. Land registration involves a process of the official recording of rights to land through deeds or titles aimed at supplying legal security to the right holders and potential buyers [56]. The sequence of the land registration process in Ghana is summarized in Figure 3 below. For details, see [57–59].

From an actor network theoretical (ANT) view point, Figure 3 below can best be understood not only based on the connection between the different divisions, but also, by the type of communication technology that connects these divisions and their work processes together. ANT helps to analyze the way in which actors (both human and non-human) build and maintain networks [43], for the purpose of achieving a goal. ANT is broadly advocated for in development research works particularly those focusing on technology. This is because, in a practical sense, 'there is ever-greater use of networks of individuals and organizations to deliver development and an ever-greater role for the material (especially technology) in development processes' [43], p. 38]. In the context of this study therefore, ANT theory gives a sound theoretical basis for understanding the different land administration processes, performed by the different land divisions, and stakeholders, and the role of blockchain technology in this relation towards achieving land administration transparency.

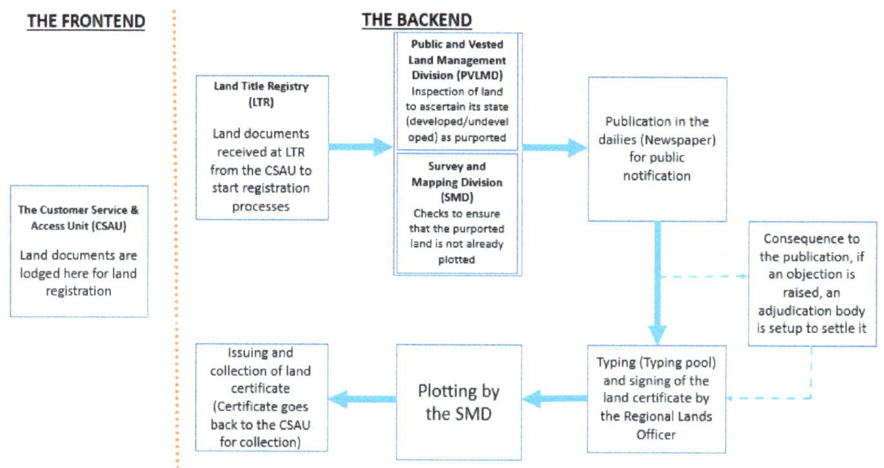

Figure 3. Land registration process. Based on [57,58].

From Figure 3 above, the frontend (Customer Service and Access Unit (CSAU)) serves as the intermediary between clients and the lands commission. Clients visit and submit their land documents, or complaints on any land process to the frontend desk of the CSAU. The CSAU, after certifying the documents, relays these to the right divisions at the backend, be it (LTR, PVLMD, or SMD). For land registration specifically, the CSAU first relays land documents to the LTR, from where it goes through all the formal processes with the different divisions until completed, brought back to the CSAU, and clients invited to pick up their certificates. Although other additional departments, such as the Land Valuation Division (LVD), are involved before land can be successfully registered, Figure 3 above is a simplified process which is understandable since the LTR is the first and last department involved in the registration process [58]. Other incidental activities include the submission to and stamping of land documents at the LVD before acceptance for registration, and also the settling of any objections that might be raised upon the publication in the dailies. However, when all documents are found correct and no objections raised, the above process should take on average 3–5 months to complete, but depending on individual cases and circumstances, certain cases could take longer [58].

Land information infrastructure on the other hand is concerned with the cadastral and topographic datasets [55].

3.2.2. Land Valuation Processes

The main processes considered here are the valuation and taxation of land and properties [55]. Valuation is an estimate or opinion of value based on expertise to meet the supply and demand under certain conditions. These conditions may be subjective or objective depending on the context of the valuation [60]. Valuation must be an unbiased estimate or opinion, a knowledgeable or learned opinion of value, and a supported estimate of a defined value. The value must represent a reasonable market value which according to the 2017 International Valuation Standards Council's (IVSC) definition. is *'the estimated amount for which an asset or liability should exchange on the valuation date between a willing buyer and a willing seller in an arm's length transaction, after proper marketing and where the parties had each acted knowledgeably, prudently and without compulsion'* [48] (p. 4), [30] (p. 4). There are five different methods for asset valuation, namely (1) the market approach or comparative method, (2) the income approach or investment method, (3) the residual approach or development method, (4) the profit method, and (5) the cost approach or contractor's method, see details in [48]. The choice of a method relies on three aspects, the nature of the asset, the basis of the valuation, and the purpose of the valuation [60]. The nature of the asset is concerned with the physical properties, characteristics and conditions of the asset. The basis of the valuation may include, market value or the market rent, worth and investment value, and fair or equitable value, while the purpose for the valuation may also include, for sale and purchase, rental, mortgage, insurance, compensation, and lease [48]. Figure 4 below shows the valuation process in Ghana.

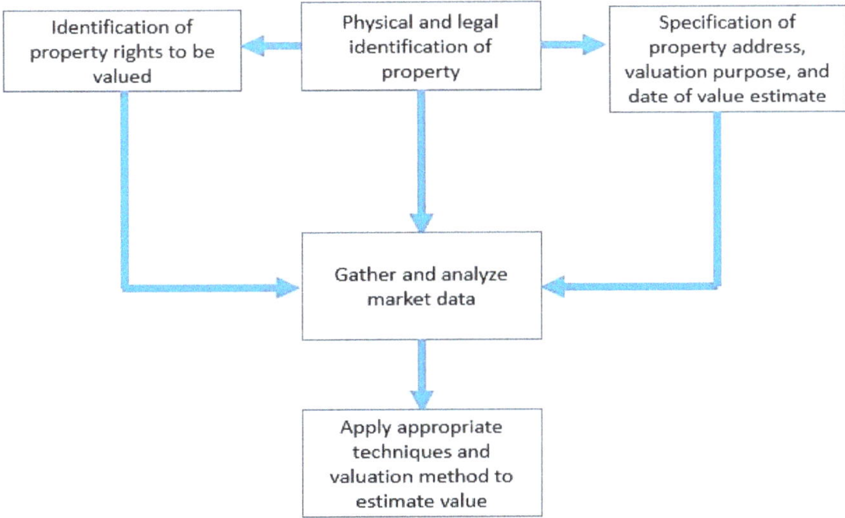

Figure 4. Land valuation process. Authors' construct.

Land taxation, currently referred to as property taxation or rating in Ghana, is one of the oldest tax forms [61]. In Ghana, this tax is paid with respect to a developed land or an immovable property [62]. Property tax differs amongst countries as it is paid in respect of; the land only in Kenya and Jamaica, buildings and improvements on land in Kosovo, and Tanzania, or to both in Canada, Germany, Japan, some parts of Australia, the United Kingdom, Indonesia, Thailand, Guinea, and Tunisia [62]. In Ghana, District Assemblies are the governmental institutions charged with the responsibility of preparing and levying property tax or rates in their areas of jurisdiction as per Section 144 of Act 936 [63]. Property taxation in Ghana is based on the replacement cost method/contractor's cost method of valuation. The tax is the replacement cost of the property after depreciation is deducted, and this should not exceed 50% of the replacement cost of an owner occupier's premises and must not be less than 75% in other cases [63]. The property taxation process in Ghana is illustrated in Figure 5 below.

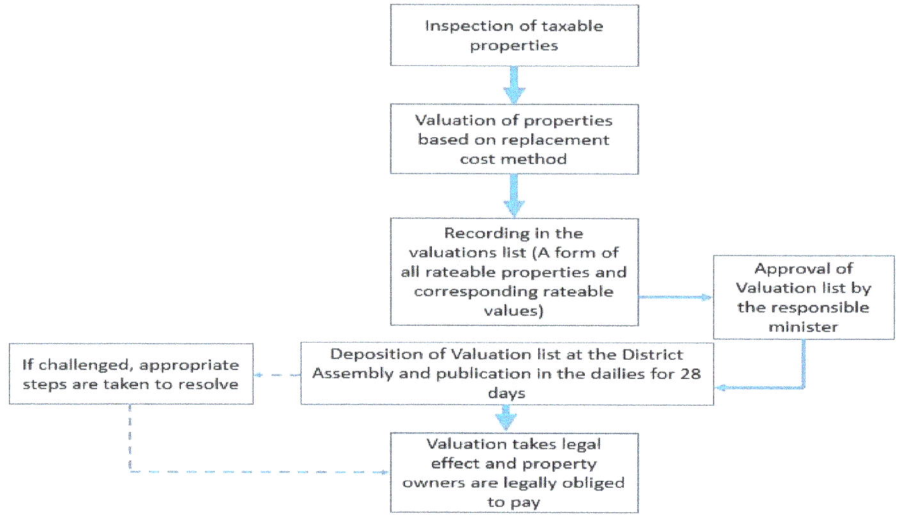

Figure 5. Property taxation process. Based on [62,63].

3.2.3. Land Use Planning and Land Development Processes

Land use planning and land development are closely linked and as such discussed together [30]. Land use planning is concerned with the planning and control of the use of land and natural resources, while the land development process is concerned with the implementation of development plans. Land development involves the building of new physical infrastructures and the implementation of construction planning and a change of land use through planning permissions and the granting of permits [55]. The designation of different land areas for different use types such as residential, commercial, recreational, and markets, and the actual carrying out of these plans based on the adoption of planning policies and land use regulations for a country, covering the national, regional to the local levels [55]. Land use planning and development in Ghana is concerned with balancing competing land uses for sustainable human settlement development [64]. The main legislation regulating land use planning in Ghana is the Town and Country Planning Ordinance, 1945 (CAP 84) [31]. Other legislations that border on physical planning in general include, National Development Planning System's Act, 1994 (Act 479), the Local Government Act, 2016 (ACT 936) and the National Building Regulations, 1996 (LI 1630) [65,66]. Land use planning in Ghana covers spatial, land use, and human settlement planning [67]. In Ghana, land use planning must ideally be based on decentralization and participatory principles [51,66,68]. Land use and development plans are prepared at the district level, forwarded and harmonized with those prepared at the regional level, and the two forwarded to the national level to the National Development Planning Commission (NDPC), where they are evaluated, and approval can be given for implementation [51,66,68].

The land use planning and development process in Ghana starts with the survey and definition of an area base map. This stage entails the collection, analysis, interpretation, and presentation, in a readily understood form, of all the data that are likely to influence the proposals which will be included in the land use plan. Here, planners with the help of local community people study the area to become conversant with all the characteristics which will help in defining the broad land use categories; residential, commercial among others [67]. Data are gathered through different survey types including a physical survey for data relating to topography, landscape, agricultural lands, and sometimes the geology of the area, a social survey gathers data on the population and its characteristics like the size, composition, structure, and housing, traffic transportation survey data includes the occupation, place of work, or school, origin and destination of work, rail and road networks, and parking facilities.

The survey stage is then followed by the planning stage. In planning, the goals and terms of reference are established as the first stage. At this stage, the planning area is defined, and all the involved people contacted. Some basic data of the area are gathered, and followed by a preliminary identification of problems and opportunities, as well as constraints to implementing improvement. The planning period is then set. The second stage is the organization of the work. This involves listing the planning tasks and activities and identifying the people or organizations responsible for these tasks or for contributing to them. Then, the needed resources are set out, and the work plan for the project as a whole is drawn up. Administrative matters and logistics are then arranged and provision is made for transport and other equipment. The third stage is a problem analysis which analyzes the causes of the problem in relation to the data already collected on the existing situation; population, land use, land resources, income, and occupation, among others. Constraints to change are then identified. The fourth stage involves identifying opportunities to change by first identifying and drafting a range of land use types that might help to achieve the goals of the plan. Generate a range of options for solving each problem in terms of opportunities; economic measures, land resources, government actions, the people, improved technology, and in terms of land use strategies; no change, maximum production, maximum conservation, etc. The fifth stage is the land suitability evaluation. At this stage, for each promising land use type, establishing the land requirements and matching these with the properties of the land to establish physical land suitability. The sixth stage comprises the appraisal of alternatives through social and economic analysis. That is, for each physically suitable combination of land use, the environmental, economic and social impacts, of the favorable and unfavorable, and of alternative courses of action, are assessed. Therefore, there should be an environmental impact assessment, financial and economic analysis, social impact assessment, and strategic planning. The seventh stage is the choosing of best options. Firstly, public and executive discussions are held on the viable options and their consequences. The comments from these discussions are then assembled and reviewed, and based on these, the necessary changes are made to the options. A decision is then made on which changes in land use should be made or worked towards. The last stage of planning is the preparation of a land use plan through zoning. This starts with the allocation or recommendation of the selected land uses for the chosen areas of land, followed by preparing the maps, the basic or master land use plan and supporting maps. After this, plans for how the selected improvements should be brought about, and how the plan is to be put into practice are made through an appropriate land management approach. A policy to guide the implementation is then drawn up, the budget is prepared and any necessary legislation drafted. It is important to mention the need for the involvement of decision-makers, sectoral agencies, and land users. The last stage of land use planning and development process is implementation. At this stage, the plan is put into action which is the responsibility of both the implementation agencies (mainly the town and country planning department) and the planning teams. During the implementation, there is the monitoring and revision of the plan in light of the goals defined at the initial stage as well as in light of the experiences that occur.

The land use planning and development process is summarized in Figure 6 below.

The next section discusses the transparency issues inherent to these different land administration processes and the role of blockchain technology to potentially resolve these.

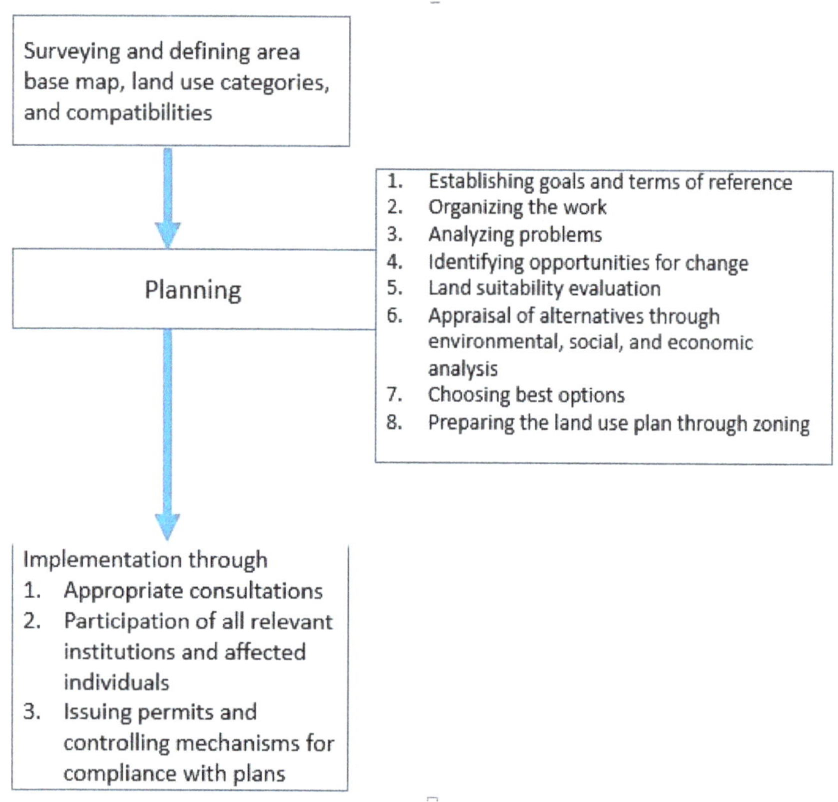

Figure 6. Land use planning and development process. Authors' construct.

4. Discussion

Transparency Issues of Land Administration Processes in Ghana and the Role of Blockchain Technology

Figure 7 below shows a four-dimensional framework for the transparency of land administration. The framework and its subsequent analysis and synthesis comprehensively capture and identify the transparency issues in the land administration processes, as highlighted under the findings, and highpoint how these processes can be made transparent, and the role of blockchain towards this. The transparency of land administration processes involves carrying out and sharing up-to-date information on ownership, value, and the use of land and all of its associated resources among related institutions, right holders and other stakeholders, including third parties, as well as, acting on the information in an open manner [30,55]. Transparency allows citizens unbridled access to land data, activities, organizations and professionals in an open and participatory manner in taking and implementing land decisions [4]. The availability/sharing of and accessibility to relevant land data, openness, and participatory processes in land administration thus underline the transparency of land administration in the context of this paper. Transparency issues appear akin across the different land administrations processes. For this reason, the discussion of blockchain's role towards addressing these issues has been integrated so as to give a better correlation and appreciation of the issues across the different land administration processes.

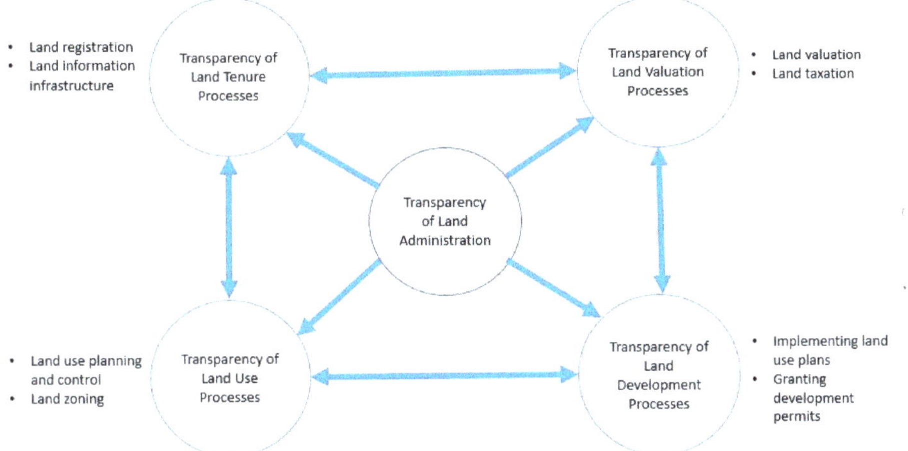

Figure 7. Transparency of the land administration framework. Source: authors' construct based on [30,55].

Taking the inventory of tenure by land registration significantly contributes to the openness, availability of information, and transparency of who owns and uses land [69]. Ensuring transparency thus depends on the establishment of land registers where they do not exist yet, making records accessible, securing transaction procedures, and documenting processed information [30]. When only regarding the legal context related to land registration in Ghana, one will expect openness and transparency in the system as outlined in our findings. However, [61] notes that in practice, the land registration procedure is cumbersome and fraught with lots of informal dealings, secrecy, bribery and corruption. In a 2016 survey, 69% and 9% of citizens that had received their registration certificates between 3–5 months and 6–8 months, respectively, indicated that they had paid bribes to middlemen or staff of the national land commission (NLC) to facilitate the process, whereas, those that had refused to pay bribes had their documents neglected, and prolonged to between 6 and 12 months to receive their land certificates [58]. This malicious delay due to the non-payment of the bribe is attributed to the lack of openness and transparency in the system which inhibits clients' ability to know the status of their registration documents in order to tell if documents are being unnecessarily delayed. The author [70] was correct to note that there is information asymmetry in Ghana's land sector, and that land information is monopolized by public land institutions. There is a lack of transparency [71] as well as information imbalance among land stakeholders, which greatly debilitates obtaining credible information due to the failure to divulge information between stakeholders, particularly to clients. This situation refutes the openness, availability and accessibility to information, and the participatory principles of land administration transparency.

On a technical level, the computerization of land registration processes can help to enhance transparency by giving citizens direct access to relevant data and also allow them to monitor process' progress [69]. Technically, introducing blockchain in an already established registration processes is feasible [2,29,44,46]. Ethereum Blockchain's smart contract for example is possible in Ghana's case [71]. Smart contract applications allow for predefined rules and requirements of the registration processes in order to be carried out successfully when these rules and requirements are met [16]. The design architecture of a public permissionless blockchain allows all transaction stakeholders, a free accessibility to information about the transaction and its processes by integrating all of them. In this way, the ability of documents to move through the stages of registration is independent of any single NLC officer or middleman, but subject to meeting all predefined conditions of the process, which every stakeholder can monitor equally. The whole registration process, from lodgment stage, through to the issuing and collection of the land certificate, thus becomes controlled by all stakeholders in the transaction due to their integration. This will not only expedite the registration process but also ensure the trust and credibility of land registration documents and processes. It is important to mention that, the manual stages of physical inspection, as well as survey and mapping services will still remain. However, these physical stages can now be brought under the complete monitoring of all concerned stakeholders, since everyone is aware of every stage of the registration. Stakeholders can therefore monitor these physical stages when they are due. This will help stakeholders to be able to validate the outputs of these physical processes as the accurate representation of the ground realities or not. Thus, in lieu of the transparency of land administration processes, a blockchain can boost land registration process by enhancing;

- Openness: through the decentralized broadcasting of transactions to all the integrated stakeholders, every decision or action can be known to all and no action can be hidden. Thus, although the different stages of the registration process involve different stakeholders, every stakeholder is aware of each stage, as well as, what, how, and when work is done on the transaction which allows for openness.

- Availability and access to land information: information imbalance obstructs accessibility to credible land data which breeds ignorance and permits fraudulent deals as some stakeholders become oblivious of other happenings in the transaction. The decentralized broadcasting of transactions and all associated information across the stakeholders, coupled with the verification and validation, as well as the hashing of new transaction blocks to historical blocks allows for easy accessibility to all relevant information (both current and historical) on land ownership, parcel, and rights, by all stakeholders at all times. This will help eliminate information asymmetry and its associated challenges of bribery and corruption.

- Participatory processes: the verification and validation through the consensus mechanism foster maximum participation in the entire registration processes from all stakeholders, since this allows the majority stakeholders to be part of transactions' decision making every time. The consensus mechanism takes place at every stage of the registration process until it is completed. Moreover, the broadcasting of process stages to all stakeholders automatically induces participation in the processes either actively or inactively. This is because everyone is aware of every happening and can give their contributions accordingly as and when necessary. That is, stakeholders are always privy to and aware of all the happenings and processes. This makes every stakeholder part of the transaction and registration processes in the participatory sense.

Regarding land information infrastructure, this typically relies on accurate and accessible cadastral and topographic datasets [55]. In Ghana, however, land information at the disposal of the different divisions of NLC is always not up-to-date because there is a lack of synchronization within the information infrastructure [71]. This challenge sometimes allows unauthorized tampering with land documents and data by some unscrupulous officials [71], across all divisions and in all the land processes. This is made possible because of the manual land administration system. This challenge can however be eliminated through the digitization and application of blockchain across the different divisions [2]. Every change on blockchain updates automatically without human efforts. This will thus provide up-to-date data at every point in time across all the divisions of the NLC and in all their processes. In this way, in addition to facilitating the data accessibility, openness and participatory processes, the blockchain will ensure up-to-date land data all the time to enhance all the land administration processes and decisions [6,55]. These potentials of blockchain if combined with the publication in the dailies stage of registration process and land taxation processes, will boost openness, transparency and participation for all citizens for transparency in the system.

Land valuation must represent an unbiased estimate, a learned opinion and a supported value estimate. Where there is no openness in the valuation processes, biases cannot be identified. Where there are difficulties in accessing market data on comparable properties, valuation will not reflect the reasonable market value. Moreover, where there is no participation in the valuation process from involved stakeholders, it will not be possible to achieve an 'arm's length transaction' since parties will have limited idea of the actual market situation. Again, maximum participation helps to avoid value conjecture on the part of some valuers who may skip some appropriate valuation steps like the physical inspection of the property and its comparables, due to the laborious and tedious nature of these valuation steps. In [30], the authors note that land value data is useful to achieve the arm's length transaction as it gives data for comparison purposes. A major challenge of valuation process in Ghana is the access to readily available and up-to-date market data on comparable properties, either from property owners or from the land institutions. This is due to the secrecy amongst land stakeholders, lack of transparency, and also information imbalance as identified in [70–72]. The same challenges that lead to land registration challenges and lengthy processing time. These greatly affect the valuation processes and the possibility of valuation results to reflect the current market situation and factors. On this basis, if registered properties and registration processes are carried out on blockchain as discussed already, the valuation processes can be linked to and carried out on this blockchain system. In so doing, since all registered properties and their data are readily available, it will facilitate access to the market data on comparables, particularly of registered properties. The openness of the system will also permit all stakeholder awareness of the valuation process to achieve the arm's length transaction [30], and a truly reflexive market value which is based on the prevailing market situations and factors. The choice of valuation method and its appropriateness can also be evaluated by stakeholders, particularly given that comparables with the same basis and purpose of valuation can easily be found from registered properties via the blockchain system. Therefore, with the secure, immutable, time-stamped, and up-to-date characteristics of the blockchain-based land administration system [24], accessibility to comparables for valuation is made easier and faster, as well as is open to the knowledge of all stakeholders. Thus, the valuation process from, identifying property owners, and comparables, through choosing a valuation technique, to actual valuation, can then be carried out via the blockchain system. This can help to eliminate value conjecture by some valuers due to the difficult accessibility to market data on comparables, as well as ignorance on the part of other stakeholders of the valuation processes. This can also address petty mistakes like wrong addresses, incorrect party details, valuation dates, as well as the exact rights to properties since other stakeholders can identify and rectify these through verification and validation.

The valuation process is similar to the taxation process. The difference is that the valuation list for taxation is, however, published in the dailies for 28 days before they can become legally binding, In property taxation, a major challenge in making property tax administration effective in Ghana is the difficulty in connecting properties to their locations, and also where transparency is lacking in the system [62,72,73]. To boost effectiveness therefore, there must be openness, as well as availability of and accessibility to property location and other information. Since the taxation process is just like the valuation processes, this can also be carried out using blockchain. Blockchain will make the identification of registered property easier for taxation as they are readily listed in the system. In this way, the process involved in levying property taxes will become open to stakeholders and give easy access to information queries and clarifications, to make the system open and participatory for transparency. Moreover, taxation records can be kept securely in this system to eliminate inherent illegalities as well as to ensure that all taxes are channeled into the right government coffers since any diversion of taxes will be known on the blockchain system.

In view of the transparency challenges of the land administration processes in the foregoing discussions, the current Ghanaian system of land use planning and development has been criticized, despite the requirement for all developments to proceed with issuing development permits [72]. The argument underlying this critique is that the system does not promote compliance [65,73]. This is a problem not only in Ghana but across the sub-Saharan Africa region. The authors in [74] noted that between 50% and 75% of all the new houses in the region's cities were developed on lands delivered through processes that do not comply with all the legal requirements. In Ghana, [74] again noted that 31% of property owner respondents had building permits while 69% had no building permits, and neither were they in the process of or taking steps with the aim of acquiring one. Of the 31% respondents that had building permits, only 23.3% had acquired permits prior to starting construction, while 76.7% did so subsequently to their building commencement [74]. The educated and formal sector employees who were aware of land use planning and development were the most that had building permits and 'there is the likelihood for such people to have connections and influence to aid their acquisition of building permit' [74] (p. 21). Without such connections and influence, a person is likely to face challenges like unnecessary delays, and the paying of illegal monies, just as was seen under the land registration process, before they can receive permits to commence developments [74]. There is therefore negative trust perceptions for land use planning and development officials, and the system [75]. In [66], the authors identified a lack of involvement and or participation and better knowledge of land use planning amongst the majority of citizens, and recommended 'the need for planning authorities to adopt participatory land use planning together with customary landholders, and educating them on the essence of comprehensive land use planning approaches' [66] (p. 4), [42]. These problems account for the low compliance with land use planning and development regulations, leading to a high rate of unauthorized developments. Land use planning and development processes need to be as open and as transparent as possible to allow for equal awareness, better knowledge, and accessibility to the system for all citizens. Adopting participatory approaches to planning by involving citizens, particularly those affected by the planning scheme, is a means to create awareness and to boost trust for the system [76]. Blockchain technology which integrates all stakeholders in a transaction and decision-making processes can facilitate the participatory planning approaches. A permissionless public blockchain (this allows all stakeholders to have open access, join, and partake in decisions without restrictions) is useful in this sense [45]. Citizens have to sign up to this permissionless blockchain via their computers or other supportive devices. They will then be assigned confidential private keys with which to sign into the system every time to be able to initiate a request or partake in discussions or transactions as seen in Figure 1 under findings. No external permission is necessary. Therefore, stakeholders can login to see all land use planning and development discussions and actions, follow it and contribute to it where necessary. To achieve this will, however, require intensive public education and awareness creation for the majority of citizens to know the use and be able to partake in the system. If this is done, it can improve more citizenry participation in processes and decisions on land use planning and development. The improved accessibility and participatory processes can consequently enhance openness, transparency, and increased trust among stakeholders. This is because it is impossible to hide decisions and processes from any stakeholder. Digitizing land use planning and development processes and data on blockchain system will therefore integrate all stakeholders. In this way, stakeholders can monitor areas for which land use permits have been granted and areas for which they have not been granted since these data will reflect on the blockchain system, and be known to all stakeholders. Citizens can then act as watchdogs, and to report on any developments that commence without the right approval. This can help to end the non-compliance to land use planning and development schemes, as well as the indiscriminate and unauthorized developments.

Improved participation through the use of a blockchain system for land transaction helps citizens to have control and security over the data. This enhances the take-up and trust in government institutions and processes to support sustainable economic growth as identified in the implementation of Georgia's blockchain land registration project [76]. A counter argument that such an improved open accessibility and participatory process can lead to opportunistic behaviors and misinformation can be made. However, blockchain's design architecture provides for systematic review and checks for all decisions and information. This is done by the majority stakeholders through the verification and validation (consensus mechanism) of data based on the good knowledge of actual grounds work, sources, history, and credibility of the stakeholder that is making or giving such decisions and data before they are accepted as true and authentic [23]. Based on the blockchain's elements of distributive decentralization (which integrates all land stakeholders), the consensus mechanism, hashing of records, immutability, and synchronization of data, land administration processes can be carried out in a way that is open to all stakeholders to ensure transparency, enhance trust amongst stakeholders, as well as achieve up-to-date data at all times for land decisions. This can be achieved by adopting a single permissionless public blockchain system for the different land administration divisions and their processes. The manual land administration functions like surveying and physical inspections will still be manual but can now be done with all other stakeholders being aware. This is because, blockchain is a decentralized technology and permits everyone on the system to know and be aware when each of the land transaction stages is due. This allows stakeholders to be able to follow, and to keep an oversight check on these processes to confirm accuracy through validation. Applying blockchain across all processes of land administration in such a simultaneous approach has synergistic effects of real-time data update, accessibility, and openness across them. This makes it easier for each stakeholder to keep-up, and to participate in decisions and transactions. It also ensures easy access to readily available land data for all interested stakeholders.

5. Conclusions

This paper aimed to identify the essential elements and relations between the blockchain technology and transparency of land administration in the existing literature, and to assess the potential of blockchain to improve the transparency of land administration processes—based on the context from Ghana. These aims were achieved through a comprehensive review of all the land administration processes in Ghana, the inherent transparency issues in them, and the possibility of blockchain to support and enhance transparency in these processes simultaneously. The paper argued and demonstrated that the completeness of land administration transparency is when transparency is achieved across all land administration processes, and stakeholders simultaneously. A single permissionless public blockchain can help achieve this. However, there is the need for the different land divisions to establish standardization in the land administration processes prior to the blockchain's application in such a compressive approach. This is because, where there is no such standardization, there is a high possibility of inconsistencies and irregularities in the processes across the different divisions which can affect the efficient working of the blockchain system across all the divisions.

This study is relevant for all land stakeholders, as it provides a better understanding, and an interpretive approach to the social and political realities of land administration in Ghana. It has also extended the discourse on the topic and offers a quick and easy reference guide for scholars, practitioners, and policy makers as hitherto, land administration processes and transparency issues in Ghana have been discussed individually, in piecemeal and scattered across different works, which hindered a better appreciation of the topic due to the polarization and different epistemological views [35,49,51–53,62,65].

As part of the policy implications for blockchain adoption, there should be a review of all paper-based land transactions for errors and corrections, and a comprehensive digitization of land administration transactions and processes in the country, in addition to the public–private partnership in the blockchain-based land administration process. Again, an intensive public education, particularly for land stakeholders, is necessary to understand the blockchain system before implementation can begin. Finally, as seen in the findings, blockchains will affect the institutional relations and shared authorities between all stakeholders which include government agencies, local chiefs and individual landowners. This is because, land decisions and associated activities are no longer dependent on a single party, but are going to be a shared effort. It is therefore important that this new decentralized and shared authority be deliberated, and negotiated to reach a consensus, particularly with the chiefs. This is because chiefs own 80% of land in Ghana and hence, there is a need for their consent and cooperation if blockchain implementation can be successful. As a consequence to establishing this, blockchain can fundamentally change the transparency variations in land administration to be more equal and homogeneous, regardless of the type of land.

The nascent nature of the topic area, and limited conceptualizations, hindered the ability to explore more options and in further detail, the blockchain architecture types that can possibly support such a comprehensive transparency of land administration as presented in this study. We therefore recommend that future research focuses on exploring this area. Moreover, since this study focused mainly on statutory land administration processes, future research works should consider the topic from the customary land administration perspective especially, given that customary lands cover 80% of land in Ghana, are governed by different indigenous customary laws, and are based on a low level of technological how-how. Finally, future researcher works, and potential institutions for blockchain adoption, should be aware of and take into consideration the technology's flaws in terms of limited storage capacity, the limitations to its scalability and speed, as well as the huge electricity consumption for its operation.

Author Contributions: Conceptualization, methodology, investigation, resources, formal analysis, validation, data curation, writing—original draft, writing—review and editing: P.D.A. and W.T.d.V.; Supervision: W.T.d.V. All authors have read and agreed to the published version of the manuscript.

Funding: This research received no external funding.

Acknowledgments: This study was carried out within the timeframe of a Ph.D. research program at the Technische Universität München (TUM), the Chair of Land Management and Land Tenure. The Ph.D. program was funded by the Katholischer Akademischer Auslander-Dienst (KAAD). We appreciate the financial support of the studies. This research received no external funding. Appreciation also goes out to the anonymous readers that read through and offered constructive comments to help improve the initial draft, as well as to the main reviewers of the paper. We say a big thank you for your valuable comments for the paper's improvement.

Conflicts of Interest: The authors declare no conflict of interest.

References

1. Vos, J.; Lemmen, C.; Beentjes, B. Blockchain-Based Land Administration Feasible, Illusory or a Panacea? In Proceedings of the Responsible Land Governance: Towards an Evidence Based Approach, Washington, DC, USA, 20–24 March 2017. [CrossRef]
2. Lemmen, C.; Vos, J.; Beentjes, B. Ongoing Development of Land Administration Standards: Blockchain in Transaction Management. *Eur. Prop. Law J.* **2017**, *4*, 478–502. [CrossRef]
3. Dawidowicz, A.; Źróbek, R. Land Administration System for Sustainable Development—Case Study of Poland. *Real Estate Manag. Valuat.* **2017**, *4*, 112–122. [CrossRef]
4. Bagdai, N.; van der Molen, P.; Tuladhar, A. Does uncertainty exist where transparency is missing? Land privatisation in Mongolia. *Land Use Policy* **2012**, *4*, 798–804. [CrossRef]
5. Williamson, I.P.; Grant, D.M. United Nations-FIG Bathurst Declaration on Land Administration for Sustainable Development: Development and Impact. In Proceedings of the XXII FIG International Congress, Washington, DC, USA, 19–26 April 2002.
6. Enemark, S. Building land information policies. In Proceedings of the Special Forum on Building Land Information Policies in the Americas, Aguascalientes, Mexico, 26 October 2004; Volume 26, p. 2004.
7. Locke, A.; Henley, G. The Possible Shape of a Land Transparency Initiative: Lessons from Other Transparency Initiatives. 2013. Available online: http://search.ebscohost.com/login.aspx?direct=true&db=lah&AN=20133405582&site=ehost-live%5Cnhttp://www.odi.org.uk/sites/odi.org.uk/files/odi-assets/publications-opinion-files/8599.pdf%5Cnemail:a.locke@odi.org.uk (accessed on 1 February 2020).
8. UN-Habitat. Tenure Responsive Land Use Planning: A Guide for Country Level Implementation. 2016. Available online: www.unhabitat.org (accessed on 18 April 2020).
9. Bell, K.C.; Bell, K.C. *Good Governance in Land Administration*; World Bank: Washington, DC, USA, 2007.
10. Phuong, T.H. Enhancing Transparency in Land Transaction Process by Reference Architecture for Workflow Management System. In Proceedings of the PACIS, Ho Chi Minh City, Vietnam, 11–15 July 2012; p. 69.
11. Anand, A.; McKibbin, M.; Pichel, F. Colored Coins: Bitcoin, Blockchain, and Land Administration. In Annual World Bank Conference on Land and Poverty. 2015. Available online: https://www.ubitquity.io/home/resources/worldbank_land_paper_ubitquity_march_2016.pdf (accessed on 8 January 2020).

12. Bagdai, N.; van der Veen, A.; van der Molen, I.P.; Tuladhar, A. Transparency as a Solution for Uncertainty in Land Privatization—A Pilot Study for Mongolia. In Proceedings of the Surveyors Key Role in Accelerated Development, Eilat, Israel, 3–8 May 2009.

13. Jaitner, A.; Caldeira, R.; Koynova, S. Transparency International-Land Corruption in Africa-Finding Evidence, Triggering Change. In Proceedings of the Annual World Bank Conference on Land and Poverty, Washington, DC, USA, 20 March 2017.

14. Spielman, A. Blockchain: Digitally Rebuilding the Real Estate Industry. Ph.D. Thesis, Massachusetts Institute of Technology, Cambridge, MA, USA, 2016.

15. Lemieux, V.L. Evaluating the Use of Blockchain in Land Transactions: An Archival Science Perspective. *Eur. Prop. Law J.* **2017**, *4*, 392–440. [CrossRef]

16. Rizal Batubara, F.; Ubacht, J.; Janssen, M. Unraveling Transparency and Accountability in Blockchain. In Proceedings of the 20th Annual International Conference on Digital Government Research, Dubai, UAE, 18–20 June 2019; pp. 204–213. [CrossRef]

17. Müller, H.; Seifert, M. Blockchain, a Feasible Technology for Land Administration? In Proceedings of the FIG Working Week, Hanoi, Vietnam, 22–24 April 2019.

18. Singh, H.; Gupta, H.; Singh, A.; Litoria, P.K. Applications of Blockchain for Land Record Management. In Proceedings of the National Conference on Role of Geospatial Technologies to Bridge the Rural and Uraban Divide, Ludhiana, India, 22–23 February 2018.

19. Eder, G. Digital Transformation: Blockchain and Land Titles. In Proceedings of the OECD Global Anti-Corruption & Integrity Forum, Paris, France, 20–21 March 2019.

20. Shang, Q.; Price, A. A Blockchain-Based Land Titling Project in the Republic of Georgia: Rebuilding Public Trust and Lessons for Future Pilot Projects. *Innov. Technol. Gov. Glob.* **2019**, *4*, 72–78. [CrossRef]

21. Bal, M. *Securing Property Rights in India Through Distributed Ledger Technology*; Observer Research Foundation: New Delhi, India, 2017.

22. Benbunan-Fich, R.; Castellanos, A. Digitalization of land records: From paper to blockchain. In Proceedings of the International Conference on Information Systems 2018, ICIS 2018, San Francisco, CA, USA, 13–16 December 2018.

23. Thakur, V.; Doja, M.N.; Dwivedi, Y.K.; Ahmad, T.; Khadanga, G. Land records on blockchain for implementation of land titling in India. *Int. J. Inf. Manag.* **2020**, *4*, 101940. [CrossRef]

24. Lazuashvili, N.; Norta, A.; Draheim, D. Integration of Blockchain Technology into a Land Registration System for Immutable Traceability: A Casestudy of Georgia. In Proceedings of the International Conference on Business Process Management, Vienna, Austria, 1–6 September 2019; Volume 361, pp. 219–233. [CrossRef]

25. Krishnapriya, S.; Sarath, G. Securing Land Registration using using Blockchain Blockchain. *Procedia Comput. Sci.* **2020**, *4*, 1708–1715. [CrossRef]

26. Vos, J. Blockchain-based land registry: Panacea illusion or something in between? In Proceedings of the IPRA/CINDER Congress, Dubai, UAE, 22–24 February 2016.

27. Kempe, M. *The Land Registry in the Blockchain—Testbed*; A development project with Lantmäteriet, Landshypotek Bank, SBAB, Telia Company, ChromaWay and Kairos Future. European Urology Supplements; Kairos Future: Stockholm, Sweden, 2017. [CrossRef]

28. Peiró, N.N.; García, E.J.M. Blockchain and Land Registration Systems. *Eur. Prop. Law J.* **2017**, *4*, 296–320. [CrossRef]

29. Yapicioglu, B.; Leshinsky, R. Blockchain as a tool for land rights: Ownership of land in Cyprus. *J. Prop. Plan. Environ. Law* **2020**. [CrossRef]

30. Yildiz, U.; Zevenbergen, J.; Todorovski, D. Exploring the Relation between Transparency of Land Administration and Land Markets: Case Study of Turkey. In Proceedings of the FIG Working Week 2020 Smart Surveyors for Land and Water Management, Amsterdam, The Netherland, 10–14 May 2020.

31. Torraco, R.J. Writing Integrative Literature Reviews: Using the Past and Present to Explore the Future. *Hum. Resour. Dev. Rev.* **2016**, *15*, 404–428. [CrossRef]

32. Snyder, H. Literature review as a research methodology: An overview and guidelines. *J. Bus. Res.* **2019**, *4*, 333–339. [CrossRef]

33. Ntihinyurwa, P.D.; de Vries, W.T. Farmland fragmentation concourse: Analysis of scenarios and research gaps. *Land Use Policy* **2020**, *4*, 104936. [CrossRef]

34. Torraco, R.J. Writing Integrative Literature Reviews: Guidelines and Examples. *Hum. Resour. Dev. Rev.* **2005**, *4*, 356–367. [CrossRef]

35. Obeng-odoom, F. Urban property taxation, revenue generation and redistribution in a frontier oil city. *Cities* **2014**, *4*, 58–64. [CrossRef]

36. Webster, J.; Watson, R.T. Analyzing the Past to Prepare for the Future: Writing a Literature Review. *MIS Q.* **2002**, *26*, xiii–xxiii.

37. Gale, N.K.; Heath, G.; Cameron, E.; Rashid, S.; Redwood, S. Using the framework method for the analysis of qualitative data in multi-disciplinary health research. *BMC Med. Res. Methodol.* **2013**, *4*, 1–8. [CrossRef]

38. Levack, W.M. The role of qualitative metasynthesis in evidence-based physical therapy. *Phys. Ther. Rev.* **2012**, *4*, 390–397. [CrossRef]

39. Karamitsos, I.; Papadaki, M.; Al Barghuthi, N.B. Design of the Blockchain Smart Contract: A Use Case for Real Estate. *J. Inf. Secur.* **2018**, *4*, 177–190. [CrossRef]

40. Natarajan, H.; Krause, S.K.; Gradstein, H.L. *Distributed Ledger Technology (DLT) and Blockchain*; FinTech note, no. 1; World Bank Group: Washington, DC, USA, 2019.

41. Themistocleous, M. Blockchain Technology and Land Registry. *Cyprus Rev.* **2018**, *4*, 195–202.

42. Shuaib, M.; Daud, S.M.; Alam, S.; Khan, W.Z. Blockchain-based framework for secure and reliable land registry system. *Telkomnika* **2020**, *4*, 2560–2571. [CrossRef]

43. Oberdorf, V. Building Blocks for Land Administration: The Potential Impact of Blockchain-Based Land Administration Platforms in Ghana. Master's Thesis, Utrecht University, Utrecht, The Netherlands, 2017.

44. Kaczorowska, M. Blockchain-based land registration: Possibilities and challenges. *Masaryk Univ. J. Law Technol.* **2019**, *4*, 339–360. [CrossRef]

45. Makala, B.; Anand, A. Blockchain and Land Administration. *J. Dev. Stud.* **2018**, *4*, 1–34.

46. Ølnes, S.; Ubacht, J.; Janssen, M. Blockchain in government: Benefits and implications of distributed ledger technology for information sharing. *Gov. Inf. Q.* **2017**, *4*, 355–364. [CrossRef]

47. Petkova, P.; Jekov, B. Blockchain in e-Governance. In *Selected and Extended Papers from X-th International Scientific Conference 'E-Governance and e-Communication'*; SSRN: Rochester, NY, USA, 2018; p. 149.

48. Shapiro, E.; Mackmin, D.; Sams, G. *Modern Methods of Valuation*; Taylor & Francis: Boca Raton, FL, USA, 2012.

49. Agyemang, F.S.K.; Morrison, N. Recognising the barriers to securing affordable housing through the land use planning system in Sub-Saharan Africa: A perspective from Ghana. *Urban Stud.* **2017**, *4*, 2640–2659. [CrossRef]

50. Stefanović, Ð.M.; Pržulj, D.; Ristic, S.; Stefanović, D. Blockchain and Land Administration: Possible applications and limitations. In Proceedings of the International Scientific Conference on Contemporary Issues in Economics Business and Management, Kragujevac, Serbia, 9–10 November 2018.

51. Fuseini, I.; Kemp, J. A review of spatial planning in Ghana's socio-economic development trajectory: A sustainable development perspective. *Land Use Policy* **2015**, *4*, 309–320. [CrossRef]

52. Kuusaana, E.D. Property rating potentials and hurdles: What can be done to boost property rating in Ghana? *Commonw. J. Local Gov.* **2015**, 204–223. [CrossRef]

53. Kleemann, J.; Inkoom, J.N.; Thiel, M.; Shankar, S.; Lautenbach, S.; Fürst, C. Peri-urban land use pattern and its relation to land use planning in Ghana, West Africa. *Landsc. Urban Plan.* **2017**, *4*, 280–294. [CrossRef]

54. Stahl, J.; Sikor, T.; Dorondel, S. Transparency in Albanian and Romanian land administration. Paper Forthcoming in the Next Issue of Cahiers Options Méditerranéennes entitled "La Question Foncière Dans les Balkans". 2008. Available online: http://www.landcoalition.org/pdf/07_paper_transparency_land.pdf (accessed on 14 August 2020).

55. Enemark, S. Understanding the land management paradigm. In Proceedings of the FIG Commission 7, Symposium on Innovative Technologies for Land Administration, Madison, WI, USA, 18–25 June 2005; pp. 19–25.

56. Zevenbergen, J. A Systems Approach to Land Registration and Cadastre. *Nord. J. Surv. Real Estate Res.* **2004**, *1*. Available online: https://journal.fi/njs/article/view/41503 (accessed on 18 August 2020).

57. Sittie, R. Land title registration. The Ghanaian experience. In Proceedings of the 23rd FIG Congress, Munich, Germany, Munich, Germany, 8–13 October 2006.

58. Ehwi, R.J.; Asante, L.A. Ex-Post Analysis of Land Title Registration in Ghana since 2008 Merger: Accra Lands Commission in Perspective. *Sage Open* **2016**, *4*, 2158244016643351. [CrossRef]

59. Mintah, K.; Baako, K.T.; Kavaarpuo, G.; Otchere, G.K. Skin lands in Ghana and application of blockchain technology for acquisition and title registration. *J. Prop. Plan. Environ. Law* **2020**. [CrossRef]

60. Asiama, K.O.; Bennett, R.; Zevenbergen, J.; Asiama, S.O. Land valuation in support of responsible land consolidation on Ghana's rural customary lands. *Surv. Rev.* **2018**, *4*, 288–300. [CrossRef]

61. Kuusaana, E.D. Property taxation and its revenue utilisation for urban infrastructure and services in Ghana: Evidence from Sekondi-Takoradi metropolis. *Prop. Manag.* **2016**, *4*, 297–315. [CrossRef]

62. Petio, M.K. Role of the Land Valuation Division in Property Rating by District Assemblies in Ghana's Upper East Region. *Commonw. J. Local Gov.* **2013**, *4*, 69–89. [CrossRef]

63. GOG. *Local Government Act*; GOG: Warsaw, Poland, 2016.

64. Yeboah, E.; Shaw, D.P. Customary land tenure practices in ghana: Examining the relationship with land-use planning delivery. *Int. Dev. Plan. Rev.* **2013**, *4*, 21. [CrossRef]

65. Awuah KG, B.; Hammond, F.N.; Lamond, J.E.; Booth, C. Benefits of urban land use planning in Ghana. *Geoforum* **2014**, *4*, 37–46. [CrossRef]

66. Kuusaana, E.D.; Eledi, J.A. Customary land allocation, urbanization and land use planning in Ghana: Implications for food systems in the Wa Municipality. *Land Use Policy* **2015**, *48*, 454–466. [CrossRef]

67. GOG. *Land Use and Spatial Planning Act*; GOG: Warsaw, Poland, 2016.

68. Owusu, G. Small towns and decentralized development in Ghana: Theory and practice. *Afr. Spectr.* **2004**, *4*, 165–195.

69. Van der Molen, P. Some Measures to Improve Transparency in land Administration. 2007. Available online: http://www.fig.net/pub/fig2007/papers/ts_1a/ts01a_05_molen_1304.pdf (accessed on 18 August 2020).

70. Adiaba, S.Y. A Framework for Land Information Management in Ghana. Available online: https://wlv.openrepository.com/bitstream/handle/2436/332138/THESIS0920118doc.pdf?sequence=1&isAllowed=y (accessed on 2 March 2020).

71. Agbesi, S.; Tahiru, F. Application of Blockchain Technology in Land Administration in Ghana. In *Cross-Industry Use of Blockchain Technology and Opportunities for the Future*; IGI Global: Hershey, PA, USA, 2020; pp. 103–116. [CrossRef]

72. Mantey, S.; Tagoe, N.D. Geo-Property Tax Information System-A Case Study of the Tarkwa Nsuaem Municipality, Ghana. In Proceedings of the FIG Working Week, Rome, Italy, 6–10 May 2012; pp. 6–10.

73. Boamah, N.A. Constraints on property rating in the Offinso South Municipality of Ghana. *Commonwealth J. Local Gov.* **2013**, 77–89. [CrossRef]

74. Awuah, K.G.B.; Hammond, F.N. Determinants of low land use planning regulation compliance rate in Ghana. *Habitat Int.* **2014**, *4*, 17–23. [CrossRef]

75. Siiba, A.; Adams, E.A.; Cobbina, P.B. Chieftaincy and sustainable urban land use planning in Yendi, Ghana: Towards congruence. *Cities* **2018**, *4*, 96–105. [CrossRef]

76. Goderdzishvili, N.; Gordadze, E.; Gagnidze, N. Georgia's Blockchain-powered Property Registration: Never blocked, Always Secured: Ownership Data Kept Best! In Proceedings of the 11th International Conference on Theory and Practice of Electronic Governance, Galway, Ireland, 4–6 April 2018; pp. 673–675. [CrossRef]

Publisher's Note: MDPI stays neutral with regard to jurisdictional claims in published maps and institutional affiliations.

 *land*

Article

Evaluation of the Objectives and Concerns of Farmers to Apply Different Agricultural Managements in Olive Groves: The Case of Estepa Region (Southern, Spain)

Antonio Alberto Rodríguez Sousa [1], Carlos Parra-López [2], Samir Sayadi-Gmada [2], Jesús M. Barandica [1] and Alejandro J. Rescia [1,*]

[1] Department of Biodiversity, Ecology and Evolution (BEE), Teaching Unit of Ecology (UDECO), Faculty of Biological Sciences, University Complutense of Madrid, 28040 Madrid, Spain; antonr05@ucm.es (A.A.R.S.); jmbarand@ucm.es (J.M.B.)

[2] Department of Agricultural Economics and Sociology, Andalusian Institute of Agricultural Research (IFAPA), Camino de Purchil s/n, 18080 Granada, Spain; carlos.parra@juntadeandalucia.es (C.P.-L.); samir.sayadi@juntadeandalucia.es (S.S.-G.)

* Correspondence: alejo296@ucm.es; Tel.: +34-91-394-47-39

Received: 2 September 2020; Accepted: 28 September 2020; Published: 1 October 2020

Abstract: Olive groves are representative of the landscape and culture of Spain. They occupy 2.5 M ha (1.5 M ha in Andalusia) and are characterised by their multifunctionality. In recent years, socio-economic and environmental factors (i.e., erosion) have compromised their sustainability, leading farmers to abandon their farms or intensify their management. The main objective/purpose of this research was to study the drivers and concerns that condition farmers' choice of a given olive grove management model. Taking the *Estepa* region as a case study (Andalusia, Spain), surveys were conducted among farmers with integrated and organic managed olive groves. The socio-economic aspects were the main objectives and concerns of the farmers with integrated olive groves. In the case of farmers with organic management, conservation objectives prevailed, and their concerns were oriented to environmental threats. The education level was a key factor in the adoption of given farm management, as it increased the level of environmental awareness. In the context of multifunctional agriculture, it would be desirable to increase this awareness of the environmental threats against olive groves, in order to provide incentives for the implementation of agri-environmental practices that would enhance the sustainability of these systems.

Keywords: farm income; landscape ecology; multifunctional agriculture; olive groves; social demands; socio-ecosystems; sustainability

1. Introduction

The agricultural systems of olive groves form multifunctional socio-ecological landscapes of notable importance in the Mediterranean basin, occupying 5 M ha of the Useful Agricultural Surface (UAS) of Europe [1–3]. Spain is the country with the largest olive-growing area in the world, reaching 2.5 M ha, 60% of which is concentrated in the Andalusia Region (southern Spain) [4]. This wide extension of olive growing in Spain gives it a high production of olives and, in particular, of olive oil. It is the first supplier of this product, with an average yield of 1.19 t year^{-1} throughout the last five collection campaigns (2012/2013–2016/2017) [5]. However, despite their wide representation and continued production, the olive grove agricultural systems present a high degree of uncertainty regarding their sustainability. The main driving factors of this situation are the rural migration and, consequently, the abandonment of agricultural land that has taken place since the 1950s, along with price volatility in agricultural markets [6,7]. In addition, the implementation, in 1957, of the Common Agricultural Policy (CAP), which originally provided incentives for the productive performance of farms, contributed negatively to the sustainability of traditional agricultural systems [8–11].

In face of the economic vulnerability of olive groves, farmers have had to opt for alternative models of agricultural management that would help ensure the persistence of these crops. In this sense, some farmers have chosen to intensify the management of their farming systems by increasing olive tree density and providing energy inputs such as fertilizers, pesticides, and the application of irrigation to maximize production yield [12–14]. However, this intensification led to some negative multidimensional externalities, such as greater soil erosion and diffused pollution derived from the indiscriminate use of agrochemicals [11,15]. In the last few years, other types of environmentally friendly management have been consolidated, such as integrated or organic farming (Table 1). In this type of agricultural system, taking into account the multifunctional agriculture framework, it is highly advisable to consider their economic, social, and environmental dimensions (i.e., Triple Bottom Line approach) in order to assess their sustainability and to promote a stable supply of ecosystem services (ES) to society [16–18].

Table 1. Main characteristics and agricultural practices carried out in the integrated and organic farming of the olive grove [14]. *Desvareto* is an agricultural practice related to the removal of stems from the olive tree.

Characteristics and Agricultural Practices	Integrated Farming	Organic Farming
Mechanisation	Allowed on slopes <20%	Allowed on slopes <20%
Water regime	Rainfed or deficit irrigation	Rainfed or deficit irrigation
Age of olive trees (years)	>25	10–25 (modern crops)
Plant density (trees ha^{-1})	100–500	100–500
Pruning	Biannual	Biannual
Waste disposal	Burning/Grinder	Grinder
Desvareto	Required	Required
Plant covers	Partial	Total
Pest management	Synthetic pesticides (chemical compounds)	Non-synthetic pesticides (organic compounds)
Fertilisation	Synthetic (foliar and soil; fertigation)	Organic (foliar and soil)
Harvesting	Manual vibrator	Manual vibrator

Although both farming models have great similarities, the integrated olive management model allows the implementation of partial plant covers, the use of chemically synthesised pesticides and fertilisers (i.e., NPK fertilisers, glyphosate) in a regulated way by external agencies, and deficit irrigation in water stress situations [14,19]. Differentially, organic olive groves are modern crops, where only the use of organic pesticides and fertilisers is allowed, and the use of irrigation is minimised. Additionally, from a legislative point of view at the Spanish level, the implementation of partial and total plant covers is mandatory in integrated and organic agriculture respectively, in order to minimise the loss of organic matter and soil fertility due to erosive processes [3,20–24].

As farming systems, the contribution of provisioning ES of olive groves is essential and, therefore, must be valued from the political dimension. The CAP, consisting of an income support pillar (Pillar 1) and a rural development support pillar (Pillar 2), with annual subsidies granted to agriculture (37.8% of the general budget of the European Union (EU)) and specific national policies such as the Law on Olive Groves or the General Plan for Olive Groves in Andalusia support farmers at different levels by improving the profitability of crops and olive groves, respectively [22,23,25,26]. Historically there has been a transition from a productivist CAP where the "single payment" was predominant in Pillar 1, to a more environmentally-friendly CAP, where 30% of the budget for direct payments is based on a "greening" regime, referring to the obligations of farmers with arable land to introduce crop rotation and diversification, and to preserve natural grasslands [27]. On the other hand, aid under Pillar 2 of the CAP favours rural development, becoming more important in olive groves [28,29]. This support is aimed at farmers who, in a non-mandatory way, adopt environmentally-friendly agricultural management models such as integrated or organic farming, where the implementation of plant cover that mitigates the consequences of erosion processes on soil degradation and its negative impact on olive productivity stands out [20,26,30]. In the current political and legislative framework, the new post-2020 CAP reforms are geared towards achieving environmental objectives such as fighting climate change and supporting European farmers in achieving a sustainable and competitive agricultural sector [31]. Thus, basic payments will continue to be dependent on the size of farms, giving greater priority to small and medium-sized farms and young farmers [32]. The new challenges proposed for the new CAP focus on promoting an intelligent, resilient, and diversified agricultural sector that guarantees food security; the emphasis on environmental care; and the strengthening of the socio-economic fabric of rural areas in order to ensure the sustainability of agriculture in Europe [33,34]. Assuming that a continuous and stable contribution by ES to society is a guarantee of the sustainability of agricultural systems, olive groves stand out for their multifunctional nature and the multiple functions they contribute to society [35,36]. In this sense, although the most valuable ES provided by olive groves corresponds to productive and supply services of olives and olive oil [14,37], "Agenda 2000" and the 2003 reforms of the CAP led to the recognition of non-productive services for these agricultural systems [38]. The olive groves contribute to regulation ES, helping to mitigate erosion and climate change because of their carbon sequestration capacity [39–41]. They also contribute to socio-cultural ES, because of the rural culture associated with these crops and their contribution to employment generation (i.e., 10% of the agricultural sector), and to agricultural income (i.e., 6% of national income in Spain) [4,17]. In addition, as components of agricultural landscapes, they constitute reservoirs of agrobiodiversity and wild biodiversity acting as transversal ES [42].

Although there are numerous comparative studies analysing the different management models applied in olive groves, quantifying their multifunctional character and evaluating their positive and negative externalities [3,11,17,20,21,43], the motivations (drivers) and concerns related to the adoption of a particular type of olive management, and the influence of soil erosion over these perceptions remains little investigated. Specifically, knowing the reasons behind farmers' choice of a given agricultural management model in olive groves is extremely important from socio-cultural, ecological, and political dimensions. This knowing allows us to understand how the cultural heritage and tradition linked to these crops influence their management and to encourage the implementation of tillage practices and subsidies that contribute to the economic and environmental stabilisation of olive groves [3,8,9,26]. On the other hand, soil erosion is one of the main threats to the sustainability of olive groves [9]; therefore, to know the ecological and economic impact of this threat on the sustainability of olive groves [18,21,43], it is necessary to understand how erosion affects farmers' perception of agricultural problems. Using a case study, the Protected Designation of Origin (PDO) *Estepa*, with 70% of its area covered by olive groves and annual benefits close to €225 M [44], the main objectives of this research were: (a) to evaluate the factors (aims/drivers and concerns of farmers) that determine the choice of a given olive management model; and (b) to analyse quantitatively the influence of the soil erosion level of the lands on the priorities of these choices. In this way, the information gathered will provide the basis for further targeted research to help ensure a fair standard of living for farmers and a stable supply of ES to society, reducing uncertainty about the sustainability of olive farming systems.

2. Material and Methods

2.1. Study Area

The olive-growing region of *Estepa* in Seville (Andalusia, Spain) was chosen as a particular case study (Figure 1), consolidated as a Protected Designation of Origin (PDO) at the European level in 2010 [45].

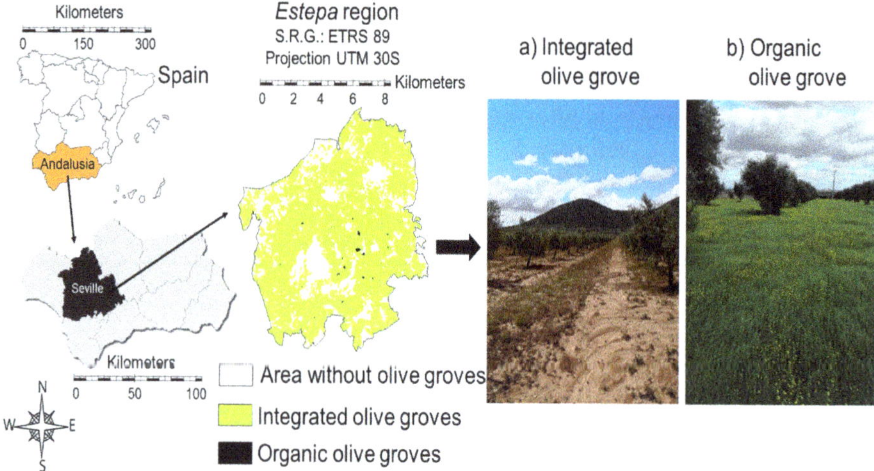

Figure 1. The geographical location of the PDO Estepa. Areas without olive groves, with integrated olive groves (**a**) and with organic olive groves (**b**) and images of the two types of management are shown.

This region, where the density of olive trees ranges from 100 to 500 trees ha^{-1} and climate is temperate Mediterranean with an annual rainfall of 400–500 mm, has 39,694 hectares of olive groves [43,46]. Olive groves are located mainly on limestone and silty soils with a depth between 30–150 cm [21]. Of the total olive-growing area in the *Estepa* region, 95% of the olive groves are managed under integrated rainfed management, with a minimum representation of plantations with deficit irrigation. In this type of management, the agricultural yield ranges between 3500–6000 kg olives ha^{-1} depending on the addition of water to the crop. It should be noted that the implementation of partial plant cover is required, and the application of chemical phytosanitary products is allowed in a regulated manner, including the possibility of adding maximum water volumes of 1500 cm^3 only during periods of water stress [14,44]. The integrated management is subject to the recommendations of the Integrated Production Associations (IPAs) and the Integrated Agricultural Treatment Groups (IATGs), whose main function is to provide technical guidance to farmers on good agricultural management practices in their crops, regulating the production and marketing of the olive oil produced [47]. In the study area, rainfed organic management is still incipient, with isolated plots with young trees covering approximately 500 ha. The agricultural yield of these farms ranges from 3500 to 5000 kg olives ha^{-1}, and it is mandatory to use total live or inert plant covers [44].

2.2. Location of Olive Groves, Erosive Levels, and Sample Design

Using official cartography and cadastral information [48,49], the olive groves belonging to the study area were geo-referenced, enabling the spatial and geographical location of olive groves managed in an integrated and organic way, as well as estimating their erosion levels using the Universal Soil Loss Equation (USLE) (Equation (1)) [50,51].

$$A = R \times K \times LS \times C \times P \tag{1}$$

where A: annual soil losses (t ha^{-1} year^{-1}); R: rain erosivity (J ha^{-1}); K: soil erodibility (Mg J^{-1}); LS: length and grade of the slope of the territory (dimensionless); C: ground cover (dimensionless); P: agricultural conservation practices (dimensionless).

Equation (1) was calibrated specifically for the study region following the study of Rodríguez Sousa et al. [21,43,46] and according to the erosive levels proposed by Moreira-Madueño (1991) [52] (Table 2). Rain erosivity (factor R), soil erodibility (factor K), and length and grade of slope (factor LS) were estimated using specific scientific-technical references for the study area and the criteria of Gisbert-Blanquer et al. (2012) [53,54]. The standards of Gómez et al. (2003) [55] were applied for C factor (tree cover) taking a value of 0.16 for integrated olive groves, considering adult olive trees with a 2.5 m canopy radius, corridors between trees of 4 m, and the presence of partial vegetation covers. For organic olive groves, the C factor took a value of 0.06 due to the presence of total vegetation covers. Finally, because all the groves presented tillage practices without mechanical erosion control, a value of 1 was assumed for P factor (agricultural conservation practices) [43].

Table 2. Estimation of soil loss (A, t ha^{-1} year^{-1}) and classification of the erosion levels of integrated and organic olive groves in the *Estepa* region according to the USLE model.

Management	Erosion Level	Factors					A
		R	K	LS	C	P	
Integrated	Null	109.7	0.82	0.00 (0%)	0.16	1	—
	Slight	109.7	0.89	0.18 (3%)	0.16	1	2.81
	Moderate	109.7	0.56	0.70 (7%)	0.16	1	6.88
	Severe	109.7	0.95	2.20 (15%)	0.16	1	36.68
Organic	Null	109.7	0.82	0.00 (0%)	0.06	1	—
	Moderate	109.7	0.56	0.70 (7%)	0.06	1	2.58

Based on this classification, while integrated olive plots were found in all erosion levels, for organic olive groves, only farms in the null and moderate erosive states were found. Through the combination of agricultural management and erosion levels, six different treatments were identified: integrated olive groves on soils with null, slight, moderate, and severe erosion; and organic olive groves on soils with null and moderate erosion. Additionally, in each treatment, a random sampling was carried out, selecting nine plots in each erosion level integrated olive groves, obtaining a sample size of $n = 9 \times 4 = 36$ plots. On the other hand, all the organic olive grove plots were selected ($n = 19$ plots), from which 9 showed null erosion and 10 moderate erosion. Finally, the overall sample size was $n = 55$ plots (Figure 2).

2.3. Surveys Implementation

In each of the selected plots (i.e., sampling points), a survey was carried out for each owner/farmer, collecting a total of $n = 55$ surveys of farmers dedicated full time to olive growing (i.e., income coming mainly from agriculture ≥80%). In this way, for each plot, the biological genus, age category, and educational level of each surveyed farmer were collected, and the agricultural management (i.e., integrated or organic) of each olive crops was checked, which verified the correct implementation of the agronomic practices required in each case (i.e., obligatory use of partial and total plant covers in integrated and organic agriculture respectively [22,23]). In addition, qualitative information was collected using a scale from 0 (not important) to 9 (very important) for the main priorities of each owner with respect to their objectives to be achieved through the agricultural management model adopted in their plot, which also evaluated the agricultural perception regarding the main concerns considered as threats to the sustainability of the olive grove over time. The objectives and concerns proposed to farmers were selected based on European criteria based on the Eurobarometer 2016 technical report [56], which combines the analysis of the socio-economic and environmental dimensions of agricultural systems. In this sense, the proposed objectives were related to ensuring good quality of life for farmers, social factors, and environmental variables that would contribute to increasing the supply of ES from olive groves to society. On the other hand, the concerns were related to the main social and environmental threats against the sustainability of olive groves. Table 3 compiles the variables proposed for the qualitative assessment of the surveys carried out by farmers.

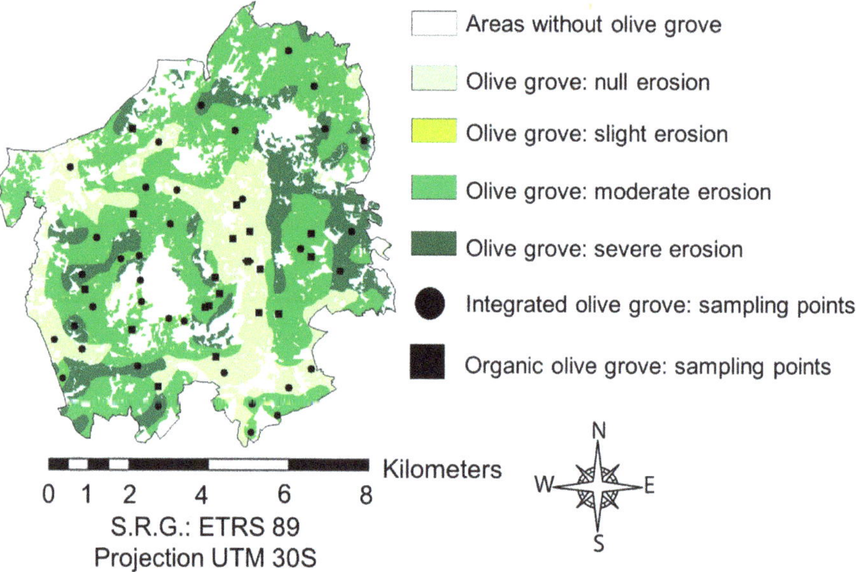

Figure 2. Sample design in the *Estepa* region, highlighting the sampling points where the surveys were carried out according to the agricultural management (i.e., integrated or organic) and the erosive state of the plots (i.e., null, slight, moderate, or severe).

Table 3. List of agricultural objectives and concerns proposed to farmers for assessment. In the "Other" section, farmers were able to suggest options not incorporated in the designed survey.

Objectives	Concerns
1. Farm income	1. Rural aging
2. Stability and economic security	2. Scarce infrastructure and public services
3. Personal reputation	3. Abandonment of rural activities
4. Respect for environment	4. Desertification and soil erosion
5. Get healthy products	5. Lack of local employment alternatives
6. Employment generation	6. Climate change
7. Generating a quality landscape and preserving the natural heritage	7. Low productivity and economic viabilit of the olive grove
8. Contributing to tourist offer	8. Loss of landscape and biodiversity of the olive grove
9. Other:	9. Other:

2.4. Statistical Analysis

In order to comparatively analyse the priorities/weights obtained for the objectives and the assumed concerns related to the olive grove, these ratings were standardised (i.e., normalised) to a range of 0 to 1 following the feature scaling or MinMax methodology (Equation (2)) as in the study by Rodríguez Sousa et al. (2019) [21]:

$$nX = (X - Xmin) \times (Xmax - Xmin)^{-1} \tag{2}$$

where nX: standardised/normalised variable (dimensionless, value ranging from 0 to 1); X: original variable; $Xmin$: minimum value of the original variable; $Xmax$: maximum value of the original variable.

A non-parametric Kruskal–Wallis test (H-test) was carried out in order to ascertain the existence of significant differences in the standardised priorities of the objectives and concerns expressed by farmers with integrated and organic olive plots, assuming the non-normal and heteroscedastic character of the results from the surveys carried out, due to working with frequency data (i.e., frequency of a given qualitative assessment for each option suggested in the surveys) [57]. In addition, having sampled plots from four different erosion levels (i.e., null, slight, moderate, or severe) to check the possible influence of erosion on the observed differences, a Tukey's post-hoc test was carried out on the integrated olive grove data. However, for organic olive groves, having only data coming from plots with two possible erosion states (i.e., null, moderate), the possible influence of the erosion processes was tested directly using an H-test without the need to perform any post-hoc test. All statistical analyses were carried out with RStudio and SPSS software, using the "car" library and "pgirmess" package and considering a level of significance of $\alpha = 0.05$ [58–60].

3. Results

3.1. Personal Information and Type of Soil Cover Applied

The farmers surveyed with integrated olive grove plots ($n = 36$) were males aged 45–54 years with an academic background lower than average university education (i.e., mandatory secondary education or high school). At the same time, of the farmers with organic olive groves ($n = 19$), 80% ($n = 15$) were men and 20% were women ($n = 4$), all of them within an age range of 35–45 years and with higher education. Specifically, 47.37% of the surveyed ($n = 9$) presented studies corresponding to professional training; 47.37% ($n = 9$) medium university studies (i.e., degree or bachelor's degree); and 5.26% ($n = 1$) higher university studies (i.e., official master's or PhD).

The checked soil cover conducted on-site during the surveys showed that some plots of olive groves presented inert covers (i.e., remains of pruning), scarce in the study area, or could be live covers. Cruciferous species, grasses, and leguminous were the most representative species: *Diplotaxis muralis* ((L.) DC., 1821), *Festuca indigesta* (Boiss., 1838), and *Vicia sativa* (L., 1753), respectively.

3.2. Assessment of Farmers' Objectives Related to Agricultural Management and Erosion Levels

Table 4 shows the standardised priorities of farmers with relation to socio-economic objectives corresponding to all erosive states of the olive-growing managements assessed.

Highly significant differences were detected with higher priorities in the integrated management of the olive grove to the objectives related to farm income, stability, and economic security. In the case of the organic olive growing, highly significant differences were detected for the objectives related to respect for the environment, the generation of a quality landscape and preservation of natural heritage, and the contribution to the tourist offer.

Tables 5 and 6 show the influence of erosive processes on the priorities of the agricultural objectives for the integrated and organic olive groves respectively.

Figure 3 compiles the standardised results of the Tukey's post-hoc test and the Kruskal–Wallis test carried out to find the existence of significant differences between the erosive states of integrated and organic olive groves, respectively.

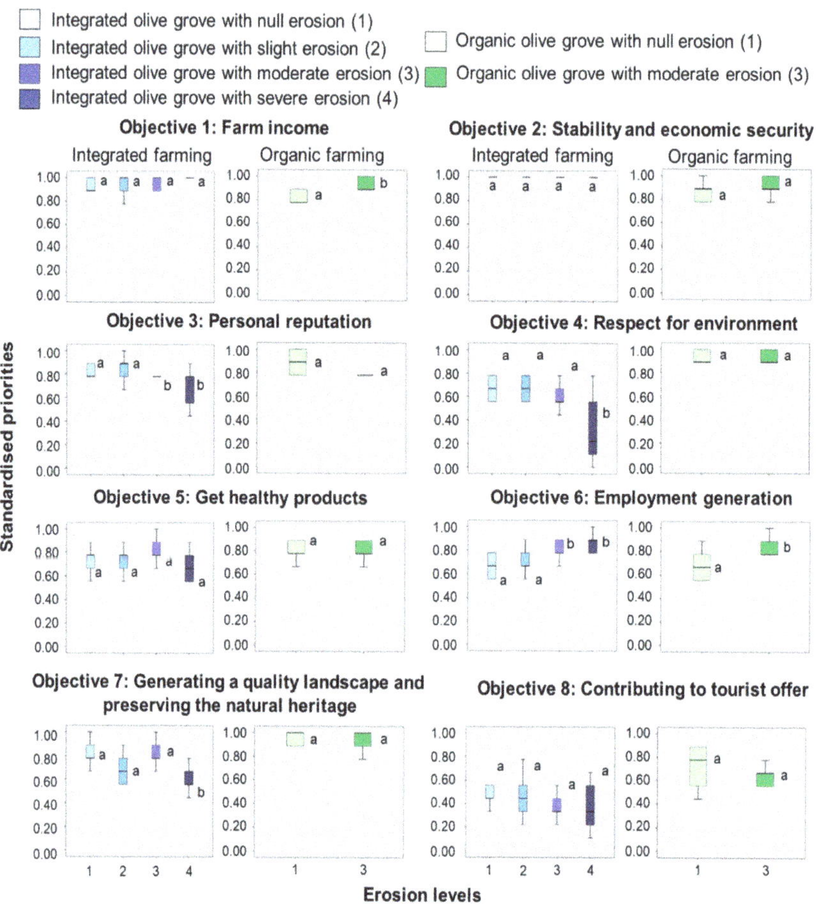

Figure 3. Boxplots showing the results of Tukey's post-hoc analysis and the Kruskal–Wallis test for each objective evaluated in the integrated and organic olive groves according to their erosion levels. The letters inside each boxplot indicate the classification groups generated to establish similar categories to each objective.

In the integrated management of the olive groves, the presence of two statistically differentiated groups was detected in the analysis of a greater personal reputation. In this sense, the plots with null and slight erosion (i.e., group a) showed a significantly higher weighting than plots with moderate and severe erosion (i.e., group b). Similarly, significant differences were also observed for the objective related to the employment generation, where plots with null and slight erosion (i.e., group a) showed significantly lower priorities than those evidenced by farmers with moderate and severe erosion plots (i.e., group b), the farmers of those plots giving greater importance to the achievement of this objective. For the objectives related to respect for the environment and the generation of a quality landscape and preservation of the natural heritage, two statistically differentiated groups were identified. This result showed that the owners of the plots with null, slight, and moderate erosion (i.e., group a) were, in a very significant way, higher than the priorities shown by the farmers with severely eroded plots (i.e., group b). For the organic management, very significant differences were observed according to the erosion level of the plots in relation to the farm income and the employment generation, with higher priorities in those plots with moderate erosion compared to plots with null erosion.

Table 4. Standardised priorities for the objectives in the integrated and organic management of the olive groves in the *Estepa* region in all erosive states assessed. Statistical values of *F* (Kruskal–Wallis test), and *p*-values; significant (<0.05 *), very significant (<0.01 **), and highly significant (<0.001 ***) are shown.

Objectives	Standardised Priorities		*F*	*p*-Value
	Integrated Olive Grove	Organic Olive Grove		
1. Farm income	0.96	0.88	19.23	<0.001 ***
2. Stability and economic security	0.99	0.90	20.23	<0.001 ***
3. Personal reputation	0.77	0.83	2.68	0.121
4. Respect for environment	0.58	0.93	94.97	<0.001 ***
5. Get healthy products	0.74	0.79	2.10	0.166
6. Employment generation	0.76	0.75	0.05	0.815
7. Generating a quality landscape and preserving the natural heritage	0.73	0.95	46.49	<0.001 ***
8. Contributing to tourist offer	0.42	0.67	33.49	<0.001 ***

Table 5. Standardised priorities for the evaluated objectives of integrated olive groves in the *Estepa* region within its erosive levels. Statistical values of *F* (Kruskal–Wallis test) and *p*-values; significant (<0.05 *), very significant (<0.01 **), and highly significant (<0.001 ***) are shown.

Objectives	Integrated Olive Grove				*F*	*p*-Value
	Erosion Levels					
	Standardised Priorities					
	Null	Slight	Moderate	Severe		
1. Farm income	0.94	0.95	0.96	1.00	1.96	0.139
2. Stability and economic security	0.96	0.99	1.00	1.00	1.45	0.245
3. Personal reputation	0.83	0.84	0.77	0.65	4.15	0.014 *
4. Respect for environment	0.67	0.68	0.62	0.35	6.97	0.001 **
5. Get healthy products	0.75	0.72	0.81	0.68	2.13	0.115
6. Employment generation	0.67	0.73	0.80	0.84	4.07	0.015 *
7. Generating a quality landscape and preserving the natural heritage	0.81	0.68	0.81	0.60	7.66	0.001 **
8. Contributing to tourist offer	0.47	0.43	0.41	0.37	0.55	0.648

Table 6. Standardised priorities for the evaluated objectives of organic olive groves in the *Estepa* region within their erosive levels. Statistical values of *F* (Kruskal–Wallis test) and *p*-values; significant (<0.05 *), very significant (<0.01 **), and highly significant (<0.001 ***) are shown.

	Organic Olive Groves			
	Erosion Levels			
Objectives	Standardised Priorities		*F*	*p*-Value
	Null	Moderate		
1. Farm income	0.83	0.93	13.63	0.002 **
2. Stability and economic security	0.88	0.93	1.60	0.224
3. Personal reputation	0.88	0.78	4.41	0.052
4. Respect for environment	0.94	0.93	0.21	0.653
5. Get healthy products	0.78	0.80	0.26	0.616
6. Employment generation	0.68	0.83	9.76	0.007 **
7. Generating a quality landscape and preserving the natural heritage	0.96	0.94	0.43	0.520
8. Contributing to tourist offer	0.70	0.64	0.80	0.382

3.3. Assessment of Farmers' Concerns Related to the Sustainability of the Agricultural Management Evaluated and Erosion Levels

Table 7 shows the standardised priorities for the main concerns that may affect agricultural sustainability in relation to the evaluated olive managements for all erosion levels. It should also be highlighted that an additional concern that was not incorporated in the pre-designed survey was detected for farmers with organic olive groves directed towards the possible threat that the use of glyphosate as a broad-spectrum herbicide may pose to food security (i.e., ninth concern). Because the use of glyphosate and its consequences was only proposed at one level of the study (i.e., organic olive grove), a statistical test comparing this concern with the integrated management model could not be performed.

The results (i.e., standardised priorities) showed the existence of significant and highly significant differences between the integrated and organic management models of olive groves with regard to concerns about rural aging, the abandonment of rural activities, and the low productivity and economic viability of the olive grove, with greater weightings in integrated olive groves. On the other hand, with higher priorities in organic farming, highly significant differences were identified between the two agricultural management models studied with regard to concerns about desertification and soil erosion, climate change, and loss of landscape and biodiversity of the olive grove.

Tables 8 and 9 show the standardised priorities of the main concerns of farmers, for the integrated and organic management of olive groves, according to the erosion levels of the plots.

Table 7. Standardised priorities for the main concerns as threats to the agricultural sustainability in the integrated and organic management of olive groves in *Estepa* region in all erosive states. Statistical values of F (Kruskal–Wallis test) and p-values; significant (<0.05 *), very significant (<0.01 **), and highly significant (<0.001 ***) are shown.

Concerns	Standardised Priorities		F	p-Value
	Integrated Olive Grove	Organic Olive Grove		
1. Rural aging	0.81	0.75	4.65	0.046 *
2. Scarce infrastructure and public services	0.70	0.72	0.56	0.464
3. Abandonment of rural activities	0.97	0.64	179.06	<0.001 ***
4. Desertification and soil erosion	0.60	0.94	338.46	<0.001 ***
5. Lack of local employment alternatives	0.63	0.59	2.54	0.130
6. Climate change	0.50	0.90	250.24	<0.001 ***
7. Low productivity and economic viability of the olive grove	0.92	0.70	42.33	<0.001 ***
8. Loss of landscape and biodiversity of the olive grove	0.40	0.86	207.97	<0.001 ***
9. Other: Use of glyphosate as a threat to food security	—	0.85	—	—

Table 8. Standardised priorities for the main concerns in the olive groves with integrated management within their erosive levels in *Estepa* region. Statistical values of F (Kruskal–Wallis test) and p-values; significant (<0.05 *), very significant (<0.01 **), and highly significant (<0.001 ***) are shown.

Concerns	Integrated Olive Grove				F	p-Value
	Erosion Levels					
	Standardised Priorities					
	Null	Slight	Moderate	Severe		
1. Rural aging	0.75	0.77	0.84	0.86	4.89	0.007 **
2. Scarce infrastructure and public services	0.62	0.63	0.73	0.83	7.80	<0.001 ***
3. Abandonment of rural activities	0.94	0.96	0.98	1.00	2.66	0.064
4. Desertification and soil erosion	0.47	0.48	0.64	0.80	23.30	<0.001 ***
5. Lack of local employment alternatives	0.60	0.62	0.63	0.65	0.59	0.621
6. Climate change	0.44	0.48	0.51	0.58	3.88	0.018 *
7. Low productivity and economic viability of the olive grove	0.81	0.90	0.96	1.00	22.19	<0.001 ***
8. Loss of landscape and biodiversity of the olive grove	0.40	0.41	0.42	0.37	0.23	0.870
9. Other: Use of glyphosate as a threat to food security	—	—	—	—	—	—

Figure 4 compiles the standardised results corresponding to the Kruskal–Wallis test and the Tukey's post-hoc test carried out to classify the erosive states of each olive management model in statistically differentiated groups for each agricultural concern studied.

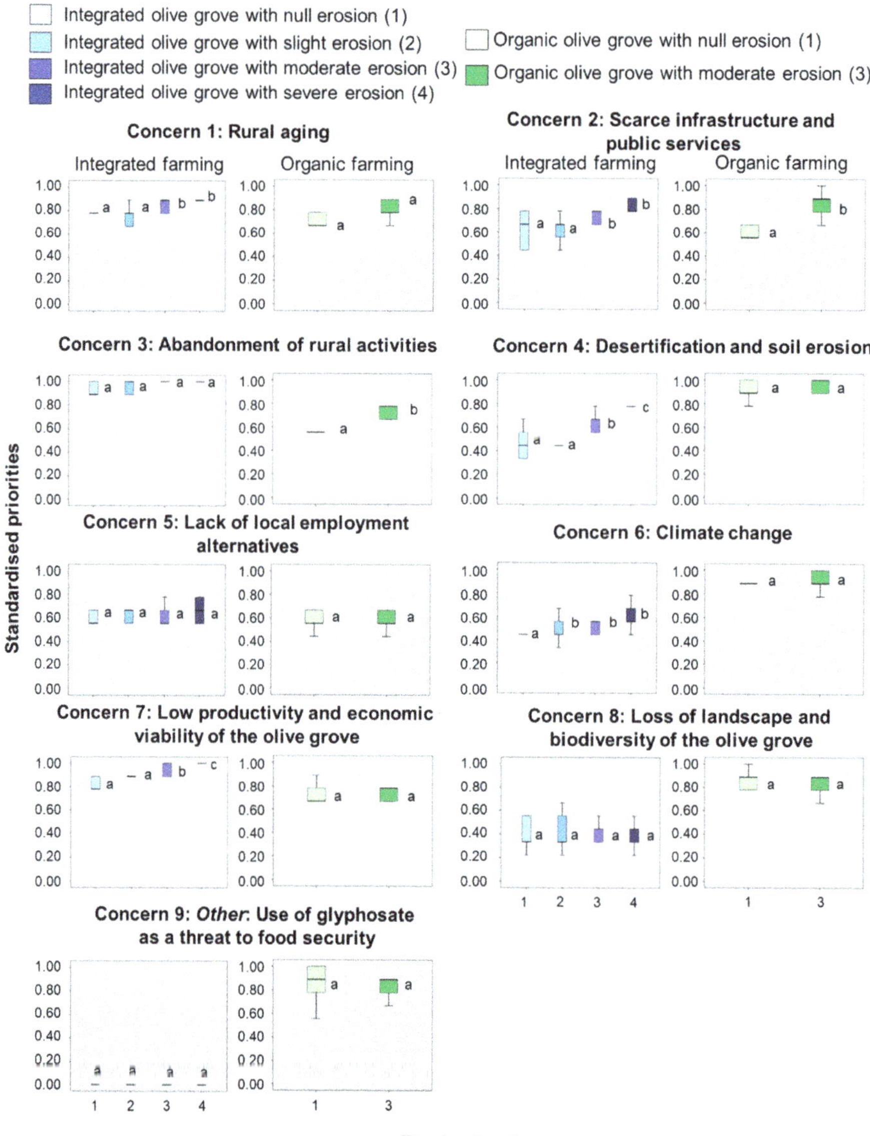

Figure 4. Boxplots showing the results of Tukey's post-hoc analysis and the Kruskal–Wallis test for each concern evaluated in the integrated and organic olive groves according to their erosion levels. The letters inside each boxplot, indicate the classification groups generated to establish similar categories to each objective.

In the integrated olive grove, there were significant differences according to the erosion level of the plots with respect to the climate change, originating two statistically differentiated groups, one formed by the plots of null erosion (i.e., group a), and another group formed by plots with slight, moderate, and severe erosion (i.e., group b), this last one showing a higher weighting on this threat. For the concern related to rural aging, two groups with very significant differences were observed. Thus, plots with null and slight erosion (i.e., group a) showed a lower degree of concern for this threat than plots with moderate and severe erosion (i.e., group b). Highly significant differences were found for different concerns of farmers: (a) for the scarcity of infrastructures and public services, two statistically differentiated groups—plots with null and slight erosion and plots with moderate and severe erosion; (b) for desertification and soil erosion, three different groups—plots with null and slight erosion, plots with moderate erosion, and plots with severe erosion; and (c) for the low productivity and economic viability of the olive grove, three groups—plots with null erosion, plots with slight erosion, and plots with moderate and serious erosion. For these last three concerns, greater weightings were observed as the erosion status of the plots increased. Finally, in organic olive groves, highly significant differences in terms of erosion were detected for the concerns related to the scarcity of infrastructures and public services and the abandonment of rural activities, showing higher priorities in the plots with higher erosion levels (i.e., moderate erosion).

Table 9. Standardised priorities for the main concerns in the olive groves with organic management within their erosive levels in *Estepa* region. Statistical values of *F* (Kruskal–Wallis test) and *p*-values; significant (<0.05 *), very significant (<0.01 **), and highly significant (<0.001 ***) are shown.

	Organic Olive Groves			
	Erosion Levels			
Concerns	**Standardised Priorities**		*F*	*p*-Value
	Null	Moderate		
1. Rural aging	0.72	0.78	2.17	0.160
2. Scarce infrastructure and public services	0.59	0.85	39.20	<0.001 ***
3. Abandonment of rural activities	0.56	0.73	21.18	<0.001 ***
4. Desertification and soil erosion	0.91	0.96	1.88	0.189
5. Lack of local employment alternatives	0.59	0.58	0.11	0.736
6. Climate change	0.89	0.91	0.37	0.550
7. Low productivity and economic viability of the olive grove	0.68	0.73	0.83	0.375
8. Loss of landscape and biodiversity of the olive grove	0.88	0.84	0.87	0.363
9. Other: Use of glyphosate as a threat to food security	0.88	0.83	0.74	0.401

4. Discussion

Assessing the sustainability of farming systems in general, and olive growing in particular, is a complex objective that requires the use of the multifunctional framework of agriculture and a joint assessment of the economic, social, and environmental dimensions of these systems [17,43,61–63]. In this sense, only an approach based on the Triple Bottom Line that balances these three dimensions can substantially contribute to the analysis of the viability of olive grove agricultural landscapes over time, ensuring the maintenance of the multiple ES that these landscapes generate for society [3,18,64–66].

The assertion made by Sgroi et al. (2015) [67] regarding economic benefit as an essential aspect in the maintenance of agricultural and, specifically, olive-growing activity was corroborated in the study carried out, with a satisfactory farm income being the main objective to be achieved by farmers with integrated olive groves. In this sense, it is worth noting that integrated olive growing is an agricultural management model that combines practices that are more respectful of the environment than intensive olive grove management models while maintaining an optimal production level capable of generating an income that provides a fair standard of living for farmers [3,25,68,69]. On the other hand, the organic olive grove is a management model where the environmental impacts derived from the different farming practices are minimised [24,70]. Additionally, the possibility of applying chemically synthesised fertilizers is eliminated [22,23]. Although the economic aspect of the crop was an important factor in the continuity of the rural activity in this type of farming, higher priorities were shown for the objectives related to the environmental dimension of the olive grove. Specifically, the most outstanding objectives were related to the implementation of agronomic practices that contribute to the conservation of the soil environment, and the generation of a quality landscape, thus preserving the natural heritage [20,71]. This last objective takes on special importance in Andalusia, where there is a deep-rooted cultural tradition linked to olive-growing activity, being historical crops and transforming elements of the landscape in the south of Spain [10,17]. The difference detected in terms of the greater relevance (i.e., higher priorities/weightings) of the economic and environmental dimensions of the olive grove between integrated and organic olive-growing managements for the study area was, in fact, closely related to the higher educational level shown by farmers with organic olive groves [72,73]. All farmers with organic olive groves presented a higher educational level (i.e., professional training or university studies), while farmers with integrated olive grove plots presented medium studies (i.e., mandatory secondary education or high school). This different academic formation generated a greater familiarity on the part of the owners of organically managed plots with the agricultural problems related to the biology of the crop and the environment, knowing in greater detail the intrinsic dynamics of olive cultivation [8,43].

Regarding the analysis of the main agricultural concerns perceived as threats to the long-term sustainability of olive growing, differences were also observed between the two agricultural approaches analysed. In this respect, farmers with integrated olive groves were extremely concerned about the social and productive aspects of the olive grove, highlighting rural aging linked to agricultural abandonment and the impact on the economic performance of farms due to insufficient production [74], with the difficulties in the continuity of agricultural activity being the main concern, with farmers having to complement their income with other lucrative activities [75]. On the other hand, environmental concerns such as erosion and climate change dominated in farmers with organic olive groves. As with the agricultural objectives, the difference observed between the highest priorities for socio-economic and environmental concerns shown by farmers with integrated and organic olive groves respectively is intrinsically related to the age and educational level of the farmers. In general terms, farmers with integrated farms showed a higher age and average educational level, engaging in agricultural activity predominantly for reasons of cultural heritage where the main concerns are to obtain a decent standard of living from their agricultural activity and to ensure its economic sustainability [76]. On the other hand, the younger age of the farmers surveyed with organic olive groves and their higher level of education gives them a perception of the agricultural system where not just its socio-economic dimension is relevant, highlighting the environmental dimension of the crop as a key factor to ensure its ecological sustainability [46,56]. Thus, environmental factors are crucial for the long-term sustainability of the crop and, from a multidimensional perspective, are the basis of the main socio-economic threats to the crops from the alteration of biological and ecological factors in the agricultural systems with a direct impact on the viability of the farms [55,77].

In the surveys of farmers applying organic management, a recurrent concern related to food security, which was not raised in the questionnaire formulated, emerged. The farmers surveyed were apprehensive about the indiscriminate use of glyphosate as a broad-spectrum herbicide, in line with the main demands of society towards agriculture in the Eurobarometer technical report published by the European Commission (2016) [56]. In this sense, concern was expressed regarding the use of glyphosate as a broad-spectrum herbicide, a practice widely used in integrated olive groves [78]. This concern deserves special attention due to the repercussions that the use of glyphosate can have on products obtained from olive groves, in addition to the consequences derived from human consumption of these products on health, as it is a product related to the development of multiple diseases [79,80].

Assuming that the education level of farmers can condition their perception about the objectives and agricultural threats to the sustainability of the olive grove, it would be highly desirable, as already proposed by Rodríguez Sousa et al. (2019) [46] and Guzmán et al. (2013) [81], to set up and consolidate active information channels from the administration at different scales to the social actors related to the olive grove on appropriate environmental management practices to ensure the sustainability of these agrosystems over time. Different institutions at different scales and with different perspectives (top-down information arising, for example, by FAO, EU, ministries, and regional governments, environmental agencies) should continuously and affordably inform farmers, and in turn, farmers should raise their demands and concerns (bottom-up information) to these institutions. In this way, fluid and relevant information channels would be established that would result in better management of agricultural systems with a cross-cutting and adaptive objective of sustainability and rural development [82].

From a legislative point of view, the results obtained suggest how the objectives of the farmers belonging to the study area are largely adapted to the structure of the current CAP (i.e., CAP 2014–2020) and to the modifications planned for the post-2020 CAP. Although the CAP seeks to promote the economic stability of agricultural holdings in general, its environmental dimension is becoming increasingly relevant, promoting the sustainability of agriculture through aid aimed at encouraging rural development and the adoption of environmentally friendly practices [26,83,84]. Obtaining an agricultural income that allows a fair standard of living, which is the main objective of farmers with olive groves under integrated management, corresponds essentially to the support provided by Pillar 1 of the CAP, based on direct aid and the "greening" regime [28,33]. The changes planned for the post-2020 CAP have three specific objectives aimed at maintaining appropriate economic stability, namely (a) ensuring fair incomes; (b) increasing the competitiveness of the agricultural sector; and (c) rebalancing equilibrium in the food chain [31]. Furthermore, the predominant environmental priorities of farmers with organic olive groves are addressed in the support provided by Pillar 2 of the CAP. From an environmental perspective, these objectives are shared by the post-2020 CAP, where four objectives are closely linked to environmental conservation and rural development: (a) fight against climate change; (b) protection of the environment; (c) conservation of landscape and biodiversity; and (d) maintenance and promotion of living rural areas [28,31]. Regarding the concerns perceived by farmers as threats to the sustainability of their crops, the results coincide with the interpretation of these threats by the CAP 2014–2020, with the challenges becoming more acute in the post-2020 CAP. Thus, the new CAP already includes measures aimed at promoting agricultural continuity and generational change in order to reduce rural aging, a concern expressed by farmers with integrated olive groves, with aid targeted at young people and small farmers [26,85]. On the other hand, the concerns shown by farmers with organic olive groves are more linked to environmental factors which, as mentioned above, are increasingly considered at the EU level, with progressively larger budget allocation to mitigate the negative effects of agriculture (i.e., diffuse pollution, erosion, greenhouse gas emissions) and to promote environmentally friendly agricultural practices [11,29,32,55,74,86,87].

Taking erosion as one of the main threats to the sustainability of olive-growing systems due to their development in Mediterranean environments and on predominantly silty soils where there are few stabilising aggregates for the soil environment [46,88], it is worth mentioning the important role played by the CAP in promoting agri-environmental practices that minimise its economic-productive and ecological consequences [26–28]. In this sense, Pillar 2 of the CAP promotes, through the corresponding subsidies, the voluntary implementation of measures such as the application of plant cover and the minimisation of tillage practices, measures which, according to multiple studies, have proven to be very effective in reducing the loss of nutrients, fertility, and soil weight [20,53,74,89]. Considering soil erosion as a possible conditioning variable of the priorities obtained in terms of the objectives and concerns evaluated, an inverse relationship between the erosion level and the environmental objectives was evident for integrated management. This observation is due to the fact that greater erosion generally requires greater application of chemical inputs to the crop and more labour to maintain an optimal level of agricultural production [8,46]. In this sense, the demand for employment is a key agricultural concern that increases with the level of erosion on agricultural farms, and the presence of labour is essential to provide a stable food supply to society [3,25]. Thus, according to Rodríguez-Pleguezuelo et al. (2018) [20], a change towards the organic management of olive groves in

Andalusia would act as a promoter of employment generation, since according to Spanish legislation this model of agricultural management is associated with the implementation and maintenance of soil cover and the use of organic fertilizers, practices that require a greater number of specialised workers [22,23]. In addition, for the integrated management of olive groves, environmental concerns obtained higher priorities as the erosion status of the plots increased, probably due to the harmful effects of erosion on soil fertility and agricultural yield [21,46,74]. Finally, for organic management, the increase in erosive processes showed a greater concern for the scarcity of infrastructure and rural abandonment, as more machinery was needed to obtain olives, and it was more likely that these exploitations would be abandoned, especially those of a marginal nature [8,43].

5. Conclusions

Olive groves form multifunctional socio-ecological systems where the economic, social, and ecological dimensions of the crop interact with each other, leading to the generation of productive and non-productive ecosystem services (ES) to society. Despite the limitations of the study conducted, where the in situ verification of agricultural management was determined through visual observation of the agronomic practices carried out by farmers and the information collected through the questionnaires was qualitative, the results showed relevant information regarding the main conditioning factors for the adoption of a given agricultural model. Thus, while the socio-economic dimension (i.e., job stability and economic benefit) was consolidated as the main axis on which the concerns and the desired objectives of the farmers with integrated olive groves, the owners with organic olive groves showed a higher degree of education level and a greater familiarity with the main biological aspects of the crop, resulting in a higher priority for the environmental dimension of the cultivation. In this sense, the greater environmental awareness of these farmers resulted in a greater weighting of the objectives related to the yield and the farm income and of the main biological and edaphic threats (i.e., soil degradation as a consequence of erosion processes, or climate change and its impact on the sustainability of the olive crop). Additionally, the existence of erosive processes led to greater environmental awareness in plots where high/severe soil loss showed its negative repercussions on agricultural sustainability, evidencing in those plots the use of more intensive farming practices in order to maintain a stable production level of the crop.

Considering that in the present study, the degree of the education level of farmers was a key factor in the adoption of the organic olive system, it would be particularly desirable to carry out several activities (i.e., training workshops) aimed at promoting greater awareness among farmers of the environmental threats that can undermine the sustainability of olive-growing systems, encouraging the implementation of agricultural practices (i.e., minimisation of tillage practices, implementation of soil vegetal covers) that help to guarantee the viability of these systems in the long term, ensuring a stable and sustainable supply of their ES.

Taking into account the political and legislative context of the current CAP and the economic-environmental trends observed for the future post-2020 CAP, it would be highly advisable to carry out research aimed at going deeper into the key factors that influence the adoption of a particular agricultural model in olive growing by farmers. In this sense, future studies should be oriented towards the achievement of the following goals: (a) carry out more generalist works on a wider geographical scale, where the validity of the results obtained is not limited to a particular region or PDO of olive oil; (b) perform experimental designs where the surveys designed and carried out on farmers are not closed or dichotomous, encouraging the participation of rural actors and the proposal of conflicting aspects in olive groves by farmers; and (c) execute more elaborate data processing, using statistical and mathematical methodologies aimed at establishing a hierarchy of the relevance of the factors studied, with emphasis on Multi-Criteria Decision-making Analysis (MCDA), highlighting methodologies as Analytic Hierarchy Process (AHP) or Analytic Network Process (ANP). Thus, through the development of these research lines, a broad knowledge can be provided aimed at evaluating and increasing the relationship between existing agricultural demands and concerns to ensure the continuity of agricultural activity in olive groves and the multidimensional objectives that are pursued through of the subsidies granted by the EU and by the CAP to the agricultural sector, being able to increase the compatibility between rural needs and demands towards agriculture by the socio-political sector.

Author Contributions: Conceptualization, A.A.R.S., C.P.-L., and S.S.-G.; Data curation, A.A.R.S., C.P.-L., S.S.-G., and J.M.B.; Formal analysis, A.A.R.S., C.P.-L., and J.M.B.; Investigation, A.A.R.S., S.S.-G., and A.J.R.; Methodology, A.A.R.S., C.P.-L., S.S.-G., and J.M.B.; Project administration, A.J.R.; Resources, J.M.B. and A.J.R.; Supervision, J.M.B. and A.J.R.; Validation, A.J.R.; Visualization, A.A.R.S.; Writing—original draft, A.A.R.S.; Writing—review and editing, A.A.R.S., C.P.-L., S.S.-G., J.M.B., and A.J.R. All authors have read and agreed to the published version of the manuscript.

Funding: This research was not funded by any public or private institution.

Acknowledgments: To Moisés Caballero, Secretary of the PDO *Estepa*, and all the owners surveyed. Special thanks to María Aurora Rodríguez Sousa. Antonio Alberto Rodríguez Sousa is a PhD student supported through a pre-doctoral contract of researcher in training (UCM-Santander scholarship) granted by University Complutense of Madrid.

Conflicts of Interest: The authors declare no conflict of interest.

References

1. Matthews, R.; Selman, P. Landscape as a Focus for Integrating Human and Environmental Processes. *J. Agric. Econ.* **2006**, *57*, 199–212. [CrossRef]
2. Eurostat. *Estadísticas Sobre Estructura de las expLotaciones Agrícolas*; Eurostat (European Statistics): Brussels, Belgium, 2018; Available online: https://ec.europa.eu/eurostat/statistics-explained/index.php?title=Farm_structure_statistics/es (accessed on 4 April 2020).
3. Sousa, A.A.R.; Parra-López, C.; Sayadi-Gmada, S.; Barandica, J.; Rescia, A. A multifunctional assessment of integrated and ecological farming in olive agroecosystems in southwestern Spain using the Analytic Hierarchy Process. *Ecol. Econ.* **2020**, *173*, 106658. [CrossRef]
4. Agriculture and Environment. Available online: http://www.ine.es (accessed on 5 April 2020).
5. Cifras Aceite de Oliva. Available online: http://www.internationaloliveoil.org (accessed on 1 April 2020).
6. Bielsa, I.; Pons, X.; Bunce, B. Agricultural Abandonment in the North Eastern Iberian Peninsula: The Use of Basic Landscape Metrics to Support Planning. *J. Environ. Plan. Manag.* **2005**, *48*, 85–102. [CrossRef]
7. Filho, W.L.; Mandel, M.; Al-Amin, A.Q.; Feher, A.; Jabbour, C.J.C. An assessment of the causes and consequences of agricultural land abandonment in Europe. *Int. J. Sustain. Dev. World Ecol.* **2016**, *24*, 554–560. [CrossRef]
8. Duarte, F.; Jones, N.; Fleskens, L. Traditional olive orchards on sloping land: Sustainability or abandonment? *J. Environ. Manag.* **2008**, *89*, 86–98. [CrossRef]
9. Spoerer, M. Agricultural protection and support in the European Economic Community, 1962–1992: Rent-seeking or welfare policy? *Eur. Rev. Econ. Hist.* **2015**, *19*, 195–214. [CrossRef]
10. Infante-Amate, J.; Villa, I.; Aguilera, E.; Torremocha, E.; Guzmán, G.; Cid, A.; Emanueli, F. The Making of Olive Landscapes in the South of Spain. A History of Continuous Expansion and Intensification. *Environ. Hist.* **2016**, *5*, 157–179. [CrossRef]
11. López-Pintor, A.; Sanz-Cañada, J.; Salas, E.; Rescia, A.J. Assessment of Agri-Environmental Externalities in Spanish Socio-Ecological Landscapes of Olive Groves. *Sustainability* **2018**, *10*, 2640. [CrossRef]
12. Caraveli, H. A comparative analysis on intensification and extensification in mediterranean agriculture: Dilemmas for LFAs policy. *J. Rural. Stud.* **2000**, *16*, 231–242. [CrossRef]
13. Flohre, A.; Fischer, C.; Aavik, T.; Bengtsson, J.; Berendse, F.; Bommarco, R.; Ceryngier, P.; Clement, L.W.; Dennis, C.; Eggers, S.; et al. Agricultural intensification and biodiversity partitioning in European landscapes comparing plants, carabids, and birds. *Ecol. Appl.* **2011**, *21*, 1772–1781. [CrossRef]
14. AEMO. *Aproximación a los Costes del Cultivo del Olivo. Cuaderno de Conclusiones del Seminario AEMO*; AEMO (Asociación Española de Municipios del Olivo/Spanish Association of Municipalities of Olive groves): Córdoba, Spain, 2012; Available online: http://www.webcitation.org/77MCvuNPx (accessed on 8 April 2020). (In Spanish)

15. Palma, P.; Alvarenga, P.; Palma, V.L.; Fernandes, R.M.; Soares, A.M.V.M.; Barbosa, I.R. Assessment of anthropogenic sources of water pollution using multivariate statistical techniques: A case study of the Alqueva's reservoir, Portugal. *Environ. Monit. Assess.* **2009**, *165*, 539–552. [CrossRef] [PubMed]
16. Randall, A. Valuing the outputs of multifunctional agriculture. *Eur. Rev. Agric. Econ.* **2002**, *29*, 289–307. [CrossRef]
17. Martínez-Sastre, R.; Ravera, F.; González, J.; Lopez, C.; Bidegain, I.; Munda, G. Mediterranean landscapes under change: Combining social multicriteria evaluation and the ecosystem services framework for land use planning. *Land Use Policy* **2017**, *67*, 472–486. [CrossRef]
18. Lampridi, M.G.; Sørensen, C.G.; Bochtis, D. Agricultural Sustainability: A Review of Concepts and Methods. *Sustainability* **2019**, *11*, 5120. [CrossRef]
19. Real Decreto 1201/2002, de 20 de Noviembre, Por el Que se Regula la Producción Integrada de Productos Agrícolas. Available online: https://www.boe.es/buscar/doc.php?id=BOE-A-2002-23340 (accessed on 18 March 2020).
20. Pleguezuelo, C.R.R.; Zuazo, V.H.D.; Martínez, J.R.F.; Peinado, F.J.M.; Martín, F.M.; García-Tejero, I.F. Organic olive farming in Andalusia, Spain. A review. *Agron. Sustain. Dev.* **2018**, *38*, 20. [CrossRef]
21. Sousa, A.A.R.; Barandica, J.M.; Rescia, A.J.; Sousa, R. Estimation of Soil Loss Tolerance in Olive Groves as an Indicator of Sustainability: The Case of the *Estepa* Region (Andalusia, Spain). *Agronomy* **2019**, *9*, 785. [CrossRef]
22. Ley 5/2011, de 6 de Octubre, del Olivar de Andalucía. Available online: https://www.boe.es/buscar/act.php?id=BOE-A-2011-17494 (accessed on 3 March 2020).
23. Plan Director del Olivar Andaluz Decreto 103/2015. Available online: http://www.webcitation.org/77MO1YwQe (accessed on 1 April 2020).
24. EUR-lex. *Regulation (EU) 2018/848 of the European Parliament and of the Council of 30 May 2018 on Organic Production and Labelling of Organic Products and Repealing Council Regulation (EC) No 834/2007*; EUR-lex (European Union Law): Brussels, Belgium, 2018; Available online: https://eur-lex.europa.eu/legal-content/EN/TXT/?uri=uriserv:OJ.L_.2018.150.01.0001.01.ENG (accessed on 25 September 2020).
25. EC. *The Attitudes of European Citizens towards Environment. Special Eurobarometer 217/Wave 62.1—TNS Opinion & Social*; EC (European Commission): Brussels, Belgium, 2005; Available online: http://ec.europa.eu/public_opinion/index_en.htm (accessed on 25 March 2020).
26. Nazzaro, C.; Marotta, G. The Common Agricultural Policy 2014–2020: Scenarios for the European agricultural and rural systems. *Agric. Food Econ.* **2016**, *4*, 89. [CrossRef]
27. Louhichi, K.; Ciaian, P.; Espinosa, M.; Perni, A.; Paloma, S.G.Y. Economic impacts of CAP greening: Application of an EU-wide individual farm model for CAP analysis (IFM-CAP). *Eur. Rev. Agric. Econ.* **2017**, *45*, 205–238. [CrossRef]
28. Dwyer, J.; Ward, N.; Lowe, P.; Baldock, D. European Rural Development under the Common Agricultural Policy's 'Second Pillar': Institutional Conservatism and Innovation. *Reg. Stud.* **2007**, *41*, 873–888. [CrossRef]
29. Ramniceanu, I.; Ackrill, R. EU rural development policy in the new member states: Promoting multifunctionality? *J. Rural Stud.* **2007**, *23*, 416–429. [CrossRef]
30. Erjavec, K.; Erjavec, E. 'Greening the CAP'—Just a fashionable justification? A discourse analysis of the 2014–2020 CAP reform documents. *Food Policy* **2015**, *51*, 53–62. [CrossRef]
31. Carey, M. The Common Agricultural Policy's New Delivery Model Post-2020: National Administration Perspective. *EuroChoices* **2019**, *18*, 11–17. [CrossRef]
32. Plan Estratégico de España Para la PAC Post 2020. Available online: https://www.mapa.gob.es/es/pac/post-2020/default.aspx (accessed on 15 September 2020).
33. Jongeneel, R.; Erjavec, E.; Azcárate, T.G.; Silvis, H. Assessment of the Common Agricultural Policy after 2020. In *Palgrave Advances in Bioeconomy: Economics and Policies*; Springer Science and Business Media LLC: Berlin, Germany, 2019; Volume 1, pp. 207–228.
34. Minotti, B.; Zagata, L. Towards Food Policy for Europe: A Comparison of the Post-2020 Common Agricultural Policy Discourses. *Eur. Countrys.* **2020**, *12*, 53–66. [CrossRef]
35. Balzan, M.V.; Sadula, R.; Scalvenzi, L. Assessing Ecosystem Services Supplied by Agroecosystems in Mediterranean Europe: A Literature Review. *Land* **2020**, *9*, 245. [CrossRef]

36. Michalopoulos, G.; Kasapi, K.A.; Koubouris, G.; Psarras, G.; Arampatzis, G.; Hatzigiannakis, E.; Kavvadias, V.; Xiloyannis, C.; Montanaro, G.; Malliaraki, S.; et al. Adaptation of Mediterranean Olive Groves to Climate Change through Sustainable Cultivation Practices. *Climate* **2020**, *8*, 54. [CrossRef]

37. Coq-Huelva, D.; García-Brenes, M.D.; Sabuco-I-Cantó, A. Commodity chains, quality conventions and the transformation of agro-ecosystems: Olive groves and olive oil production in two Andalusian case studies. *Eur. Urban Reg. Stud.* **2012**, *19*, 77–91. [CrossRef]

38. López, C.P.; Calatrava-Requena, J.; De-Haro-Giménez, T. A systemic comparative assessment of the multifunctional performance of alternative olive systems in Spain within an AHP-extended framework. *Ecol. Econ.* **2008**, *64*, 820–834. [CrossRef]

39. Lal, R. Soil Carbon Sequestration Impacts on Global Climate Change and Food Security. *Science* **2004**, *304*, 1623–1627. [CrossRef] [PubMed]

40. Rodríguez-Entrena, M.; Barreiro-Hurle, J.; Gómez-Limón, J.A.; Espinosa, M.; Castro-Rodríguez, J. Evaluating the demand for carbon sequestration in olive grove soils as a strategy toward mitigating climate change. *J. Environ. Manag.* **2012**, *112*, 368–376. [CrossRef]

41. Aguilera, E.; Lassaletta, L.; Gattinger, A.; Gimeno, B.S. Managing soil carbon for climate change mitigation and adaptation in Mediterranean cropping systems: A meta-analysis. *Agric. Ecosyst. Environ.* **2013**, *168*, 25–36. [CrossRef]

42. Solomou, A.; Sfougaris, A. Comparing conventional and organic olive groves in central Greece: Plant and bird diversity and abundance. *Renew. Agric. Food Syst.* **2011**, *26*, 297–316. [CrossRef]

43. Sousa, A.A.R.; Barandica, J.M.; Sanz-Cañada, J.; Rescia, A.J. Application of a dynamic model using agronomic and economic data to evaluate the sustainability of the olive grove landscape of *Estepa* (Andalusia, Spain). *Landsc. Ecol.* **2019**, *34*, 1547–1563. [CrossRef]

44. Specifications for Protected Designation of Origin Estepa. Ministry of Agriculture, Fisheries and Rural Development. Available online: http://www.juntadeandalucia.es/export/drupaljda/PliegoEstepamodificado.pdf (accessed on 31 March 2020).

45. Cañada, J.S.; Vázquez, A.M. Quality certification, institutions and innovation in local agro-food systems: Protected designations of origin of olive oil in Spain. *J. Rural Stud.* **2005**, *21*, 475–486. [CrossRef]

46. Sousa, A.A.R.; Barandica, J.M.; Rescia, A.J. Ecological and Economic Sustainability in Olive Groves with Different Irrigation Management and Levels of Erosion: A Case Study. *Sustainability* **2019**, *11*, 4681. [CrossRef]

47. Sánchez-Escobar, F.; Coq-Huelva, D.; Sanz-Cañada, J. Measurement of sustainable intensification by the integrated analysis of energy and economic flows: Case study of the olive-oil agricultural system of *Estepa*, Spain. *J. Clean. Prod.* **2018**, *201*, 463–470. [CrossRef]

48. Plan Nacional para la Observación del Territorio: Sistema de Información Sobre Ocupación del Suelo de España. Available online: www.siose.es (accessed on 11 April 2020).

49. Datos Espaciales de Referencia de Andalucía (DERA): G17 Divisiones Administrativas. Available online: http://www.webcitation.org/77MQd2rHN (accessed on 2 April 2020).

50. Wischmeier, W.H.; Smith, D.D. *A Universal Soil-Loss Equation to Guide Conservation Farm Planning*, 1st ed.; International Society of Soil Science: Madison, WI, USA, 1961; pp. 418–425.

51. Benavidez, R.; Jackson, B.; Maxwell, D.; Norton, K. A review of the (Revised) Universal Soil Loss Equation ((R)USLE): With a view to increasing its global applicability and improving soil loss estimates. *Hydrol. Earth Syst. Sci.* **2018**, *22*, 6059–6086. [CrossRef]

52. Moreira-Madueño, J.M. *Capacidad de Uso y Erosión de Suelos. Una Aproximación a la Evaluación de Tierras en Andalucía*; Junta de Andalucía, Agencia del Medio Ambiente: Sevilla, Spain, 1991; Available online: http://www.juntadeandalucia.es/medioambiente/web/Red_informacion_ambiental/productos/Publicaciones/articulos/articulos_pdf/Paralelo.PDF (accessed on 7 April 2020). (In Spanish)

53. Gómez-Calero, J.A. *Sostenibilidad de la producción de olivar en Andalucía. Instituto de Agricultura Sostenible. Instituto de Agricultura Sostenible*; CSIC (Centro Superior de Investigaciones Científicas/Spanish National Research Council): Córdoba, Spain, 2010; Available online: https://www.ias.csic.es/sostenibilidad_olivar/Sost_2009/Sostenibilidad_de_la_Producci%F3n_de_Olivar_en_Andaluc%EDa3.pdf (accessed on 29 March 2020). (In Spanish)

54. El factor K de la ecuación universal de pérdidas de suelo (USLE). Available online: http://hdl.handle.net/10251/16850 (accessed on 21 March 2020).

55. Gómez, J.; Battany, M.; Renschler, C.S.; Fereres, E. Evaluating the impact of soil management on soil loss in olive orchards. *Soil Use Manag.* **2006**, *19*, 127–134. [CrossRef]

56. EC. *Europeans, Agriculture and the CAP. TNS Opinion & Social. Special Eurobarometer 440*; EC (European Commission): Brussels, Belgium, 2016; Available online: http://data.europa.eu/euodp/en/data/dataset/S2087_84_2_440_ENG (accessed on 4 April 2020).

57. González, C.G.; Lise, A.V.; Felpeto, A.B. *Tratamiento de Datos con R, Statistica y SPSS*, 1st ed.; Ediciones Díaz de Santos: Madrid, Spain, 2013; pp. 217–415. (In Spanish)

58. SPSS. *IBM Statistical Package for the Social Sciences for Windows*, SPSS version 21.0; IBM Corp: Armonk, NY, USA, 2012.

59. RStudio. *Open Source and Enterprise-Ready Professional Software for R*, RStudio version 0.98.1102; RStudio Inc.: Boston, MA, USA, 2009–2014; Available online: https://www.rstudio.com/products/RStudio/ (accessed on 11 April 2020).

60. Gandrud, C. *Reproducible Research with R and R Studio*, 2nd ed.; Chapman and Hall/CRC: New York, NY, USA, 2016; pp. 29–78.

61. Rescia, A.J.; Sanz-Cañada, J.; Del Bosque-González, I. A new mechanism based on landscape diversity for funding farmer subsidies. *Agron. Sustain. Dev.* **2017**, *37*, 182. [CrossRef]

62. Cappelletti, G.M.; Grilli, L.; Nicoletti, G.M.; Russo, C. Innovations in the olive oil sector: A fuzzy multicriteria approach. *J. Clean. Prod.* **2017**, *159*, 95–105. [CrossRef]

63. De Luca, A.I.; Falcone, G.; Stillitano, T.; Iofrida, N.; Strano, A.; Gulisano, G. Evaluation of sustainable innovations in olive growing systems: A Life Cycle Sustainability Assessment case study in southern Italy. *J. Clean. Prod.* **2018**, *171*, 1187–1202. [CrossRef]

64. Parras-Alcántara, L.; Lozano-García, B.; Keesstra, S.D.; Cerda, A.; Brevik, E.C. Long-term effects of soil management on ecosystem services and soil loss estimation in olive grove top soils. *Sci. Total. Environ.* **2016**, *571*, 498–506. [CrossRef] [PubMed]

65. Torres-Miralles, M.; Grammatikopoulou, I.; Rescia, A. Employing contingent and inferred valuation methods to evaluate the conservation of olive groves and associated ecosystem services in Andalusia (Spain). *Ecosyst. Serv.* **2017**, *26*, 258–269. [CrossRef]

66. Montanaro, G.; Xiloyannis, C.; Nuzzo, V.; Dichio, B. Orchard management, soil organic carbon and ecosystem services in Mediterranean fruit tree crops. *Sci. Hortic.* **2017**, *217*, 92–101. [CrossRef]

67. Sgroi, F.; Foderà, M.; Di Trapani, A.M.; Tudisca, S.; Testa, R. Cost-benefit analysis: A comparison between conventional and organic olive growing in the Mediterranean Area. *Ecol. Eng.* **2015**, *82*, 542–546. [CrossRef]

68. Morris, C.; Winter, M. Integrated farming systems: The third way for European agriculture? *Land Use Policy* **1999**, *16*, 193–205. [CrossRef]

69. Nayak, P.; Nayak, A.; Panda, B.; Lal, B.; Gautam, P.; Poonam, A.; Shahid, M.; Tripathi, R.; Kumar, U.; Mohapatra, S.; et al. Ecological mechanism and diversity in rice based integrated farming system. *Ecol. Indic.* **2018**, *91*, 359–375. [CrossRef]

70. López, C.P.; De-Haro-Giménez, T.; Calatrava-Requena, J. Diffusion and Adoption of Organic Farming in the Southern Spanish Olive Groves. *J. Sustain. Agric.* **2007**, *30*, 105–151. [CrossRef]

71. Stobbelaar, D.J.; Kuiper, J.; Van Mansvelt, J.D.; Kabourakis, E. Landscape quality on organic farms in the Messara valley, Crete. *Agric. Ecosyst. Environ.* **2000**, *77*, 79–93. [CrossRef]

72. Phillips, J.M. Farmer Education and Farmer Efficiency: A Meta-Analysis. *Econ. Dev. Cult. Change* **1994**, *43*, 149–165. [CrossRef]

73. Bernal-Jurado, E.; Mozas-Moral, A.; Fernández-Uclés, D.; Medina-Viruel, M.J. Determining Factors for Economic Efficiency in the Organic Olive Oil Sector. *Sustainability* **2017**, *9*, 784. [CrossRef]

74. Gómez, J.A.; Infante-Amate, J.; De Molina, M.G.; Vanwalleghem, T.; Taguas, E.V.; Lorite, I. Olive Cultivation, its Impact on Soil Erosion and its Progression into Yield Impacts in Southern Spain in the Past as a Key to a Future of Increasing Climate Uncertainty. *Agriculture* **2014**, *4*, 170–198. [CrossRef]

75. Kristensen, S.B.P.; Busck, A.G.; Van Der Sluis, T.; Gaube, V. Patterns and drivers of farm-level land use change in selected European rural landscapes. *Land Use Policy* **2016**, *57*, 786–799. [CrossRef]

76. Fleskens, L.; Duarte, F.; Eicher, I. A conceptual framework for the assessment of multiple functions of agro-ecosystems: A case study of Trás-os-Montes olive groves. *J. Rural. Stud.* **2009**, *25*, 141–155. [CrossRef]

77. Ropero, R.F.; Rumí, R.; Aguilera, P.A. Bayesian networks for evaluating climate change influence in olive crops in Andalusia, Spain. *Nat. Resour. Model.* **2018**, *32*, e12169. [CrossRef]

78. Resolución de la dirección general de sanidad de la producción agraria. Available online: https://www.mapa.gob.es/agricultura/pags/fitos/registro/productos/pdf/22992.pdf (accessed on 5 April 2020).

79. Gasnier, C.; Dumont, C.; Benachour, N.; Clair, E.; Chagnon, M.-C.; Séralini, G.-E. Glyphosate-based herbicides are toxic and endocrine disruptors in human cell lines. *Toxicology* **2009**, *262*, 184–191. [CrossRef]

80. Bai, S.H.; Ogbourne, S.M. Glyphosate: Environmental contamination, toxicity and potential risks to human health via food contamination. *Environ. Sci. Pollut. Res.* **2016**, *23*, 18988–19001. [CrossRef] [PubMed]

81. Guzmán, G.; López, D.; Román, L.; Alonso, A.M. Participatory Action Research in Agroecology: Building Local Organic Food Networks in Spain. *J. Sustain. Agric.* **2012**, *37*, 127–146. [CrossRef]

82. Rodríguez-Cohard, J.C.; Sánchez-Martínez, J.D.; Gallego-Simón, V.J. Olive crops and rural development: Capital, knowledge and tradition. *Reg. Sci. Policy Pract.* **2018**, *11*, 935–949. [CrossRef]

83. Matthews, A.; Salvatici, L.; Scoppola, M. Trade Impacts of Agricultural Support in the EU. *Res. Agric. Appl. Econ.* **2017**, 120p. [CrossRef]

84. Rizov, M. Rural development under the European CAP: The role of diversity. *Soc. Sci. J.* **2005**, *42*, 621–628. [CrossRef]

85. Manos, B.; Bournaris, T.; Chatzinikolaou, P. Impact assessment of CAP policies on social sustainability in rural areas: An application in Northern Greece. *Oper. Res.* **2010**, *11*, 77–92. [CrossRef]

86. Mäder, P.; Berner, A. Development of reduced tillage systems in organic farming in Europe. *Renew. Agric. Food Syst.* **2011**, *27*, 7–11. [CrossRef]

87. West, T.O.; Marland, G. Net carbon flux from agricultural ecosystems: Methodology for full carbon cycle analyses. *Environ. Pollut.* **2002**, *116*, 439–444. [CrossRef]

88. Loumou, A.; Giourga, C. Olive groves: The life and identity of the Mediterranean. *Agric. Hum. Values* **2003**, *20*, 87–95. [CrossRef]

89. Durán-Zuazo, V.H.; Pleguezuelo, C.R.R. Soil-Erosion and Runoff Prevention by Plant Covers: A Review. *Sustain. Agric.* **2009**, *28*, 785–811. [CrossRef]

 *land*

Article

Could Mapping Initiatives Catalyze the Interpretation of Customary Land Rights in Ways that Secure Women's Land Rights?

Gaynor Paradza [1,*] , Lebogang Mokwena [2] and Walter Musakwa [3]

1   Public Affairs Research Institute, Forest Town, Johannesburg 2193, South Africa
2   New School for Social Research, Department of Sociology, New York, NY 10011, USA;
    mokwg590@newschool.edu
3   Future Earth and Ecosystems Research Group, Department of Urban and Regional Planning,
    University of Johannesburg, Johannesburg 2028, South Africa; wmusakwa@uj.ac.za
*   Correspondence: gaynorp@pari.org.za or ggparadza@gmail.com; Tel.: +27-748-220-487

Received: 31 July 2020; Accepted: 17 September 2020; Published: 23 September 2020

Abstract: Although land forms the basis for marginal livelihoods in Sub-Saharan Africa, the asset is more strategic for women as they usually hold derived and dependent rights to land in customary tenure areas. Initiatives to secure women's land tenure in customary areas are undermined by the social embeddedness of the rights, patriarchy, lack of awareness by the communities, legal pluralism, and challenges of recording the rights. As pressure on customary land tenure increases due to foreign and local land-based investment interests, land titling initiatives, tourism, and mineral resources exploration, communities and women within them are at real risk of losing their land, the basis of their livelihoods. Women stand to lose more as they hold tenuous land rights in customary land tenure areas. Accordingly, this study analyzes case studies of selected mapping initiatives in Sub-Saharan Africa to interrogate the extent to which mapping both as a cadastral exercise and emerging practice in the initiation of participatory land governance initiatives, catalyze the transmission of customary land rights in ways that have a positive impact on women's access to land in customary land tenure areas. The results indicate that mapping initiatives generate opportunities, innovations, and novel spaces for securing women's access to land in customary tenure areas which include catalyzing legislative changes and facilitating technology transfer, increasing awareness of women's interests, providing opportunities for women to participate in decision-making forums, providing a basis for securing statutory recognition for their land rights, and improving natural resource stewardship. The potential challenges include the community's capacity to sustain the initiatives, the expense of the technology and software, widespread illiteracy of women, power asymmetries and bias of the mapping experts, increased vulnerability of mapped land to exploitation, the legal status of the maps in the host community and /or country, compatibility with existing land recording systems, statutory bias in recording land rights and the potential of mapping initiatives to unearth existing land boundary conflicts. These challenges can be mediated by sensitive planning and management to ensure real and sustainable land tenure security for women. The paper contributes to debates around customary land tenure dynamics, specifically the issues pertaining to registration of primary and derived customary rights to land. These includes policy debates and choices to be made about how best to secure tenuous customary land rights of women and other vulnerable people. The paper also contributes to our understanding of what instruments in land registration toolkits might strengthen women's land rights and the conditions under which this could be done.

Keywords: customary; land; tenure; women; mapping; Sub-Saharan Africa

1. Introduction

Land in Sub-Saharan Africa is held under various tenure systems which include freehold, leasehold, and customary. Customary land tenure is the majority of land holding in Africa, accounting for 78% of the land holding [1]. The relative abundance of land held under customary tenure means that it is a critical asset for the poor who rely on subsistence for their survival [2]. The abundance of land under customary tenure coupled with the fact that it is allocated in some circumstances for free as birthright or upon marriage marital render it one of the most accessible land for women who are overrepresented among the poor. However, these land rights are increasingly vulnerable to conflict, increasing land demand fuelled by rapidly changing land markets, urbanization, and large scale land-based investments [3,4]. This leaves women vulnerable to lose the land at a time when land rights are increasingly contested. The insecurity occasioned by the instability of marriage—(which has far-reaching implications at both the household and at the community levels)—has introduced new stresses to women's sustained land access as these key institutions through which women negotiate access to customary land are increasingly in a state of flux [5,6]. The impact on women's land rights under customary tenure land is that they find themselves dependent on ever shrinking, less fertile, and increasingly expensive pieces of land. Among the factors that account for the vulnerability of women's land rights in customary tenure areas, has been women's invisibility or secondary tenure status [5]. This secondary tenure status arises out of their securing land on the back of primary land rights, which, in patriarchal customary institutions, are held by men. A review of land mapping initiatives in patrilineal systems illustrates that these could potentially provide some relief and generate opportunities for women to secure their customary land rights by increasing the visibility of derived land rights, generating opportunities for women to participate in land decision-making fora, and generate opportunities for women to articulate their land needs in land governance institutions.

Mapping of land using geospatial information technologies such as global positioning systems (GPS) are some of the increasingly popular innovations that are used by diverse constituencies working to record and secure customary land tenure rights. Mapping can also be in the form of mental mapping, ground mapping, participatory sketch mapping, transact mapping, and participatory 3-dimensional modeling [7–11]. Combining geospatial information technology and social science mapping often yield holistic mapping results [4,7,8]. These methods have been used by international and multilateral donor community, philanthropic organizations, local and transnational non-governmental organizations (NGOs) and communities, customary authorities, the national government and local government [4,7,8]. Participatory land and natural resources mapping initiatives have emerged for a variety of reasons. These include recording land rights, capturing land use and land cover, boundary mapping, and indigenous knowledge, biodiversity mapping and conflict resolution [9,11]. As a people-centered approach to improving (and making more transparent) overall land governance, participatory mapping initiatives have indeed emerged as one of the key ways to amplify the voices of usually rural communities and those who are most vulnerable within them vis-à-vis land tenure and general natural resource access and use. Participatory mapping consists of facilitating a community-wide discussion about and agreement on the boundaries of the land and natural resources that are communally owned or used. It is also about the various uses of the land and natural resources that are available within the communal area whose boundaries are being mapped [8,9]. Participatory mapping initiatives have become one of the experimental ways through which alternative processes for enhancing inclusive, gender-sensitive, sustainable, and informed land governance in Sub-Saharan Africa has been pursued. Participatory mapping improves access to spatial information [8]. Mapping exercises have entailed or have at least aimed to inform the conversion of the community participatory maps into formal, location-based, geographic maps for incorporation into or for the revision of existing cadastral databases. In customary land tenure areas, these initiatives have also been deployed to facilitate land claims and land stock taking [10].

Women and Land Tenure Vulnerability in Customary Tenure Areas

Women need secure access to land because they are highly dependent on the resource for their welfare, productivity, and empowerment [5,12,13]. Women need land to fulfill their productive and reproductive responsibilities. When women control land, they have increased bargaining power in their communities and households [12,13]. Increased women's access to land has a positive impact on children's welfare and education in Sub-Saharan Africa [2,14,15]. Securing access to land in a customary land tenure area provides a platform through which women can negotiate access to community membership, employment opportunities, assistance with childcare. Women also use customary land as a fallback insurance if they lose their residence on the dissolution of marriage [5]. The terms on which customary land rights are secured render this the most accessible land tenure to women in Sub-Saharan Africa [16].

Customary tenure is a term to describes the complex, dynamic, and evolving multi-faceted and transformational ideas and practices around land tenure [17,18]. The nature of customary land rights is that they are not registered [16]. Under customary tenure, land is held under overlapping tenures of individual household and common property [5,16]. The rights are guaranteed by diverse regulatory regimes [19,20]. Customary lands are held under customary laws which is one of a host of other laws and regulations that include statutory and non-statutory regimes. As a result of the existence of diverse laws and regulations, there is ambiguity, contradiction and loopholes in the application and interpretation of laws in the governance of customary lands [5,21]. Although customary laws and rules are not consistent across diverse groups, they are generally controlled by male biased traditional leaders who control the land on behalf of the communities [5,22,23]. Although the term is contested, [21] identified various common features of customary land tenure. These include overlapping rights, shared and inclusive rights, social and politically stable resource boundaries and controlled through guaranteeing of rights, enforcing of rights, and dispute resolution [4,18,24]. Although internal mechanisms for allocating land are diverse, people gain access to land by membership of a community through birth, marriage, or bush clearing [10]. Land beneficiaries have conditions attached which differ between men and women.

Women's land access under customary tenure varies from place to place with substantial differences between patrilineal and matrilineal societies although male kin control land allocation in both systems [2,16,25]. Although the processes are not homogenous, in Zambia, for example, in matrilineal societies, descent and inheritance follows the mother's line, while in patrilineal systems, descent follows the father's line [16,25]. Land allocation is made to male members by lineage in patrilineal societies [16]. This publication focuses on patrilineal systems, where women's access to land is informed by their relationship between men and women as spouses, siblings, fathers, daughters, sons, and brothers [26]. Women hold secondary and derived land rights. This secondary tenure status arises out of their securing land on the back of primary land rights, which, in patriarchal customary institutions, are held by men [6]. Although traditional leaders may allocate individual land rights to single women—they rarely allocate land to married women in their own right [5,16]. Customary law seems to have few provisions for divorced women and even fewer for single women [5,27]. Divorced women are vulnerable to lose their marital customary land rights upon the dissolution of the marriage [5,16,27]. Inheritance is also gender related with women's rights to inherit land disproportionately limited [5,27]. The ambiguity in the interpretation of laws and statutes is viewed as consistently discriminating against women in the arena of customary land claims by allowing those with power to oppress the powerless. Examples include the vulnerability of widows to land dispossession through unequal and inconsistent application of the law [16,25,28,29] and use of the law to discredit women involved in disputes over customary land. Another characteristic of customary land tenure is that community members often rely upon common resources such as forests, grazing lands, and water sources for their livelihoods and daily needs. Women also rely on these common resources for firewood and medicinal plants. Community members are generally considered the "co-owners" or rightful users of such land [4]. The situation has been changing as labor and monetary markets also mediate women's access to land in customary land tenure areas [5,16]. The stock of land held under customary tenure has been declining. This is because of conversion to leasehold and freehold, climate change losses, increasing land demand fuelled by rapidly changing land markets, urbanization, and large-scale land investments, globalization, structural adjustment, and perceptions that other tenure forms are more secure [3,5,18,30–32]. The pressures to formalize or register customary land tenure is driven by those who believe that this would modernize the tenure and attract investment, make it easier to record the land rights, incentivize users to take better care of resources and use the land as collateral [4,33,34]. The pressures on customary land tenure are inducing shifts in women's land rights and undermining their fragile tenure [6]. Market pressures undermine women's customary land rights as women generally lack resources to secure land on the market. The market pressures also affect common lands that women rely on for firewood as these are vulnerable to disposal on the land markets. The market pressures also result in individualization and registration of land. Women's tenure is compromised as family land is usually registered in the name of a single individual in the household who is usually male. The pressures on customary land arising from land scarcity, rapid population growth, and increasing land individualization erode the customary safeguards in place to secure women's land tenure security [35–38]. Women's customary tenure is compromised when disputes arise in land negotiations because they have limited access to decision-making forums and institutions of justice. [6,27,39]. The invisibility, secondary tenure status, marginalization of women from customary land governance forums and challenges of representing customary land rights effectively at law increase the vulnerability of women's customary tenure land rights [40–42]. As a result, when land is transferred to large scale investors, the attendant decision-making, apportionment of costs and benefits marginalizes women [30,43]. This is because women are excluded from decision-making and their invisible land claims are not acknowledged in the calculation of losses for compensation purposes. In addition, marriage, one of the institutions through which women gain and maintain access to customary land, has been undermined by divorce, increased mobility, death, and economic independence of women. The insecurity occasioned by the instability of marriage introduces new stresses to women's sustained land access as these key institutions through which women negotiate access to customary land are increasingly in a state of flux [5].

The transformation and sometimes privatization of common land (like forests, grazing areas) arrangements undermine women's land claims in customary tenure areas by undermining existing structures and arrangements for managing the vulnerability of people and women who depend on these. This is through, for example, opening up land for speculation by outsiders and increasing inequality among the different groups [31,44].

The impact on women who depend on customary tenure land is that they find themselves dependent on ever shrinking, less fertile, and increasingly expensive pieces of land. Mapping is an increasingly popular mechanism for demarcating land for purposes of transfer, valuation, and registration. This review of land mapping initiatives illustrates that these could provide some relief and generate opportunities for women to secure their customary land rights by increasing the visibility of derived land rights, generating opportunities for women to participate in land decision-making fora, and generate opportunities for women to articulate their land needs in land governance institutions.

2. Materials and Methods

The paper draws on secondary data. This consists of secondary analysis of published case studies of work that was carried out in Benin, Liberia, Lesotho, Democratic Republic of Congo, South Africa, Rwanda, Madagascar, Uganda, and Mozambique. Table 1 summarizes the cases, the nature of intervention, and sources available on the intervention. The cases were purposively selected to reflect common themes of mapping on land under customary tenure, diversity across the Sub-Saharan Africa, and cases that provided a reflection and/or impact on women's land rights. Table 1 shows that mapping interventions were undertaken as a major or minor part of other interventions to record and delineate customary land tenure. This was in Mozambique, Uganda, Madagascar, Democratic Republic of Congo,

Benin, and Lesotho. In all the cases, mapping was used to contribute to clarify land rights of land under customary tenure. The sources include inception reports, self–authored reports by implementing institutions, technical reports, and peer reviewed publications.

Table 1. Case study summaries.

Country	Nature of Intervention	The way Mapping was Part of Intervention	Sources Available on the Intervention
Liberia International Development Law Organization (IDLO)	Gather evidence relating to the type and level of support that communities require to successfully complete community land titling.	Mapping was a major part to facilitate the protection of vulnerable groups' land rights in the context of decentralized land management and administration.	Inception report Reviewed publication
Lesotho Millennium Challenge Corporation (MCC)	Review of Lesotho land administration to assess the land administration situation and to propose areas for legal reform.	Mapping was used in resettlement, compensation, verification and adjudication of land rights. Gender considerations were included from the outset of the land-related portion of the project.	End of Project Technical Report
Benin MCC	Project aimed to create secure land tenure and effective and transparent governance of land.	Mapping was an integral part of a project that aimed at improving women's access to land and the security of women's land tenure.	
Democratic People's Republic of Congo REDD+	Community involvement in mapping for conservation, planning, and management of land tenure.	Major part of the intervention. Women included in awareness sessions to ensure a participatory process. Women targeted in training in cartography.	Self-reporting by the Project Management Team Research report
South Africa	Digital mapping in marginalized communities to inform land-use decision-making.	Major part of co-constructing knowledge with communities.	2 Peer reviewed articles
Rwanda International Fund for Agriculture Development (IFAD) supported	Mapping land and natural resource right, use management of natural resources.	Strengthening of land tenure security through the facilitation of statutory land registration. Major to strengthen land use planning. Strengthening women's land rights was a major theme of the intervention.	Project self-report /learning note Conference report Peer reviewed Project evaluation report
Mozambique MCC IDLO	Support communities to acquire documentation to secure their communal lands. Gather evidence relating to the type and level of support that communities require to successfully complete community land titling.	Provide support to communities in the process of mapping their lands.	Peer reviewed publication Technical report Inception report
Uganda IFAD supported International Land Coalition supported	Community mapping to ensure tenure security customary land rights.	Major precondition to secure statutory registration of communal land. Women inclusion precondition for registration.	Best Practice self- report by NGO 3 peer reviewed journal articles
Madagascar ILC Supported	Mapping land use and natural resource management to facilitate customary land registration.	Major intervention. Strengthening women's access to land a major theme.	Learning note Conference report Best practice report Peer reviewed report

The cases were analyzed using thematic analysis. The review of the selected case studies followed a two-step thematic analysis. The first step analysis focused generally on mapping in customary area lands. The purpose was to capture the context, reasons, and institutional arrangements for mapping customary tenure land. The findings of this analysis are described in Sections 3.1 and 3.2 of the paper. The secondary analysis focused on data in the selected cases related to consideration of women's land rights in the mapping initiatives. These findings are the focus of Section 3.4. The primary focus of this study was to interrogate the potential of customary land mapping initiatives to strengthen women's land tenure in Sub-Saharan Africa. The point being to explore the potential of the intervention so that the findings of the exploratory work can used to inform policy making around strengthening women's land rights in customary land tenure areas. The limitations of the methodology include the inconsistency of sources that authors relied on which range from self-reported to peer reviewed publications. This inconsistency posed challenges in the consistency and depth of information available. The use of existing data to fit new research questions focusing on (a) the impact of mapping on customary land and (b) the potential of the intervention to increase tenure security for women's land rights was limited by the fact that the authors had no other knowledge of these cases outside the written information. For example, none of the cases had information on the gender disaggregated indicators, long term impact of the initiative on women's land tenure rights in customary tenure areas, or the strategies for securing women's land tenure status outside the locality. The authors drew on their own experiences in mapping and land tenure security to strengthen their insight and analysis. Although the article would have benefitted from more information on the process, strengths, and challenges of the primary research, the authors believe that the diversity of cases analyzed generate rich data, diverse contexts, and local experiences of customary land mapping in Sub-Saharan Africa. The authors believe that the selection of cases and the data can be reliably used as a basis for critically analyzing the potential of mapping as a tool for increasing customary land tenure security for women in Sub-Saharan Africa.

3. Results and Discussion

3.1. Overall Justification for Mapping Initiatives in Customary Land Areas

Mapping initiatives are closely tied to a broader process of land regularization or documentation. Maps are an effective medium for a large variety of development projects. Maps enhance stakeholders' ability to visualize space, how it is used, land use and land cover changes, and how the demarcated space relates to adjacent parcels of land and resources. Maps provide a platform for stakeholders to gain an appreciation through a visual—even if rudimentary—instrument, the nature of the resource- and land-related problems that the community may be facing. The International Development Law Organization (IDLO) implemented the *Community Land Titling Initiative* in Liberia, Uganda, and Mozambique to experiment with different models through which to support communities in the acquisition of knowledge regarding the legislative requirements and process for securing government documentation or titles for their communal lands [4,9].

Mapping can enable stakeholders to visualize the spatial distribution of complex problems [8,10]. For example, Mozambique's land delineation process enabled communities to define their land claims in customary land [8]. Mapping can help stakeholders to understand the relations over land and natural resources and illustrate important social, traditional, and historical as well as indigenous knowledge [10,45]. Mapping is also used to carve out the socio-political environment, especially institutional structures that govern land [46], while also creating the possibility for good land governance by enhancing transparency. Mapping empowers communities by providing them with opportunities to convert tacit indigenous knowledge embedded in people's memories—especially those of elders'– into external usable knowledge [10]. Mapping is also used to highlight environmental vulnerability by marking coastal erosion and riverine flood lines. In Kenya, GIS and remote sensing techniques where used for natural resource management and for project monitoring [44]. Mapping increases stakeholders' awareness of, for example, land use, land demand, and or population density. In the Democratic Republic of Congo, mapping was used to devolve central control over resources as mapping of customary land boundaries by local authorities and traditional leaders and chiefs enabled these lower levels of governance to claim autonomy in decision-making over land from the central government. The initiative also enabled communities to gain knowledge of the contractual issues that could affect their land rights and capacity to participate in decision-making processes. The intervention also facilitated the management and use of village land and natural resources, monitoring of loss of forest cover, and helped communities identify land rights and confirm land limits [7]. The recording of land claims on a map increases the visibility of the claims and improves communities and women's chances of securing their tenure as these maps become the documentary evidence of communities' claim to the land. Within communities, participatory mapping can generate discussions and lead to the resolution of inter- and intra-community disputes over natural resources [9,10]. In terms of land use, structured participatory mapping processes can provide communities with an opportunity to develop land use plans to inform primary and secondary land rights. This will enable them to monitor changes in the environment and resource distribution [7]. In uSuthu, Swaziland, the International Land Coalition supported—IFAD mapping intervention enabled communities to plot water sources and soil types. Through this mapping exercise, extension staff could advise traditional leaders on land designation as well as generate guidelines for future land and resource allocation. Mapping is commonly used to generate local land occupation plans, support the adjudication of land claims and the issuance of legal certificates of occupation by people with unrecorded customary land rights [44]. Mapping was used in Mozambique to enable communities to claim and register interest in land, establish the spatial extent of the *Direito de Uso e Aproveitamento dos Terras* (DUAT) or interest in the land or 'the right of use and benefit of customary land' [8,39,46], and mitigate the risk of boundary contestation with neighboring communities.

In Mozambique, the Millennium Challenge Corporation (MCC) funded mapping enabled communities to realize the extent of their land, thereby making it possible for them to make informed decisions about land use and enhance their capacity for informed, evidence-based engagements, and negotiation with potential investors [46]. Mapping can support the recording and intergenerational transfer of knowledge which sustains traditional indigenous knowledge and promotes ecosystem resilience [10]. Mapping can facilitate negotiation among community members regarding resource priorities. Mapping can be used to inform advocacy and enhance community cohesion in the face of land-related challenges. In South Africa, where spatial injustice is represented by skewed distribution of land, mapping became a political exercise to represent diverse narratives of land history. These narratives are used to support restitution land claims by communities [47].

In Madagascar, IFAD worked with rural communities who were previously barred from owning land. In 2005, the government introduced a land policy to improve land tenure for these groupings, thereby enabling rural communities to formalize their land rights and garner tenure security. As a result of the government's new land policy, communities were issued with certificates and existing titled ownership, parks, and reserves were mapped using existing records that were in turn validated by the communities. Customary rights were also identified and mapped using satellite images of the communes. This was complimented with multi stakeholder land tenure diagnosis, planning, and participatory community land use maps [10]. Maps can be used to represent a first attempt at a geographical representation of natural resources and therefore, a document that can be used to negotiate rights to land and natural resources. In Uganda, pressure over land and natural resources exposed customary land users to violation of their land rights [47–49]. The situation is especially bad in Karamoja, an area richly endowed with mineral resources where investors are rushing to claims concessions. Customary land rights held by pastoralists remained vulnerable because the communities lacked resources and mechanisms to demarcate their land. The communities' lack of awareness of their rights, coupled with the government's delays to implement legal provisions aimed at increasing the community's control over customary tenure land [49], compounded the situation. In order to obtain a certificate of community ownership or a freehold title over communal land in Uganda, a community has to establish a Community Land Association (CLA), the first step of which requires mapping and boundary harmonization [47,48]. The community map is a precondition for the registration of the Communal Land Association. The Uganda Land Alliance, with support from Dan Church Aid and the Ford Foundation, carried out community mapping of communal land resources to support the registration of the Communal Land Associations. The mapping resulted in the identification of customary lands in the area, boundaries, grazing lands, watering points, areas for gathering firewood, and shrines. The mapping showed that customary land was vulnerable to land grabbing [48]. Finally, on the back of the recognition of immense agricultural land tenure pressures in Rwanda and their negative impact on food security, the International Fund for Agricultural Development (IFAD) provided support for the Kirehe Community-Based Watershed Management Project (KWAMP). While the objective of KWAMP was not directly about enhancing participatory land governance through community land and resource mapping, a key component of the project necessitated this approach. In particular, under the *Water and land use management* component of the project, the intention was addressing challenges and weaknesses in the existing regulatory framework while also granting enhanced tenure security for farmers by registering their land rights. The government developed simple methods by which people could map their own boundaries using satellite images and aerial photography [44]. These images were then used to inform the parceling out and titling of land to those farmers receiving poverty reduction interventions.

3.2. Institutional Arrangements for Mapping Initiatives

Mapping initiatives that have been undertaken in Liberia, Mozambique, Benin, Lesotho, and Uganda are multi-stakeholder in their design. Diverse actors who range from community-based organizations to development partners initiate the process. These stakeholders play differing roles each of which contributes to increased land tenure security of women in customary lands. Development partners integrate gender into their strategies, governance, metrics, organizational structure, and budgets [46]. In the case of Liberia, Mozambique, and Uganda, the International Development Law Organization (IDLO) was the central initiating actor with funding from the United Kingdom Department of International Aid (UKAid), Australian Aid (AUSAid), as well as from the Bill and Melinda Gates Foundation and the Open Society Foundation [4,8]. In the case of Benin and Lesotho, the Millennium Challenge Corporation (MCC) has played the leading role [46]. The World Wide Fund for Nature (WWF) has similarly played a strategic role in the Democratic Republic of Congo [7] while IFAD provided land officers to support the districts in Rwanda and also engaged in community sensitization as part of broader land demarcation pilots in Rwanda. In Madagascar, IFAD supported and funded the project to support development in the Menabe and Melkay regions [10]. In Benin, local governments play an important role in registration and are responsible for issuing certificates and upholding land information [44]. The WWF in the DRC worked systematically to include customary land authorities in the mapping exercises. This was through their inclusion at inception and validation activities. The WWF provided computers and training for young men and women on using the mapping technology [7]. The MCC facilitated women's land tenure security by insisting on the revision of gender blind and discriminatory legislation as a pre-condition of their involvement. This precipitated the revision of discriminatory statutory laws that undermined women's land tenure status in Lesotho [46]. In Liberia, Mozambique, and Uganda, the traditional leaders were responsible for identifying sites that are of cultural significance. They also assisted the technicians to identify and verify all community boundaries [8]. Traditional leaders also supported the mapping by participating in the discussions and facilitating the participation of their communities.

The government of Rwanda initiated the 2008 National Land Centre that was responsible for overall land management and mapping [10]. The central government and its agencies were also responsible for implementing the Land law and formulating Land Policies that framed the mapping initiatives. The gender sensitive Land Policy and the Organic Land Law No. 08/200515, which is the basis of a legal framework for land management and administration in the country, prohibits any discrimination based on sex in matters relating to ownership or possession of rights over the land, as the wife and the husband have equal rights over the land [50]. This legislation provided a gender sensitive policy and legislative environment for grounding the customary land tenure mapping initiatives. The Ministry of Lands, Environment, Forestry, Water, and Mines in Rwanda developed an implementation program for the land policy and law. The registrar of land titles maintained the registers and record. The district registrar convened community meetings and certified the legal documents to endorse the mapping in Mozambique [8]. They also formalized community groups. In Rwanda, the Commissions scrutinized disputes arising in the regularization process and ensured the participation of communities. The district institution in Rwanda maintained the land records and archives and in some cases was responsible for the authorization and approval of the plans submitted by communities in respect of their land governance and use [50]. The government of Uganda made it mandatory for women to be included in the customary land mapping decision-making processes [49]. The formal recognition of women's land claims and use of common lands in customary mapping exercises increased the potential for women to secure customary land tenure status by capturing and increasing the visibility of these claims. This made it easier for women to prove and defend their claims on customary lands in the various decision-making and registration forums. All the cited governments mobilized political will—an often overlooked but strategic precondition for realizing women's land tenure security. Civil society, women's groups, and community organizations provide expertise and local knowledge; support the change once the intervention has ended [46]. Local Civil Society Organizations (CSOs) often facilitate the mapping exercise and acted as mediating agents between the community and the mapping technicians, either from the private or public sectors. In Rwanda, the CSOs that have participated in the implementation of mapping initiatives have been responsible for enhancing the conflict resolution capacity of the traditional leaders and raising women's awareness about land rights. In Uganda, the Uganda Land Alliance, a civil society organization supported the organization and registration of local communities to enable them to have legal entities which they could use to register their land claims. The Uganda Land Alliance presided over election of the management committees and ensured that each management committee of nine had at least three women members [49]. Other civil society organizations supported raising communities' awareness of gender issues and providing training for the central and local government stakeholders. In Rwanda, the Rwanda Initiative for Sustainable Development (RISD), a local Civil society organization ensured that women's views were heard by soliciting the views of the grassroots women to be considered [50].

In Kiepersol, South Africa, researchers worked with communities to explore the dynamics in a land claim case where community members mapped their heritage claims. The communities led the research team to their former homelands and identified significant places like graveyards and places they played when they were younger [47]. Similarly, the village level committees in Tanzania have been responsible for managing land and establishing and administering local registers of communal land rights, issuing certificates of customary land rights, and managing the land. Communities may delineate land according to customary use including forests, habitats, and sites of cultural and historical importance. Communities also have a role in the validation of boundaries occupation and land use [8].

3.3. Legal Status of the Mapping Outputs

The legal status of the maps and outputs of the interventions is important as it informs the extent to which the outcomes of these innovative interventions can be used to secure women's land tenure in a customary land tenure area. In Rwanda, the information was entered into the Land Tenure Regularization Support System [33] and used for titling purposes. In Madagascar, the information generated was used to develop land occupation plans and grant certificates of occupation [33]. By contrast, despite the immense effort that was undertaken by IDLO, SDI, CTV, and LEMU in implementing the *Community Lands Titling Initiative* in Liberia, Mozambique, and Uganda, at the time of project report publication in 2012, none of the participating communities had managed to receive formal documentation in recognition of their communal territorial claims [35]. These ambiguous project outcomes notwithstanding, however, in Uganda financial institutions are required by law to accept the legitimacy of the Certificate of Communal Ownership (CCOs) created from community mapping exercises as valid titles and as such, those communities that had managed to complete and submit their applications for a CCO or for freehold titles stood a better chance of leveraging their ownership for land investment purposes or for the facilitation of sale. In order to increase the legitimacy of maps, the maps must be prepared to government standards and guidelines. At the time of the publication of the project in 2015, 52 communities whose lands and mapping were ready were still waiting for the registration certificate from the government of Uganda [41].

In Uganda, communal area land governance is fragmented among the various ministries and managed by various often overlapping land governance institutions. These include the district land boards, land tribunals, local land courts, and customary institutions [41,46]. The 1988 Land Act which conferred legal rights to customary tenure land rights and the 2013 Land policy that recognized customary land tenure as being at par with other land tenure systems in Uganda, increased the opportunity for communities and women within them to secure their customary land tenure rights. The consolidation of the customary and statutory land governance systems potentially removed loopholes between the two systems which are exploited by the powerful to undermine women's land tenure in customary land tenure areas.

In the reducing emissions from deforestation and forest degradation (REDD) initiative in the DRC, the maps were officially recognized by the relevant authorities. The WWF worked with community partners and administrative authorities to facilitate the process required to obtain official recognition of community maps. This process of recognition begins with the Local Territory Administration and continues through to the District Commissioner, Provincial Interior Ministry and, finally, the National Interior Ministry. The maps and numerical data gathered have been shared with institutions, including *Institute Géographique du Congo* and *Institut National de la Statistique* at the national and provincial levels. Printed maps are distributed to communities, customary authorities, including land chiefs, and the territory administrators [34]. Beyond the mapping exercise undertaken by IDLO and CTV in Mozambique, other donor-sponsored community land ownership regularization initiatives have, following the delineation, produced a map of the community DUAT with any other information, such as rights-of-way, which has then been registered in the Cadastral Atlas, culminating with the issuing of a Certificate of Delimitation in the name of the community [33]. In Rwanda, the maps generated were recognized at law as legal documents [10].

3.4. Impact of Customary Land Mapping on Women's Customary Land Rights

The mapping exercises increased communities' awareness of women's claims and rights to land and related resources. The mapping enables communities to identify and increase awareness of the way in which women exploit natural resources like rangelands, water, and forests. This, for example, included the mapping of water routes which are often overlooked when large scale land resources are alienated and highlights the extra burden that women had to walk longer distances to secure water for domestic use. The maps also capture the women's dependence and use of forest produce and resources in ways that male biased mapping would not. The integration of women and communities into land use planning [8] enabled women to participate and increased their visibility. The recording of women's land rights on the maps and claims and routes to access water and forest produce increased the visibility of women's claims on land. The increased visibility means that women's concerns become represented in various forums.

The use of resource mapping techniques and community validation enabled women to highlight the vulnerability of the resources they use in relation to competing interest (for example, large scale land investors, the government, and powerful individuals in the communities). The Rwanda initiative used maps to identify winners and losers in the implementation of irrigation projects and in the implementation of irrigation schemes. This development made it possible for the project process to identify women and other vulnerable groups who often lose out of the large land-based investments like irrigation schemes where men are often better placed to capture the benefits from employment and subcontracting arrangements [50]. This increased the chances of women's issues and claims being taken into consideration in the assessment and mitigation of impacts and changes in land tenure and use.

The inclusion of women in the mapping process by granting equal recognition to women as community members, as legitimate members of local land governance structures, and as resource users, entrenched their interests in relation to land, water, and other natural resources. The protocols for mapping land and resource rights provide opportunities for women to participate in decision-making. In Mozambique, the *Lei de Terras (1997)* law makes it mandatory to include women in every step of registering community land. The *Lei de Terras* explicitly grants women equal rights in community property and their participation in every component of community land-related governance [8,39]. In Uganda, the 1988 Land Act, which provides for the establishment of Communal Land Associations for the purposes of managing communal land resources, specifies that a third of the Communal Land Association (CLA) executive committee members must be women. This provision has paved the way for the participation of women in the management of communal lands. The CLA is the legal entity that can map and register land on behalf of the community and as such, by making women's proportional representation a requirement, the Land Act has gone a significant way towards ensuring women's tenure security [49]. In Rwanda, the data and information collected for maps and spatial is gender disaggregated [50] to facilitate monitoring. The recognition of women as rightful actors and participants in decision-making processes goes some way towards increasing their security as they are consulted in decisions to alienate land, plan land use, allocate land, and or negotiate and enter into commercial agreements with outsiders. This recognition potentially enables women to benefit equally in compensation that may be paid out for the alienation of land. The other ways in which women's inclusion in decision-making in land matters included discussions with elders, raising community awareness, and training women to participate in the decision-making fora and following up on them [10].

Mapping initiatives potentially strengthen women's land rights in a marital union; the strategic, but unfortunately, increasingly unstable institution through which women traditionally gain access to customary land. The Benin MCC report recommended that the customary obligation of husbands to provide land to wives be captured at an individual level and other secondary rights, like communal rights to harvest, be captured in rural landholding plans (*Plan Foncier Rural* (PFR)). The report also recommended that the law and regulations, at a minimum, presume co-ownership of land between spouses and that a communication, education, and training plan be adopted to raise community's awareness and appreciation of these recommended developments [46].

Mapping initiatives such as the community land delimitation in Mozambique and community land registration in Uganda enable women to make significant progress towards gaining statutory recognition of customary land rights. In Mozambique and Uganda where women's land rights in customary areas were weak, the successful implementation of the mapping initiatives provided women with a form of statutorily recognized land rights and records. Participatory land and resource mapping can strengthen women's land rights in customary tenure areas by facilitating the recording of secondary land rights, which would otherwise remain invisible and ignored during transactions with outsiders and powerful institutions with an interest in the appropriation or acquisition of communal land and attendant resources. Where these derived rights to land are not made visible nor recognized as legitimate claims, the result in many instances is that women lose out on compensation and beneficiation claims. While not the panacea for enhanced tenure security for women, the recording of women's secondary land rights is a fundamental first step towards bringing the customary land rights into the public domain where other institutions (such as advocacy groupings and even judicial courts) can participate in the mediation of these rights [46].

The improved land and resource governance that resulted from the transparent and democratic processes that land governance institutions were subjected to increased women's land tenure security in customary areas as they were less vulnerable to corruption and gender-based violence. Widows and single women benefited from the transparency as their land rights were recorded in a medium other than people's memories. This provided them with an alternative platform for securing and defending their land rights. In Benin, the mapping went further by drawing specific maps of secondary rights and vulnerable groups and making a plan to protect and enhance their access to land [46].

In Rwanda, gender considerations are taken into account at every step of the mapping and registration process. This means that men and women are included in awareness campaigns and specific aspects related to women's land rights are spelled out; the data are disaggregated by sex; both men and women are involved in the identification of parcels and boundaries; the delineation team includes both men and women and all the data related to all the components of the family are registered including wives and daughters [50]. In the context of a delimited community, civil law provides a useful legal safeguard tool in this regard, since the ownership of the use right is shared by every single member of the community, meaning that decisions must legally be taken by women as well as by men, and cannot be mediated through households, traditional authorities, or other 'representative' bodies (unless specifically mandated to do so). In reality, of course, there are challenges in the application of this tool [7].

In Benin and Lesotho, mainstreaming gender and amending gender-blind legislation was a precondition for the Millennium Challenge Account signing a compact with the respective countries. This catalyzed women's land tenure security to the top of the agenda, and it formed an integral part of all the developments during the projects. In Benin, this included mapping initiatives. As a result, women were included in decision-making. Experts were hired with the specific role of safeguarding women's interests in the planning phase of mapping programmers, ensuring their participation in these initiatives, and overseeing the development and incorporation of gender-sensitive indicators as part of the matrices for monitoring and measuring the success of outcomes. As a result, the stakeholders have had to commit to closing the gender gap in land access [46].

The formalizing of women's rights as part of the delimitation of community rights generated opportunities for women to negotiate and shift attitudes and approaches without coming into direct conflict with the status quo. Subjecting the maps to community validation at various stages also ensured that women's concerns are highlighted, and their views inform the mapping conceptualization and development process.

The arrangements made to transfer technology and skills potentially transfer important skills and competencies linked to land-use management to local communities [7,8]. Women and or women only groups were targeted in these capacity building initiatives [7,44]. This includes training of cartographers. This makes it possible for local communities to gain intimate knowledge of the techniques, technologies, and processes necessary to promote and promote their tenure land tenure. The IDLO initiatives transferred skills to community-based para-legals, capacitating them with the knowledge of relevant legislative provisions for instituting participatory processes for formal recognition of communal land rights. The para-legal—and communities, more generally—also obtained conflict resolution skills, which were critical for managing intra- and inter-community disputes over land boundaries and resource use arrangements. Overall, there was increased capacity on the part of the communities to make decisions and participate in natural resource governance structures and negotiations. Through the Millennium Challenge Account-funded project in Benin, the number of para-legal professionals in the mayoral administrations in twenty communes was increased, significantly improving the local governments' capacity to assist citizens, including women's groups, to make effective use of the maps, to negotiate, and conclude agreements for use and occupancy of lands of new proprietors or of communal reserves [46].

3.5. Limits of Mapping for Securing Women's Land Tenure in Customary Areas

Analyses of these initiatives for the reform of customary land management demonstrated their scope and importance but also their limits. Mapping initiatives, while encouraging, must be subjected to empirical scrutiny to assess the extent to which they have actually translated into real security of tenure beyond the lifespan of outsider-driven or facilitated mapping initiatives. Mapping initiatives are not only difficult to launch but necessarily involve processes that can prove costly. The costs, which include equipment and software licensing, are beyond the reach of poor communities and women who have limited economic opportunities to earn money. This can be mitigated to an extent by open source technology, the involvement of funding partners, and documenting or registering the community land as a meta-unit [8]. Mapping initiatives can potentially create new and or unearth long standing boundary conflicts within the community or between a community and its neighbors, which can increase pressure on tenuous women's communal land tenure rights. Participatory appraisal and boundary mapping can mediate boundary conflicts. This includes the inclusion of local community information on history, culture, social organization, spatial occupation, land use, population dynamics, and possible conflicts and their resolution in the mapping [4]. Since mapping initiatives and the space they open up for women to secure land rights in customary land tenure areas threaten power interests, they may be subject to resistance by those who are threatened who include patriarchal gate keepers.

This is mediated by expressing the importance of gender issues in the early stages of project design and developing a plan to communicate to difference target audiences [4,46].

The supply driven nature of the initiatives undermine sustainability of the project in the long term beyond the project period. It is important to monitor the project using gender disaggregated data and adjust methods and targets if necessary [46,50]. There is no guarantee that the local land administration institutions (the municipalities and the provincial cadastral services) will have the skills, tools, and capacity to maintain the cadastral registers [10]. In that regard, it is important to build capacity of local level institutions. This is done though providing technical tools and training to the decentralized institutions. This can be complimented by increasing communities' capacities to use the mapping processes to maintain the cadastral registers. Women should be integrated into these initiatives [4,7,10]. The real changes associated with the paper-based and paper-generating initiatives will remain beyond the reach of many women unless they are supported by interventions to support the largely illiterate rural women to secure the necessary documentation. This is mediated by simplifying and streamlining land titling processes and including knowledgeable women in the project [46] as well as securing enforcement mechanisms. There is a risk that land institutions will not recognize the maps and or the rights and claims represented on them. The risk is higher if the maps are not presented in a format that is not consistent with existing land right recording systems. In order to mitigate this risk, mapping initiatives should study and take into account existing systems so that the new information is compatible with existing records. For example, in the DRC, the team worked to ensure that all information and data from the participatory mapping exercises was incorporated into national databases [7]. Since the mapping processes and attendant conversations involve choices and discussions, there is a risk that women and women's interests may be overlooked or set aside by those who have power over the mapping process [4]. The mandatory representation of women in decision-making fora and entities formed to facilitate the mapping, implementing a women's empowerment/participation strategy and working to ensure women's full involvement in all community land documentation activities are all strategies to address this challenge [8,46,48].

In the communities represented, women form diverse age groups class and positions of power [27,51–53]. The differentiation of women and communities, if not taken into account during the mapping, can create challenges. For example, negotiating land rights under customary law can put certain community members at a significant disadvantage, particularly for certain categories of women such as widows and divorcees that lack the social power and support to successfully assert their interests. The process can also potentially perpetuate existing inequalities by favoring those with power [10,52] Mapping initiatives should identify vulnerable groups and put in place mechanisms to protect them as well as the acknowledgement of rights holders secondary and derived rights to land [46]. The choice by the mapping authors can also undermine women's land claims. For example, state law bias in recording marriage can marginalize those whose marriages are not recorded in Statute. This marginalizes the majority of women who are married under customary law and other informal conjugal unions. In order to mitigate the risk of disenfranchisement of people who held derived land rights, implementers can develop contract forms and processes to document secondary rights. This includes amending gender-blind legislation and recording and registration of women's secondary rights to land, which include the right to use their husband's land and customary obligation of husbands to provide land to wives be captured as a real right to land [46]. Mapping of community resources may expose vulnerable resources and communities to exploitation. As a result, the communities and women within them will be vulnerable to displacement and loss of land to commercial interests. In order to protect women within communities from marginalization during registration, communities can register lands as a collective. The mapping would reference customary boundaries and empower communities to control and regulate intracommunity land holdings and use to protect the vulnerable women [4,8]. The risk of women losing communal resource rights in the registration of mapping outcomes can be addressed through registration of women's rights to harvest produce from communal resources [46].

4. Recommendations for Improving on the Practice and Scaling-Up of Mapping Initiatives

Although mapping initiatives potentially catalyze the transmission of women's land rights, there is need for a longitudinal study, and impact-monitoring impacts are hard to gauge over the duration of a short-term project. The extended multi-year monitoring would require a different project timeline and funding model. The development partners' pilot and project approaches make it difficult to consolidate and entrench processes and practices. For sustainability: internally-driven participatory governance ethos and practice, it is important to work both at the level of the community (raising awareness about legislative framework, supporting community discussion of and implementation of actual mapping processes in a gender sensitive way) as well as at the level of the government. The latter is necessary not only to ensure that the practices of government officials on land governance become more inclusive and gender sensitive, but also to support the state itself in the timeous and efficient discharge of its functions in respect of issuing the communal/household land titles or other formal recognition of tenure [50]. The involvement of women may have to be negotiated at length with gate keepers and community elders. This is important in communities where women are traditionally excluded from decision-making on land matters. It is also important to train and build capacity of women to ensure they play an active role in these community structures [49]. Land mapping procedures and outcomes should take note of, and where possible, preserve and build upon the aspects of untitled customary tenure that already favor women, given that they are already accepted as legitimate in communities. Land tenure mapping initiatives in customary areas can work better for women's land rights if they also include community-specific gender analysis and crafting strategies to address gender disparities into their programs.

5. Conclusions

The paper, which draws from case studies in selected countries in Sub- Saharan Africa, has shown that customary land mapping initiatives are an innovation that, given certain conditions and resources, can potentially catalyze the transmission of women's customary land tenure rights in ways that increase tenure security. This is through providing women access to decision-making forums and facilitating women's participation in land registration processes. The initiatives that recognize women's derived land rights precipitate change of the terms on which women negotiate access to resources in communal areas, providing alternative identity to marriage and kinship-based systems, open up novel spaces between custom and statute, which potentially enable women to negotiate for customary land, and increasing the visibility of women's land interests and narratives in customary land tenure areas. However, mapping, if not handled carefully, can potentially undermine women's fragile land rights in customary land tenure areas by entrenching existing power asymmetries, complicating existing land recording systems, increasing the cost of securing land by disbursing costly technology, unearthing long standing conflicts which can increase pressure on fragile land claims. There is a need for mapping initiatives to address issues of sustainability, localization and scaling, and affordability to ensure that the intervention brings about demonstrated changes to women's land tenure status in customary land tenure areas.

Author Contributions: Conceptualization G.P. and L.M. Methodology G.P. and W.M., formal analysis, G.P. writing L.M. and G.P., review and editing G.P. and W.M. All authors have read and agreed to the published version of the manuscript.

Funding: This research received no external funding.

Conflicts of Interest: The authors declare no conflict of interest.

References

1. Alden Wily, L. Collective Land Ownership in the 21st Century: Overview of Global Trends. *Land* **2018**, *7*, 68. [CrossRef]

2.   Cotulla, L.; Toulmin, C.; Quan, J. *Better Land Access for the Rural Poor. Lessons from Experience and Challenges Ahead*; IIED: Hertfordshire, UK, 2006.

3.   Paradza, G.; Sulle, E. Agrarian Struggles in Mozambique: Insights from Sugar Plantations. In *Africa's Land Rush Rural Livelihoods and Agrarian Change*; Hall, R., Scoones, I., Tsikata, D., Eds.; James Currey: Suffolk, UK, 2015; pp. 150–162.

4.   Knight, R. Best Practices in Community Titling. 2010. Available online: https://www.files.ethz.ch/isn/139540/Land_InceptionPaper.pdf (accessed on 30 July 2020).

5.   Makura-Paradza, G. *Single Women, Land and Livelihood Vulnerability in a Communal Area in Zimbabwe*; Wageningen Academic Publishers: Wageningen, The Netherlands, 2010.

6.   Walker, C. Piety in the Sky? Gender Policy and Land Reform in South Africa. *J. Agrar. Chang.* **2003**, *3*, 113–148. [CrossRef]

7.   World Wildlife Foundation. *Strengthening Land Tenure Through Participatory Land Use Mapping in the Democratic Republic of Congo*; World Wildlife Foundation Fact Sheet; World Wildlife Foundation: Gland, Switzerland, 2013.

8.   Knight, R.; Adoko, J.; Auma, T.; Kaba, A.; Salomao, A.; Siakor, S.; Tankar, I. *Protecting Community Lands and Resources*; International Development Law Organization (IDLO): Rome, Italy, 2012.

9.   Weyer, D.; Bezerra, C.; Vos, A. Participatory Mapping in a Developing Country Context: Lessons from South Africa. *Land* **2019**, *8*, 134. [CrossRef]

10.   Di Gessa, S. *Participatory Mapping as a Tool for Empowerment. Experiences and Lessons Learned from the ILC Network*; International Land Coalition: Rome, Italy, 2008; Available online: https://www.participatorymethods.org/sites/participatorymethods.org/files/particpatory%20mapping%20as%20a%20tool%20for%20empowerment.pdf (accessed on 15 June 2020).

11.   Van Den Brink, R.J.E. *Land Reform in Mozambique*; Agric and Rural Development Notes; The World Bank: Washington, DC, USA, 2008; pp. 2–4.

12.   Agarwal, B. Gender, resistance and land: Interlinked struggles over resources and meanings in South Asia. *J. Peasant Stud.* **1994**, *22*, 81–125. [CrossRef]

13.   Whitehead, A.; Tsikata, D. Policy Discourses on Women's Land Rights in Sub-Saharan Africa: The Implications of the Re-turn to the Customary. *J. Agrar. Chang.* **2003**, *3*, 67–112. [CrossRef]

14.   Kachingwe, N. From Under Their Feet: A Think Piece on the Gender Dimensions of Land Grab in Africa. 2012. Available online: https://landportal.org/resource/actionaid/under-their-feet (accessed on 30 July 2020).

15.   Cooper, E. Women and Inheritance in 5 Sub-Saharan African Countries: Opportunities and Challenges for Policy and Practice Changes. Ph.D. Thesis, University of Oxford, Oxford, UK, 2010.

16.   Veit, P. Focus on Land Brief: Custom, Law and Women's Land Rights in Zambian Land. 2012. Available online: http://www.focusonland.com/fola/en/countries/brief-custom-law-and-womens-land-rights-in-zambia/ (accessed on 30 July 2020).

17.   Peters, P. "Our daughters inherit our land, but our sons use their wives' fields": Matrilineal-matrilocal land tenure and the New Land Policy in Malawi. *J. East. Afr. Stud.* **2010**, *4*, 179–199. [CrossRef]

18.   Cousins, B. Potentials and Pitfalls of "communal" land tenure reform: Experience in Africa and implications for South Africa. In Proceedings of the World Bank Conference on Land Governance in Support of the MDGs, Washington, DC, USA, 9–10 March 2009.

19.   Berry, S. Hegemony on a shoestring: Indirect rule and access to agricultural land. *Africa* **1992**, *62*, 327–355. [CrossRef]

20.   Lavigne-Delville, P. *Harmonising Formal Law and Customary Land Rights in French-Speaking West Africa, London*; IIED Issue Paper 86; IIED: London, UK, 1999.

21.   Cousins, B. Debating communal tenure in Zimbabwe. *J. Contemp. Afr. Stud.* **1993**, *12*, 29–39. [CrossRef]

22.   Spichiger, R.; Kabal, E. Gender Equality and Land Administration: The Case of Zambia. Available online: https://www.diis.dk/files/media/publications/import/extra/wp2014_gender-land-zambia_rachel-spichiger_edna-kabal_web.pdf (accessed on 31 July 2020).

23.   Women for Women International. *"Women Inherit Wrappers, Men Inherit Fields" The Problem of Women's Access to Land in South. Kivu, Democratic Republic of Congo*; Women for Women International: Washington, DC, USA, 2014; Available online: https://www.landportal.org/fr/library/resources/mokoro7093/%E2%80%98women-inherit-wrappers-men-inherit-fields%E2%80%99-problem-women%E2%80%99s-access-land (accessed on 20 July 2020).

24. Du Plessis, W. African Indigenous Land Rights in a Private Ownership Paradigm. *Potchefstroom Electron. Law J. Potchefstroomse Elektron. Regsblad* **2012**, *14*, 45–69. [CrossRef]

25. Van Asperen, P.; Mulolwa, A. *Improvement of Customary Tenure Security as Pro-Poor-Tool for Land Development—A Zambian Case Study*; International Federation of Surveyors (FIG): Copenhagen, Denmark, 2006.

26. Mvududu, S.; McFadden, P. *Reconceptualizing the Family in a Changing Southern African Environment*; Women and Law in Southern Africa Research Trust: Harare, Zimbabwe, 2001.

27. Kameri-Mbote, P. Gender Issues in Land Tenure under Customary Law. Available online: http://www.ielrc.org/content/w0509.pdf (accessed on 31 July 2020).

28. Haddad, L.; Gillespie, S. Effective food and nutrition policy responses to HIV/AIDS: What we know and what we need to know. *J. Int. Dev.* **2001**, *13*, 487–511. [CrossRef]

29. Whitehead, A. Women, Men and African Agriculture. *IDS Bull.* **2003**.

30. Chinsinga, B.; Chasukwa, M. Trapped Between Farm Input Subsidy Programme and the Green Belt Initiative: Malawi's Contemporary Agrarian Political Economy. In *Africa's Land Rush Rural Livelihoods and Agrarian Change*; Hall, R., Scoones, I., Tsikata, D., Eds.; James Currey: Suffolk, UK, 2015; pp. 132–149.

31. Chimhowu, A.; Woodhouse, P. Customary vs Private Property Rights? Dynamics and Trajectories of Vernacular Land Markets in Sub-Saharan Africa. *J. Agrar. Chang.* **2006**, *6*, 346–371. [CrossRef]

32. Peters, P. Inequality and Social Conflict Over Land in Africa. *J. Agrar. Chang.* **2004**, *4*, 269–314. [CrossRef]

33. Bruce, J.; Migot-Adholla, S. *Searching for Land Tenure Security in Africa*; Kendall/Hunt: Dubuque, IA, USA, 1994.

34. Deininger, K.; Binswanger, H. *The Evolution of the World Banks' Land Policy*; World Bank: Washington, DC, USA, 1999.

35. Berry, S. *No Condition is Permanent*; University of Wisconsin Press: Madison, WI, USA, 1993.

36. Meinzen-Dick, R.; Mwangi, E. Cutting the web of interests: Pitfalls of formalizing property rights. *Land Use Policy* **2009**, *26*, 36–43. [CrossRef]

37. Daley, E.; Englert, B. Securing land rights for women. *J. East. Afr. Stud.* **2010**, *4*, 91–113. [CrossRef]

38. Garber, B. Women's land rights and tenure security in Uganda: Experiences from Mbale, Apac and Ntugamo. *Sociol. J. Afr. Stud.* **2013**, *13*, 1–32.

39. Norfolk, S.; Turner, C. Improving Tenure Security for the Rural Poor Mozambique Case Study. Available online: http://www.fao.org/3/a-k0786e.pdf (accessed on 31 July 2020).

40. Yngstrom, I. Women, Wives and Land Rights in Africa: Situating Gender Beyond the Household in the Debate Over Land Policy and Changing Tenure Systems. *Oxf. Dev. Stud.* **2002**, *30*, 21–40. [CrossRef]

41. International Land Coalition. *Gendered Impacts of Commercial Pressures on Land*; International Land Coalition: Rome, Italy, 2011.

42. Hall, R.; Paradza, G. Pressures on Land in Sub-Saharan Africa: Social Differentiation and Societal Responses. Available online: https://www.academia.edu/2128927/ (accessed on 30 July 2020).

43. Joireman, S. The Mystery of Capital Formation in Sub-Saharan Africa: Women, Property Rights and Customary Law. *World Dev.* **2008**, *36*, 1233–1246. [CrossRef]

44. UN-Habitat/IFAD/GLTN. *Using Approaches and Technologies for Mapping Land and Natural Resource Use and Rights*; Tenure Security Learning Initiative for East and Southern Africa; IFAD: Rome, Italy, 2012; Available online: https://issuu.com/landgltn/docs/land_and_natural_resources_tenure_s/20 (accessed on 7 March 2020).

45. Tripathi, N.; Bhattarya, S. Integrating Indigenous Knowledge and GIS for Participatory Natural Resource Management:State-of-the-Practice. *Electron. J. Inf. Syst. Dev. Ctries.* **2004**, *17*, 1–13. [CrossRef]

46. Scalise, E.; Giovarelli, R.; Hannay, L.; Richardson, A. Gender and Land: Good Practices and Lessons from Four Millennium Challenge Compact Funded Projects. 2014. Available online: http://www.wocan.org/resources/gender-and-land-good-practices-and-lessons-learned-four-millennium-challenge (accessed on 10 June 2015).

47. Rugadya, M.; Kamusiime, H. Tenure in Mystery: The Status of Land Under Wildlife, Forestry and Mining Concessions in Karamoja Region, Uganda. *Nomadic Peoples* **2013**, *17*, 33–65. [CrossRef]

48. Namulondo, P.; Paradza, G.; Cherlet, J. *Communal Land Associations Claim Compensation for Investments in Their Territories, Karamoja, Uganda. Case Study*; International Land Coalition Database of Good Practices: Rome, Italy, 2015.

49. Weiner, D.T.A.; Harris, T.M.; Levin, R.M. Apartheid Representations in a Digital Land Scape: GIs, Remote Sensing and Local Knowledge in Kiepersol, South Africa. *Cartog. Geogr. Inf. Systt.* **1995**, *22*, 30–44.

50. Carpano, F. Strengthening Women's Access to Land into IFAD Projects: The Rwanda Experience. 2011. Available online: https://www.issuelab.org/resources/21153/21153.pdf (accessed on 2 February 2015).

51. Rugadya, M. Titling of Customary Tenure is Not a Fix for Women's Land Right: A Review of Evidence and Practice; Washington DC, USA. Available online: https://www.academia.edu/43491123/Titling_of_Customary_Tenure_is_not_a_fix_for_Womens_Land_Right_a_review_of_Evidence_and_Practice (accessed on 13 July 2020).

52. Chigbu, U.; Paradza, G.; Dachaga, W. Differentiations in Women's Land Tenure Experiences: Implications for Women's Land Access and Tenure Security in Sub-Saharan Africa. *Land* **2019**, *8*, 22. [CrossRef]

53. Cousins, B.; Winer, D.; Amin, N. Social Differentiation in the Communal Lands of Zimbabwe. *Rev. Afr. Political Econ.* **1992**, *19*, 5–24. [CrossRef]

 *land*

Article

Evaluation of the Completeness of Spatial Data Infrastructure in the Context of Cadastral Data Sharing

Agnieszka Trystuła *, Małgorzata Dudzińska and Ryszard Źróbek

Department of Land Management and Geographic Information Systems, Faculty of Geoengineering,
University of Warmia and Mazury in Olsztyn, 10-720 Olsztyn, Poland;
gosiadudzi@uwm.edu.pl (M.D.); zrobek@uwm.edu.pl (R.Z.)
* Correspondence: agnieszka.trystula@uwm.edu.pl

Received: 3 June 2020; Accepted: 12 August 2020; Published: 14 August 2020

Abstract: The idea behind the Infrastructure for Spatial Information in Europe (INSPIRE) project was to provide EU citizens with access to various types of information, including environmental protection and spatial management data. These resources can be viewed (Web Map Service—WMS) and downloaded (Web Feature Service—WFS) online. Cadastral datasets represent one of the 34 spatial data themes in the spatial data infrastructure (SDI). The functionality of the SDI has not yet been fully achieved due to the failure of the WMS and WFS network services. The aim of this article was to assess the completeness of the SDI containing cadastral datasets. The present study has practical implications. The proposed diagnostic tool supports an assessment of the completeness of SDI resources in seven diagnostic groups (technical and legal identifiers, the cadastral information profile, the WMS network service, the WFS network service, source cadastral databases, data validity, and WMS and WFS standardization). The developed assessment methodology enables the identification of websites that publish cadastral data through INSPIRE network services, as well as problematic websites, and it has high development potential. The results of the assessment should be used in the ongoing construction of the SDI. They can also be used to improve the quality of network services and their availability for end users.

Keywords: cadastral data; spatial data infrastructure; websites publishing cadastral data; INSPIRE network services; evaluation

---

1. Introduction

Infrastructure for Spatial Information in Europe (INSPIRE) network services are widely applied in decision-making processes relating to responsible spatial management. Network services such as the Web Map Service (WMS), Web Map Tile Service (WMTS), Web Feature Service (WFS), Web Coverage Service (WCS) and Catalogue Service for Web (CSW) have been developed by the Open Geospatial Consortium (OGC), and they can be applied to develop dispersed systems and web applications that communicate across the network through appropriate HTTP protocols [1]. Network services speed up access to spatial data dispersed across multiple databases if they have been developed in accordance with OGC standards that guarantee proper service operation. WMS and WFS standards are used to develop spatial data infrastructure (SDI) according to selected EU and international standards [2–6].

The SDI was developed to facilitate the implementation of EU environmental policies and activities. The primary tasks of the SDI are to enable the exchange of spatial data between public sector organizations and to facilitate access to these data across the EU [7]. In every European country, the SDI can be implemented in a manner that promotes the development and improves the quality of European initiatives such as e-administration and the European Interoperability Framework. One of the greatest advantages of the European SDI is that it improves the functioning of public administration at all levels by facilitating access to geospatial information [8]. Under the INSPIRE Directive, the public administration authorities of the EU member states are under obligation to integrate data from various thematic fields and to provide access to such information through web service modules that support the online viewing, searching, and downloading of spatial data.

The cadaster is one of the sources of spatial information for developing thematic datasets in line with the INSPIRE Directive. The cadaster aggregates spatial information that significantly influences economic processes and economic growth. The current status and functionality of the cadaster have been shaped by historical, political, and legal factors, as well as the dynamics of Poland's socioeconomic development. The cadastral system provides access to information on land parcels, buildings, premises, and entities who hold various legal titles to the listed property. Contemporary cadastral databases should be simple, effective, and reliable [9] in order to improve the functioning of organizations that are responsible for real estate management. Cadastral data are used to resolve decision-making problems in the process of achieving environmental, social, economic, legal, and tax policy objectives [10–13]. These objectives cannot be achieved without access to cadastral data. Cadastral data should be made available through network services [14–16] in line with national regulations on open access to public data. Public access to data is an essential instrument of social control over state administration, and it increases the responsibility and transparency of government activities. The relevant data are provided in the form of cadastral maps by the Head Office of Geodesy and Cartography [17].

The aim of this article was to assess the progress made in the development of the SDI, which is based on websites publishing cadastral data that have been submitted to the register of spatial datasets and services under the SDI. The present study has practical implications, and it proposes a tool for validating the progress in SDI development in seven diagnostic groups: technical and legal identifiers, the cadastral information profile, the WMS network service, the WFS network service, INSPIRE theme 1.6 and 3.2 source databases, data validity, and WMS and WFS standardization. The aim of the assessment was to diagnose the current status of SDI development, to identify the strengths and weaknesses related to the quality of publicly available cadastral data, and to formulate recommendations for further activities with the aim of improving their effectiveness in various decision-making processes.

Nearly 400 websites publishing cadastral data (county (powiat) cadastral databases) in Poland need to be consolidated, and the relevant data resources have to be standardized. The article evaluates websites publishing cadastral data to assess the progress made in the implementation of network service solutions as one of the key features of the SDI. The proposed procedure for evaluating websites publishing cadastral data involved the following stages:

- The determination of the main objective of service evaluation.
- The description of the criteria for diagnosing the functionality of websites publishing cadastral data and access to cadastral data (legal, organizational, and technical aspects).
- The development of indicators for evaluating selected diagnostic criteria.
- The interpretation of the results of cadastral service evaluation.

2. Background

The aim of the INSPIRE concept was to establish a framework for improving the availability, relevance and interoperability of spatial data for environmental policy-making and activities that exert a direct or an indirect impact on the environment [2,18–21]. An interoperable SDI is an institutional concept that aims to better respond to the public demand for geographic data in a wide range of thematic domains. This concept continues to evolve, and it has emerged as the main SDI that supports social and economic policy-making around the world [8,22]. The purpose of the SDI is to store, share, and maintain spatial data and metadata at an appropriate level. The main features of the SDI are presented in Figure 1.

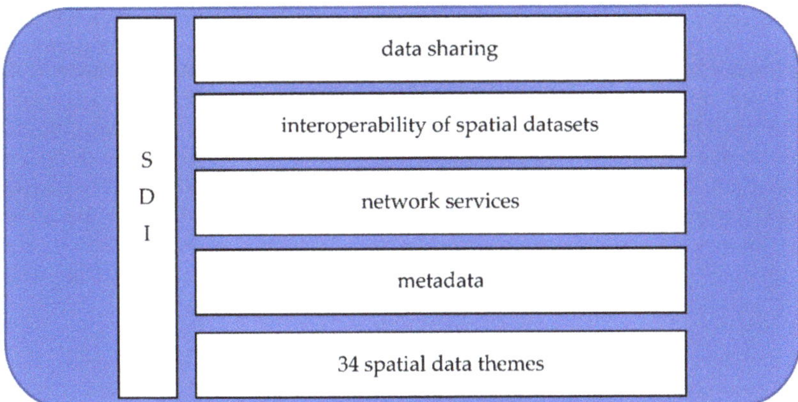

Figure 1. Spatial data infrastructure (SDI) features. Source: [2].

Network services pose one of the greatest challenges to the development of the SDI in Europe. These services create access to spatial data, including cadastral parcels, at both the national and European levels [2,23,24]. The spatial data themes referred to in the INSPIRE Directive include cadastral parcels and buildings [2]. Cadastral parcels are listed in Annex I, and they are considered reference data, i.e., data that constitute the spatial framework for linking and/or identifying other types of information in various thematic fields, such as buildings, the environment, soil use, and land use. The INSPIRE Directive focuses on the geographic attributes of cadastral data. In the context of the INSPIRE Directive, cadastral parcels mainly serve as locators of general geo-information, including environmental data. According to the technical standards laid down by the INSPIRE Directive [25], cadastral parcels fall under the scope of one or more INSPIRE themes if they are defined by cadastral or equivalent registers, as well as if they have uniform legal status and are available as vector data. From the perspective of the implemented directive, the INSPIRE model of cadastral data only covers the geometric part of the cadastral system. Legal aspects and ownership data are not taken into consideration even if they are part of the dataset because the member states have the right to limit public access to spatial data and services [2]. In view of the above, the cadastral data model is simple and highly compatible with other INSPIRE databases, such as databases of buildings whose specifications are based on geographic location in line with the developed guidelines [26].

The technical implementation of network services falls subject to the technical specifications developed by the OGC. The member states are under obligation to establish and operate a network of the following spatial data services:

- Discovery services that support the search for spatial datasets and services based on the content of the corresponding metadata, as well as enabling users to display metadata content.
- View services that, as a minimum, enable users to display, navigate, zoom in/out, pan, or overlay viewable spatial datasets, as well as to display legend information and any relevant metadata content.
- Download services that enable users to copy, download, and, where practicable, directly access spatial datasets or parts of such sets.
- Transformation services that enable users to transform spatial datasets with a view to achieving interoperability.
- Services that enable users to invoke spatial data services.

These services have to account for specific user requirements, and they have to be easy to use, available to the public, and accessible via the internet or any other appropriate means of telecommunication [2]. Cadastral data published via network services can be viewed (OGC WMS) and downloaded (OGC WFS) [25]. In line with the INSPIRE Directive, the INSPIRE geoportal is the main European spatial database that integrates spatial data resources and enables the member states to view (OGC WMS) and download (OGC WFS) spatial data. The geoportal also supports measures aiming to monitor the entire INSPIRE data collection. The WMS is based on the HTTP interface, and it enables users to view and integrate maps with other spatial data from the INSPIRE geoportal. Three functions have been identified in line with OGC standards: GetCapabilities for acquiring detailed descriptions of maps available on the server, GetMap for downloading maps, and GetFeatureInfo for requesting information about the objects displayed on the map. GetCapabilities and GetMap are obligatory functions that have to be implemented in every WMS [4]. The WFS is an internet service that provides access to geographic objects and enables users to download and edit objects in the database. The service also contains tools for creating, storing, and parameterizing server queries [5].

The INSPIRE concept promotes access to knowledge about European resources at the national, regional, and local levels. Modern societies have a vast need for a broad spectrum of information relating to environmental protection, cultural heritage protection, spatial management, investments, internal and external security, the development of a knowledge-based economy, e-administration, e-society, and, consequently, civil society [27]. Poland has developed the relevant legal tools [28] for implementing the provisions of the INSPIRE Directive. According to [1,29], the cadaster plays an important role in the Polish SDI as a reference for other spatial data themes covered by the INSPIRE Directive.

The cadaster provides access to the spatial data themes referred to in the INSPIRE cadastral model. The data pertaining to cadastral parcels (Annex I to the INSPIRE Directive, theme 6) can be compatible with other INSPIRE spatial data themes, such as buildings (Annex III, theme 2). The geoportal.gov.pl web service is being developed with the use of the open source technology, and it is operated by the Chief Surveyor General of Poland (CSG) in line with EU and national regulations [25,28] to provide access to SDI resources in Poland. The CSG is also responsible for 15 INSPIRE data themes, including cadastral parcels and buildings. The data relating to cadastral parcels (Annex I, theme 6) and buildings (Annex III, theme 2) are published by two groups of network services. The first group is based on the WMS, and it enables users to view data layers relating to cadastral parcel boundaries, parcel numbers, and buildings. In Poland, access to cadastral data was created by harmonizing the resources of the Land Parcel Identification System (LPIS), which supports direct payments to farmers under the Common Agricultural Policy. The boundaries of cadastral parcels are determined based on cadastral system data. The second group of web services involves the WMS and the WFS, which publish cadastral data that are available in county centers for geodetic and cartographic documentation (county cadastral databases) and have been previously notified in the geoportal's service repository.

According to [14], the WMS specifications for the cadastral data distributed by Polish counties include the following functions:

1. GetMap—for viewing cadastral maps in the PNG format.
2. GetFeatureInfo—for accessing information such as cadastral parcel ID, parcel number, the territorial unit for which the cadastral database is kept, the number of the land and mortgage register, and the date on which cadastral data were last updated.
3. GetCapabilities for accessing data layers via the WMS and basic layer parameters such as coordinate systems, graphic formats, and accessible data themes.

A cadastral parcel can be localized (its geometric parameters can be downloaded) using a service based on the OGC WFS standard. All WFS-based applications should have the following functionalities:

1. GetCapabilities, which returns metadata.
2. DescribeFeatureType, which returns a description of feature types from the cadastral parcel layer.

3.   GetFeature, which returns the cadastral parcel, its geometry and features based on the legal definition, parcel identification data, or coordinates.

3. Materials and Methods

3.1. Study Area

In the geographic sense, the analyzed county cadastral databases are located in the Eastern Poland macroregion, which is one of the least economically developed regions of the EU [30]. This fact was one of the key arguments for selecting the study area. Eastern Poland is a peripheral macroregion that occupies an area of nearly 99,000 km^2 and accounts for 32% of Poland's territory. Its eastern border marks the eastern border of Poland and the eastern border of the EU (Figure 2). Websites that are tasked with providing valid cadastral data as reference data for the SDI under the Act on Spatial Data Infrastructure were analyzed and evaluated in 14 counties of Świętokrzyskie Voivodeship [28].

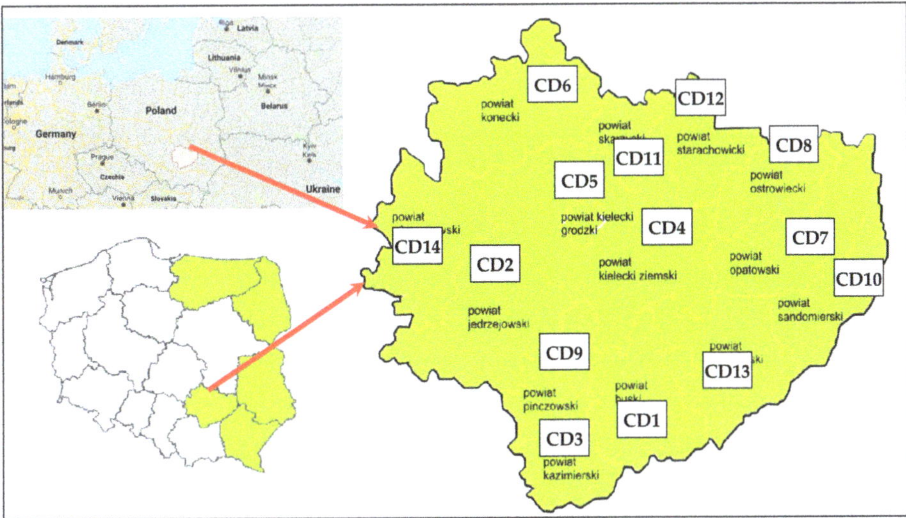

Figure 2. General location of Świętokrzyskie Voivodeship. Source: Own study.

County governors are responsible for maintaining and updating Polish cadastral systems. The analyzed county databases provide access to cadastral data, and they are partly integrated with the National Integration of Land Registers (NILR) service that groups the county WMS under a single URL address. The NILR is a tool that supports the national geoportal by facilitating the presentation of cadastral data directly acquired from public organizations that are responsible for integrating and updating cadastral data. Counties that are only partly integrated with the NILR rely on network services based on the cadastral resources of the LPIS that are not regularly updated. According to [31], county databases differ in the accuracy with which the boundary points of cadastral parcels have been mapped, and they contain discrepant information on the area of cadastral parcels, as well as errors relating to the classification of land-use types. Regardless of the manner in which spatial infrastructure nodes at the county level have been integrated with the NILR, cadastral data should be prepared in line with the EU model described in [25]. In the period covered by this study, the largest number of counties that published cadastral data based on LPIS resources were located in Świętokrzyskie Voivodeship (Figure 3).

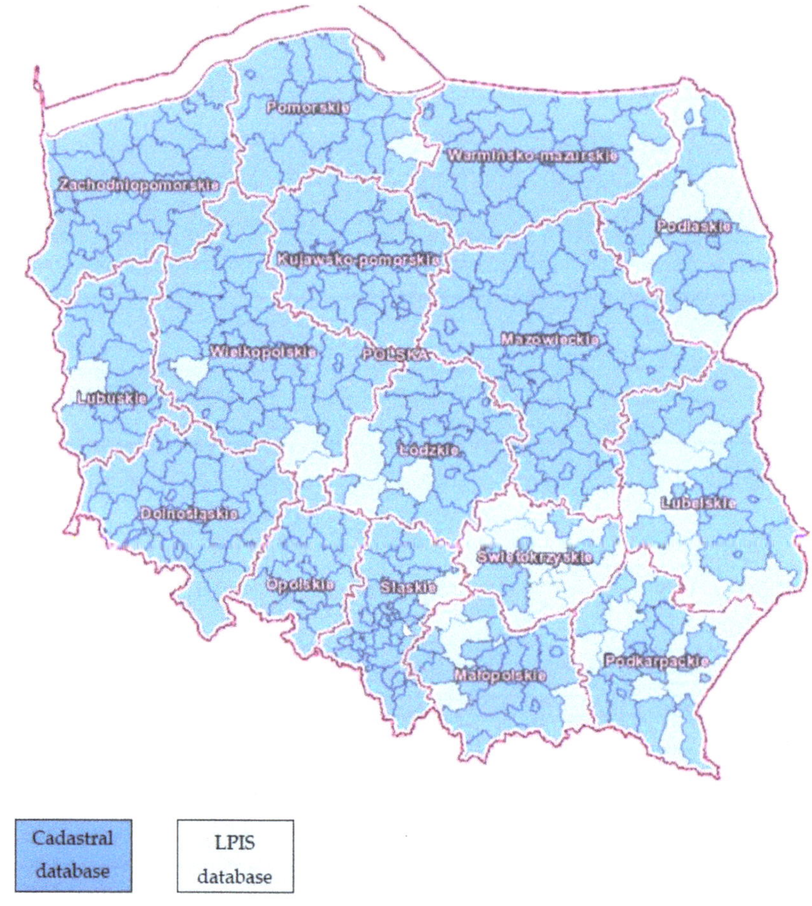

Figure 3. County infrastructure nodes in the National Integration of Land Registers (NILR) service. Source: https://geoportal.gov.pl/.

Nearly 90% of Polish counties are fully integrated with the NILR service (cadastral data are regularly updated). In Świętokrzyskie Voivodeship, cadastral databases were fully integrated with the NILR in 5 counties, i.e., in 36% of public administration units in that voivodeship. The remaining nine counties published cadastral data based on LPIS resources. As a result, Świętokrzyskie Voivodeship ranks last on the list of Polish voivodeships that publish valid cadastral data. This study covered all counties of Świętokrzyskie Voivodeship regardless of their integration status with the NILR

3.2. Methodology

According to [32], an evaluation is a process of collecting and analyzing data to identify the strengths and weaknesses of programs, policies, and organizations with the aim of improving their effectiveness. Evaluations have three objectives, which are to measure impacts, understand the causal path, and engage stakeholders in learning processes. The present study evaluated the progress made in the harmonization of Polish legal acts relating to the development of the SDI at the local level with the EU regulations.

The deployed methodology involved three research stages (Figure 4) that were developed by merging several approaches [33], including:

1. A review of the literature addressing the problem, with the main focus on selected legal, organizational, and technical aspects related to the publication of cadastral data using network services such as the WMS and the WFS.
2. Research into the usability of websites publishing cadastral data combined with an expert interview.
3. An analysis of the register of spatial datasets and services with the aim of exploring its structure and operating principles.
4. Evaluations involving the identification of success or failure criteria for the online publication of cadastral data.
5. Inference aiming to formulate, in a clear and unambiguous manner, the crucial results of the evaluation of selected websites that publish cadastral data.

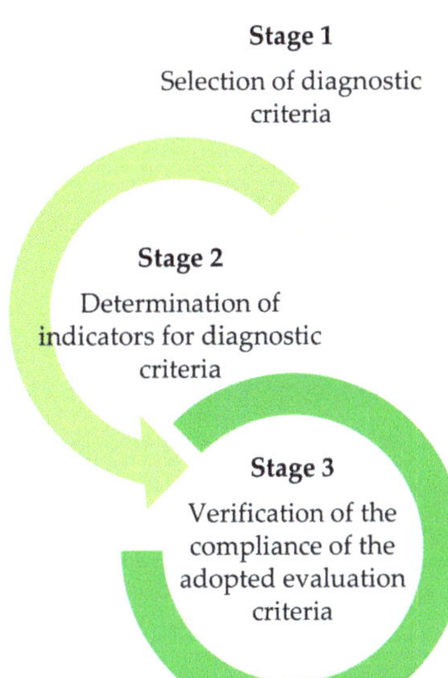

Figure 4. Research plan. Source: own study.

The adopted approach supported an analysis and evaluation of selected processes relating to the development of cadastral systems in line with the EU solutions.

The analyzed websites that publish cadastral databases should make the collected data available through network services. To verify the websites' compliance with the provisions of the INSPIRE Directive, complex phenomena with varied origin were analyzed with the use of a qualitative method.

Diagnostic criteria for evaluating cadastral systems that publish cadastral data via network services (the WMS and the WFS) and that determine the overall validity of the performed analysis were selected in the first stage of research. In the second stage, indicators were assigned to selected diagnostic criteria for evaluating network services. The anticipated compliance of the assigned indicators was determined.

Stage one: The identification of the diagnostic criteria for evaluating websites publishing cadastral data via the WMS and WFS services

The developed indicators were used to monitor the development of cadastral systems [11,34–36]. Diagnostic criteria and the relevant indicators were identified based on the provisions of:

1. Directive 2007/2/EC of the European Parliament and of the Council of 14 March 2007 Establishing an Infrastructure for Spatial Information in the European Community [2].
2. Act on Spatial Data Infrastructure of 4 March 2010 [28].
3. Geodetic and Cartographic Law of 17 May 1989 [37].
4. Regulation of the Minister of Regional Development and Construction of 29 March 2001 on land and building registers [38].
5. ISO 19128 [4].
6. ISO 19142 [5].
7. Technical specifications for county-level WMSs relating to land and building registers.
8. Interviews with the experts employed by the County Centre for Geodetic and Cadastral Documentation in Świętokrzyskie Voivodeship.
9. Analysis of the structure and operating principles of the register of spatial datasets and spatial data services kept by the Head Office of Geodesy and Cartography (HOGC).

The adopted indicators for evaluating cadastral systems that publish spatial data combine the diagnostic criteria associated with the studied objects, including spatial data themes, cadastral parcel identifiers, the availability and standardization of network services, and the validity of published cadastral data.

Network services are among the identifiable features of the SDI. The criteria for evaluating systems that publish information about cadastral parcels and buildings via network services were selected based on European and Polish trends that account for local environmental needs. Seven diagnostic criteria (A–G) for evaluating equivalent cadastral systems are presented in Table 1. In systems that meet all criteria, SDIs were regarded as complete at both the local and national levels.

Table 1. Selected Diagnostic Criteria for Evaluating Cadastral Systems. WMS: Web Map Service; WFS: Web Feature Service.

Symbol	Diagnostic Criteria
A	Technical and legal identifiers
B	Cadastral data profile
C	WMS network service
D	WFS network service
E	Sources of data for databases of cadastral parcels (Annex I, theme 1.6) and buildings (Annex III, theme 3.2)
F	Standardization of WMS and WFS
G	Data validity

Source: Own study.

Stage two: The description of the diagnostic features of selected evaluation criteria, including the degree of criteria fulfilment

The indicators assigned to each diagnostic criterion for evaluating cadastral systems are presented in Tables 2–8.

Table 2. Technical and Legal Identifiers.

Diagnostic Criterion	Indicator	Symbol	Validity (0–1)
	Ordinal number of cadastral dataset	A_1	
	Publication date of cadastral dataset	A_2	
	Notification date of cadastral dataset	A_3	
	Notifying entity	A_4	
A	Identifier of cadastral dataset	A_5	0–1
	Name of cadastral dataset	A_6	
	Code of cadastral dataset	A_7	
	Legal regulations	A_8	

Source: own study.

Table 3. Cadastral Data Profile.

Diagnostic Criterion	Indicator	Symbol	Validity (0–1)
B	Annex I, theme1.6 * ° cadastral parcels ° cadastral parcel labels Annex III, theme 3.2 * ° buildings ° building labels	B_1 B_2	0–1

*—The criterion is fulfilled when all data are visible or when all data have been returned by the function. Source: own study.

Table 4. WMS Network Service.

Diagnostic Criterion	Indicator	Symbol	Validity (0–1)
C	WMS availability WMS address	C_1 C_2	0–1
	WMS indicator at county level: $P_{WMS} = \frac{N_{WMS}}{N_{us}} \cdot 100\%$	C_3	N_{WMS} (0–1) N_{us} (0–5) ** 0–100% 0—When 0% 1—When 20–100%

**—The indicated range covers 5 INSPIRE network services (WMS, WFS, Catalogue Service for Web (CSW), Web Coverage Service (WCS), and WCTS). Source: Own study.

Table 5. WFS Network Service.

Diagnostic Criterion	Indicator	Symbol	Validity (0–1)
D	WFS availability WFS address	D_1 D_2	0—Not available 1—Available
	WFS indicator at county level: $P_{WFS} = \frac{N_{WFS}}{N_{us}} \cdot 100\%$	D_3	N_{WFS} (0–1) N_{us} (0–5) 0–100% 0—When 0% 1—When 20–100%

Source: own study.

Table 6. Sources of Data for Databases of Cadastral Parcels and Buildings. LPIS: Land Parcel Identification System.

Diagnostic Criterion.	Indicator	Symbol	Validity (0–1)
E	Land and building register	E_1	0—LPIS 1—Cadastral database

Source: own study.

Table 7. Standardization of WMS and WFS.

Diagnostic Criterion	Indicator	Symbol	Validity (0–1)
F	GetMap function	F_1	0–1
	GetFeatureInfo function *:		
	° cadastral parcel ID		
	° cadastral parcel number	F_2	
	° territorial unit for which the cadastral database is kept		
	° number of the land and mortgage register		
	GetCapabilities function *:		
	° cadastral parcel layer		
	° number of cadastral parcel	F_3	
	° building layer		
	HTTP protocol	F_4	
	GetFeature function based on parcel ID	F_5	
	Function based on x and y coordinates	F_6	

*—The criterion is fulfilled when all data are visible or when all data have been returned by the function. Source: own study.

Table 8. Data Validity.

Diagnostic Criterion	Indicator	Symbol	Validity (0–1)
G	Date of last cadastral data update	G_1	0—No date 1—Date of last update

Source: own study.

Stage three: The evaluation of selected systems publishing cadastral data based on the indicated criteria and their diagnostic features

Selected cadastral systems that publish spatial data based on the selected diagnostic criteria and indicators were evaluated in the third stage of the study. The criteria responsible for the success or failure of network services that publish cadastral data were identified. The results of the evaluation were used to formulate clear conclusions regarding the analyzed network systems that publish cadastral data. The trends and prospects relating to the development of cadastral systems that publish data via INSPIRE network services were verified based on the extent to which the selected equivalent systems met the diagnostic criteria. Each of the seven diagnostic criteria were evaluated on a two-point grading scale: 0 for when at least one diagnostic criterion was not met, and 1 for when all diagnostic criteria were met. The following key was used to evaluate the completeness of the Polish SDI based on the available network services, the associated spatial data themes, and their variability over time (data validity):

- Excellent (EXC)—100% of possible points for every adopted criterion; the evaluated SDI is fully complete.
- Above Average (AAVG)—More than 60% of possible points for every adopted criterion; the evaluated SDI is characterized by above-average completeness.
- Average (AVG)—More than 40% of possible points for every adopted criterion; the evaluated SDI is characterized by average completeness.
- Below Average (BAVG)—More than 20% of possible points for every adopted criterion; the evaluated SDI is characterized by below-average completeness.
- Negative (NEG)—0–20% of possible points for every adopted criterion; the evaluated SDI is characterized by critical-level completeness.

The proposed diagnostic criteria (A–G) for evaluating cadastral systems that publish data via network services were verified based on HOGC data [39]. The results of the verification process are presented in Tables 9–15.

Table 9. Evaluation of Cadastral Databases Based on Criterion A.

Diagnostic Criterion	Indicator	Cadastral Database (CD)														Fulfilment of Criterion (Counties)	Fulfilment of Criterion (Voivodeship)
		CD1	CD2	CD3	CD4	CD5	CD6	CD7	CD8	CD9	CD10	CD11	CD12	CD13	CD14		
		Technical and legal identifiers															
A	A_1	1	1	1	1	1	1	1	1	1	1	1	1	1	0	93%	93%
	A_2	1	1	1	1	1	1	1	1	1	1	1	1	1	0	93%	
	A_3	1	1	1	1	1	1	1	1	1	1	1	1	1	0	93%	
	A_4	1	1	1	1	1	1	1	1	1	1	1	1	1	0	93%	
	A_5	1	1	1	1	1	1	1	1	1	1	1	1	1	0	93%	
	A_6	1	1	1	1	1	1	1	1	1	1	1	1	1	0	93%	
	A_7	1	1	1	1	1	1	1	1	1	1	1	1	1	0	93%	
	A_8	1	1	1	1	1	1	1	1	1	1	1	1	1	0	93%	

Source: own study.

Table 10. Evaluation of Cadastral Databases Based on Criterion B.

Diagnostic Criterion	Indicator	Cadastral Database (CD)														Fulfilment of Criterion (Counties)	Fulfilment of Criterion (Voivodeship)
		CD1	CD2	CD3	CD4	CD5	CD6	CD7	CD8	CD9	CD10	CD11	CD12	CD13	CD14		
		Cadastral data profile															
B	B_1	1	1	1	1	1	1	1	1	1	1	1	1	1	0	93%	21%
	B_2	0	0	1	0	1	0	0	0	1	0	0	0	0	0	21%	

Source: own study.

Table 11. Evaluation of Cadastral Databases Based on Criterion C.

Diagnostic Criterion	Indicator	Cadastral Database (CD)														Fulfilment of Criterion (Counties)	Fulfilment of Criterion (Voivodeship)
		CD1	CD2	CD3	CD4	CD5	CD6	CD7	CD8	CD9	CD10	CD11	CD12	CD13	CD14		
		WMS network service															
C	C_1	1	1	1	1	1	0	1	1	1	0	0	1	0	0	64%	57%
	C_2	1	1	1	1	1	0	1	1	0	0	0	1	0	0	57%	
	C_3	1	1	1	1	1	0	1	1	1	0	0	1	0	0	64%	

Source: own study.

Table 12. Evaluation of Cadastral Databases Based on Criterion D.

Diagnostic Criterion	Indicator	Cadastral Database (CD)														Fulfilment of Criterion (Counties)	Fulfilment of Criterion (Voivodeship)
		CD1	CD2	CD3	CD4	CD5	CD6	CD7	CD8	CD9	CD10	CD11	CD12	CD13	CD14		
		WFS network service															
D	D_1	0	0	0	0	0	0	0	0	0	0	0	0	0	0	0%	0%
	D_2	0	0	0	0	0	0	0	0	0	0	0	0	0	0	0%	
	D_3	0	0	0	0	0	0	0	0	0	0	0	0	0	0	0%	

Source: own study.

Table 13. Evaluation of Cadastral Databases Based on Criterion E.

Diagnostic Criterion	Indicator	Cadastral Database (CD)														Fulfilment of Criterion (Counties)	Fulfilment of Criterion (Voivodeship)
		CD1	CD2	CD3	CD4	CD5	CD6	CD7	CD8	CD9	CD10	CD11	CD12	CD13	CD14		
		Sources of data for databases of cadastral parcels and buildings															
E	E_1	0	1	0	0	1	0	1	0	1	0	0	1	0	0	36%	36%

Source: own study.

Table 14. Evaluation of Cadastral Databases Based on Criterion F.

Diagnostic Criterion	Indicator	Cadastral Database (CD)														Fulfilment of Criterion (Counties)	Fulfilment of Criterion (Voivodeship)
		CD1	CD2	CD3	CD4	CD5	CD6	CD7	CD8	CD9	CD10	CD11	CD12	CD13	CD14		
F	F_1	0	1	1	1	0	0	1	1	0	0	0	1	0	0	43%	36%
	F_2	0	1	0	1	0	0	1	1	0	0	0	1	0	0	36%	
	F_3	0	1	1	1	1	0	1	1	0	0	0	1	0	0	50%	
	F_4	0	1	0	1	0	0	1	1	0	0	0	1	0	0	36%	
	F_5	0	1	0	1	0	0	1	1	0	0	0	1	0	0	36%	
	F_6	0	1	0	1	0	0	1	1	0	0	0	1	0	0	36%	

Source: own study.

Table 15. Evaluation of Cadastral Databases Based on Criterion G.

Diagnostic Criterion	Indicator	Cadastral Database (CD)														Fulfilment of Criterion (Counties)	Fulfilment of Criterion (Voivodeship)
		CD1	CD2	CD3	CD4	CD5	CD6	CD7	CD8	CD9	CD10	CD11	CD12	CD13	CD14		
		Data validity															
G	G_1	0	1	0	1	0	0	1	1	0	0	0	1	0	0	36%	36%

Source: own study.

The criterion denoting compliance with technical and legal indicators (diagnostic criterion A) was met by 13 out of the 14 analyzed cadastral systems in the evaluated area. One cadastral system was not notified to the register of datasets and data services, and it was classified as not fulfilling criterion A.

The spatial data themes (diagnostic criterion B) relating to cadastral parcels and buildings were present in three databases. The remaining 11 databases had an incomplete data profile. One database with a complete data profile contained additional INSPIRE themes such as soil (Annex III, theme 3.3) and addresses (Annex I, theme 1.5).

The WMS (diagnostic criterion C) was evaluated based on the availability of an HTTP address. Nine of the analyzed databases published data via the WMS, and the HTTP address of one database was not available.

The WFS (diagnostic criterion D) was not available in any of the examined cadastral databases. This non-public service can only be accessed by authorized users, but this fact did not influence the evaluation results.

Two sources of cadastral data themes were evaluated (diagnostic criterion E). Only five databases contained cadastral data themes that were acquired from the cadaster.

The technical specifications relating to the publication of cadastral data via the WMS and the WFS were evaluated based on standards [4,5]. The WMS specifications were fully compliant in five databases (diagnostic criterion F). Validity was defined as data compliance with the present status of cadastral objects. This criterion is significantly influenced by time, which induces various changes in cadastral parcels and buildings. Criterion G denoted the date of the last cadastral data update, and it was fulfilled by five databases that publish cadastral data via network services.

A ranking of the examined databases based on the total number of scored points is presented in Figure 5.

The results of the evaluation based on the adopted diagnostic criteria were used to analyze the current status of the SDI, with special emphasis on the WMS and the WFS that publish cadastral data. In the EU, numerous legal, administrative, and technical obstacles had to be overcome in the process of SDI implementation [22,34,40]. The SDI was not complete in any of the examined cadastral databases, but infrastructure completeness was above average in 21% of the analyzed territorial units. An average completeness was noted in 36% of the studied cases, and a below-average completeness was found in one database (7% of the analyzed cases). Infrastructure completeness did not exceed 20% in 36% of the studied objects.

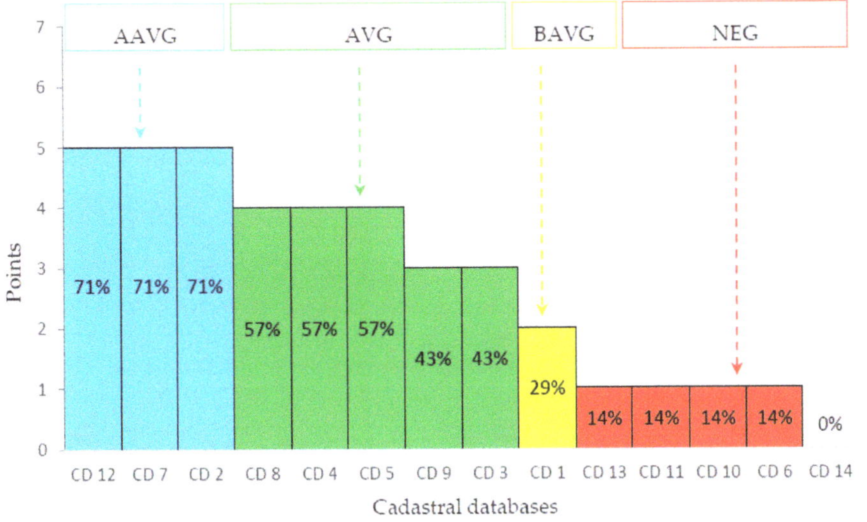

Figure 5. Total number of points scored by local cadastral databases. Source: own study.

4. Discussion

The proposed tool for verifying the completeness of the SDI supports an evaluation of websites that publish cadastral data in seven diagnostic groups to determine whether, or to what extent, this goal has been met (technical and legal indicators; cadastral data profile; the WMS; the WFS; sources of data for spatial data themes indicated in Annex I, theme 6; Annex III, theme 2; data validity; and the standardization of the WMS and the WFS). The adopted criteria support an evaluation of the factors and variables that play a key role in SDI development, and they reflect the strengths of the developed infrastructures.

The described methodology can be used to identify both cadastral web applications with high development potential (36% of the evaluated databases) and problematic services (64% of the examined cases). The analyzed territorial units differed in the level of SDI development. In view of previous studies that investigated the evolution of the SDI based on the availability of network services in Poland [14,29,41] and the EU [8,18,42–44], the evaluated territorial units have made strong and continued progress towards the achievement of a robust SDI.

In the presented evaluation, the main emphasis was placed on legal and technical aspects, the scope of cadastral data, the WMS, the WFS, and the standardization of network services. The fulfilment of seven diagnostic criteria based on the relevant indicator values is presented in Figure 6.

Diagnostic criterion A was the only parameter where the relevant indicator was fulfilled in more than 90% in the analyzed territorial units. This result validates the results of Izdebski [1], who observed that the fundamental sets of cadastral data had not been fully implemented and were not fully operational in Poland despite the fact that the SDI should be developed in line with the roadmap accepted by all EU member states. The above observation also indicates that not all cadastral datasets that are nearly fully compliant with Polish and EU regulations are fully operational. The developed cadastral system is theoretically compatible with EU requirements, but its operability continues to be limited in practice. However, the existing obstacles will most likely be overcome in the near future due either to support from EU funds that promote the implementation of central and local government initiatives in the field of the SDI or the dynamic development and dissemination of technologies for the acquisition, processing, and use of spatial data [28].

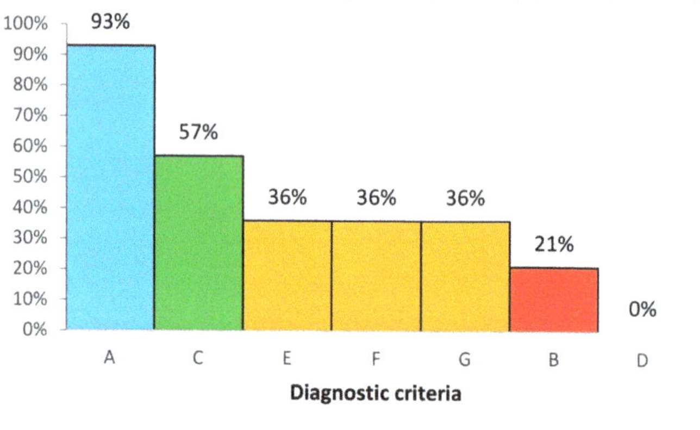

Figure 6. The fulfilment of diagnostic criteria based on the relevant indicator values in the analyzed territorial units. Source: own study.

Cadastral databases and the associated public (WMS) and non-public (WFS) services have been modernized with the support of EU funds. As demonstrated by several researchers, including those of [45], these databases have also been upgraded for compliance with Polish and EU standards. Polish counties have undertaken collaborative measures to develop and implement regional and local spatial databases as elements of the SDI, and these efforts promote problem solving, the sharing of experiences, and the achievement of strategic goals of the INSPIRE Directive.

In line with the provisions of the INSPIRE Directive, websites that publish cadastral data do not only have to be complete (diagnostic criterion B); they should also meet user expectations. Spatial data themes relating to cadastral parcels and buildings play a pivotal role in the SDI. Cadastral parcel identifiers contain information about land ownership, and cadastral parcels are also among the key reference objects for localizing other objects in spatial databases [14,25,46]. Buildings are equally important objects in cadasters, and they are linked with cadastral parcels by virtue of their legal status, attributes, and condition. Building identifiers are always linked to cadastral parcels. According to the roadmap for SDI implementation, datasets of cadastral parcels (Annex I, theme 6) should be implemented before datasets of buildings. The implementation of building data should be completed by the end of October 2020. The results of the presented analysis and previous research findings have indicated that the spatial data theme relating to cadastral parcels has been fully implemented in all datasets notified to the registers of spatial data that are covered by the Polish SDI.

The viewing of WMS data (diagnostic criterion C) and the downloading of WFS data (diagnostic criterion D) were evaluated based on the availability of these network services. The relevant criteria were not met when service addresses had not been notified or were absent. More than 64% of the examined cadastral web applications in Świętokrzyskie Voivodeship notified their data to the NILR service via the WMS. An analysis of the HOGC data archive for 2014 revealed that the relevant result had improved by more than 100% in the last five years. If the current growth trend is maintained, the WMS should be implemented in all of the examined databases in the next five years. The WFS had not been notified to the register of spatial datasets and spatial services by any of the analyzed territorial units that keep cadastral databases, which constitutes a breach of the respective legal provisions [28]. In Poland, the availability of the WFS is generally low. Only 6% of 380 county cadastral databases publish their data via the WFS, which stands in violation of the SDI strategy in the INSPIRE Directive, in particular in the context of obligatory network services. The above could be partly attributed to the misconception that data can be downloaded without transfer fees or authorization. In practice, the operators who publish cadastral data via the WFS monitor the users of data and the purpose for which the published data are used, and they set transfer fees for private users and public administration.

The responsible entities should develop principles for modelling network service processes that are compliant with SDI objectives. A dedicated data management module and mechanisms protecting datasets against unauthorized access and modification should be implemented.

The cadaster is the key component of the SDI in many EU countries [22,47]. The cadaster was not the primary source of cadastral data for the implemented SDI (diagnostic criterion E) in 64% of the analyzed databases. Cadastral parcel identifiers constitute the main reference data for many objects in INSPIRE datasets. However, temporary datasets that rely on other sources, such as the NILR service where data are not regularly updated, can be created in the process of SDI development.

The evaluation of the availability of network services in territorial units that keep cadasters in Świętokrzyskie Voivodeship (diagnostic criterion F) revealed that both the WMS (graphic presentation of cadastral data) and the WFS (search for and identification of cadastral parcels) should be harmonized with ISO standards. Only 36% of the analyzed county nodes published cadastral data via network services that were compliant with the EU standards. The first standardization efforts were undertaken in Poland already in 2007, and they led to the development of guidelines for the graphic presentation of thematic data layers in the WMS. Thematic data layers developed in county cadasters at the time were based on NILR data that were largely invalid, incomplete, and unfit for practical use, which was one of the main obstacles to the effective integration of county network services. Central administration authorities manage nearly 400 county cadastral databases with the involvement of diverse technical and organizational solutions, as well as various data visualization methods, a process that also obstructs the publication of cadastral data via network services. Before 2017, 30% of county cadastral nodes published data via WMSs. The implementation of the NILR service has radically improved the availability of cadastral data via network services [14]. The vast majority of the analyzed county databases that are fully integrated with the NILR publish cadastral data via WMSs that meet the requirements of the INSPIRE Directive and are compliant with ISO standards.

The validity of cadastral data (diagnostic criterion G) was largely determined by the register publishing such data. The above can be attributed to staffing shortages and a lack of adequate financial resources in Świętokrzyskie Voivodeship, which is one of the least economically developed regions in the EU. These problems could be resolved through financial aid from the state budget, support for human resources, the exchange of experiences, effective information flow, and assistance in SDI development.

5. Conclusions

The results of the evaluation of local network services that publish cadastral data at the county level were analyzed and interpreted to determine the progress made in the development of the Polish SDI based on a set of diagnostic criteria compliant with INSPIRE standards. This study demonstrated that the Polish SDI has been designed in line with the EU requirements, but it has not yet achieved full functionality. Considerable progress has been made since the INSPIRE Directive was transposed into Polish law, but the development of the SDI continues to face numerous obstacles. The implementation of the Polish SDI is delayed by the economic disparities between Polish regions and the existence of hundreds of county databases that publish cadastral data via network services such as the WMS and the WFS that are not always fully compliant with EU standards. The strengths and weaknesses of legal, organizational, and technical solutions adopted during the evolution of the Polish SDI were identified in the present study. The results of the evaluation constitute valuable inputs for developing the Polish SDI and network services. These results can also be used to improve the quality of the implemented network services and their availability for end users.

Territorial units, in particular counties, participate in the development of the Polish SDI pursuant to the provisions of the Act on Spatial Data Infrastructure [28]. These units are tasked with harmonizing cadastral data and ensuring the interoperability of datasets and infrastructure services. Therefore, further research is needed to identify the most effective technical solutions and legal instruments for adapting the existing spatial databases to INSPIRE requirements and other challenges of the modern world. The resulting measures would substantially support and accelerate the development of local SDIs.

The INSPIRE Directive does not cover all spatial data themes that play a very important role in local, regional, and national development. Regional geoportals rely on own guidelines and technical solutions to publish data that are not addressed by INSPIRE themes, which runs counter to the objectives of the INSPIRE Directive. In many cases, data are acquired from reliable state-run databases, but not all of these sources comply with EU requirements. Therefore, the possibility of expanding the thematic scope of European SDIs should be further investigated to guarantee that the adopted solutions promote effective spatial management.

Spatial data infrastructures will be fully compliant with the provisions of the INSPIRE Directive when the responsible entities at every level of governance actively participate in the process of SDI development. Financial support from the state and the EU is also needed to speed up the implementation of INSPIRE solutions in regions where the development of the SDI is delayed due to a lack of tools with the required functionality.

Author Contributions: Conceptualization, A.T., M.D., and R.Ż.; methodology, A.T.; formal analysis, A.T.; investigation, A.T.; resources, A.T.; writing—Original draft preparation, A.T.; writing—Review and editing, A.T., M.D., and R.Ż; supervision, R.Ż.; project administration, R.Ż. All authors have read and agreed to the published version of the manuscript. Authorship was limited to those who have contributed substantially to the work reported.

Funding: This research received no external funding.

Conflicts of Interest: The authors declare no conflict of interest. The funders had no role in the design of the study; in the collection, analyses, or interpretation of data; in the writing of the manuscript, or in the decision to publish the results.

References

1. Izdebski, W. Analysis of the cadastral data published in the Polish Spatial Data Infrastructure. *Geod. Cartogr.* **2017**, *66*, 227–240. [CrossRef]
2. Directive 2007/2/EC of the European Parliament and of the Council of 14 March 2007 Establishing an Infrastructure for Spatial Information in the European Community. Available online: http://data.europa.eu/eli/dir/2007/2/oj (accessed on 12 January 2020).
3. Open GIS Consortium Inc. *OpenGIS Web Map Server Implementation Specification*. 2006. Available online: https://www.ogc.org/standards/wms (accessed on 13 January 2020).
4. ISO 19128: 2005. Geographic Information—Web Map Server Interface. Available online: https://www.iso.org/standard/32546.html (accessed on 15 January 2020).
5. ISO 19142: 2011. Geographic Information—Web Feature Service. Available online: https://www.iso.org/standard/42136.html (accessed on 16 January 2020).
6. INSPIRE Network Services. Available online: https://inspire.ec.europa.eu/network-services/41 (accessed on 17 June 2020).
7. About INSPIRE. Available online: https://inspire.ec.europa.eu/about-inspire/563 (accessed on 18 June 2020).
8. Pashova, L.; Bandrova, T. A brief overview of current status of European spatial data infrastructures—Relevant developments and perspectives for Bulgaria. *Geo. Spat. Inf. Sci.* **2017**, *20*, 97–108. [CrossRef]
9. Kaufmann, J.; Steudler, D. *Cadastre 2014 a Vision for a Future Cadastral System*; International Federation of Surveyors (FIG): Copenhagen, Denmark, 1998.
10. Przulj, D.; Radaković, N.; Sladić, D.; Radulović, A.; Govedarica, M. Domain model for cadastral systems with land use component. *Surv. Rev.* **2019**, *51*, 135–146. [CrossRef]
11. Dawidowicz, A.; Źróbek, R. A methodological evaluation of the Polish cadastral system based on the global cadastral model. *Land Use Policy* **2018**, *73*, 59–72. [CrossRef]
12. Mika, M. An analysis of possibilities for the establishment of a multipurpose and multidimensional cadaster in Poland. *Land Use Policy* **2018**, *77*, 446–453. [CrossRef]
13. Hull, S.; Whittal, J. Human rights in tension: Guiding cadastral systems development in customary land rights contexts. *Surv. Rev.* **2019**, *51*, 97–113. [CrossRef]

14. Izdebski, W. Good practices of participation of communes and districts in creating spatial data infrastructure in Poland. Available online: http://www.izdebski.edu.pl/index.php?akcja=publikacje&kat=23 (accessed on 27 July 2019).

15. Krigsholm, P.; Zavialova, S.; Riekkinen, K.; Ståhle, K.; Viitanen, K. Understanding the future of the Finnish cadastral system—A Delphi study. *Land Use Policy* **2017**, *68*, 133–140. [CrossRef]

16. Thompson, R.J. A model for the creation and progressive improvement of a digital cadastral data base. *Land Use Policy* **2015**, *49*, 565–576. [CrossRef]

17. Otwarte Dane. Ewidencja gruntów i budynków. Available online: https://dane.gov.pl/dataset/925,dziaki-katastralne-inspire-cp (accessed on 17 June 2020).

18. Craglia, M. INSPIRE: Towards a Participatory Digital Earth. Geospatial World. 2014. Available online: https://www.geospatialworld.net/article/inspire-towards-a-participatory-digital-earth/ (accessed on 12 August 2019).

19. Kotsev, A.; Peeters, O.; Smits, P.; Grothe, M. Building bridges: Experiences and lessons learned from the implementation of INSPIRE and e-reporting of air quality data in Europe. *Earth Sci. Inform.* **2014**, *8*, 353–365. [CrossRef]

20. Bydłosz, J. The application of the Land Administration Domain Model in building a country profile for the Polish cadaster. *Land Use Policy* **2015**, *49*, 598–605. [CrossRef]

21. Cagdas, A.C.; Bovkir, R. Generic land registry and cadaster data model supporting interoperability based on international standards for Turkey. *Land Use Policy* **2017**, *68*, 59–71. [CrossRef]

22. Williamson, I.P.; Rajabifard, A.; Feeney, M.E.F. *Developing Spatial Data Infrastructures, from Concept to Reality*; Taylor and Francis Group: London, UK; New York, NY, USA, 2003; pp. 3–16.

23. Kliment, T.; Granell, C.; Cetl, V.; Kliment, M. Publishing OGC resources discovered on the mainstream web in an SDI catalogue. In Proceedings of the 16th AGILE International Conference on Geographic Information Science, Leuven, Belgium, 14–17 May 2013; pp. 1–6. Available online: https://agile-online.org/conference_paper/cds/agile_2013/short_papers/sp_s5.3_kliment.pdf (accessed on 10 August 2019).

24. Stella, G.; Coli, R.; Maurizi, A.; Famiani, F.; Castellini, C.; Pauselli, M.; Tosti, G.; Menconi, M.E. Towards a National Food Sovereignty Plan: Application of a new Decision Support System for food planning and governance. *Land Use Policy* **2019**, *89*, 104216. [CrossRef]

25. European Commission Joint Research Centre. D2.8.I.6 Data Specification on Cadastral Parcels—Technical Guidelines. 2014. Available online: https://inspire.ec.europa.eu/documents/Data_Specifications/INSPIRE_DataSpecification_CP_v3.1.pdf (accessed on 15 August 2019).

26. European Commission Joint Research Centre. D2.8.III.2 Data Specification on Buildings—Draft Technical Guidelines. 2013. Available online: https://inspire.ec.europa.eu/documents/Data_Specifications/INSPIRE_DataSpecification_BU_v3.0rc3.pdf (accessed on 15 August 2019).

27. Head Office of Geodesy and Cartography. Available online: http://www.gugik.gov.pl/bip/inspire (accessed on 2 August 2019).

28. Law on Spatial Data Infrastructure. 2010. Available online: http://isap.sejm.gov.pl/isap.nsf/download.xsp/WDU20100760489/O/D20100489.pdf (accessed on 11 September 2019).

29. Gaździcki, J. Geospatial information in Poland: Development and new challenges. Polish Association for Spatial Information. *Ann. Geomat.* 2017, *2*, 139–145. Available online: http://rg.ptip.org.pl/index.php/rg/article/view/RG2017-2_Gazdzicki/1712 (accessed on 23 February 2020).

30. Eurostat Regional Yearbook 2018 Edition. Available online: https://ec.europa.eu/eurostat/web/products-statistical-books/-/KS-HA-18-001 (accessed on 4 September 2019).

31. Kocur-Bera, K. Data compatibility between the Land and Building Cadaster (LBC) and the Land Parcel Identification System (LPIS) in the context of area-based payments: A case study in the Polish Region of Warmia and Mazury. *Land Use Policy* **2019**, *80*, 370–379. [CrossRef]

32. Berriet-Solliec, M.; Labarthe, P.; Laurent, C. Goals of evaluation and types of evidence. *Evaluation* **2014**, *20*, 195–213. [CrossRef]

33. Johnson, R.B.; Onwuegbuzie, A.J. Mixed methods research: A research paradigm whose time has come. *Educat. Res.* **2004**, *33*, 14–26. [CrossRef]

34. Kaufmann, J.; Kaul, C. Assessment of the core cadastral domain model from a cadaster point of view. In *FIG 2004: Standardization in the Cadastral Domain: Proceedings of the Workshop Standardization in the Cadastral Domain, Bamberg, Germany, 9–10 December 2004*; FIG: Copenhagen, Denmark, 2004.

35. Rajabifard, A.; Williamson, I.; Steudler, D.; Binns, A.; King, M. Assessing the worldwide comparison of cadastral systems. *Land Use Policy* **2007**, *24*, 275–288. [CrossRef]

36. Bennett, R.M.; Pickering, M.; Sargent, J. Transformations, transitions, or tall tales? A global review of the uptake and impact of NoSQL, blockchain, and big data analytics on the land administration sector. *Land Use Policy* **2019**, *83*, 435–448. [CrossRef]

37. Geodetic and Cartographic Law of 17 May 1989. Available online: http://isap.sejm.gov.pl/isap.nsf/download.xsp/WDU20190000725/U/D20190725Lj.pdf (accessed on 24 September 2019).

38. Regulation of the Minister of Regional Development and Construction of 29 March 2001 on Land and Building Registers. Available online: http://isap.sejm.gov.pl/isap.nsf/download.xsp/WDU20160001034/O/D20161034.pdf (accessed on 24 September 2019).

39. Register of Spatial Datasets and Services. Available online: https://www.geoportal.gov.pl/ewidencja-zbiorow-i-uslug (accessed on 12 December 2019).

40. Thellufsen, C.; Rajabifard, A.; Enemark, S.; Williamson, I. Awareness as a foundation for developing effective spatial data infrastructures. *Land Use Policy* **2009**, *26*, 254–261. [CrossRef]

41. Trystuła, A. Cadastral databases in web map service supporting rural development policies. *Infrastruct. Ecol. Rural Areas* **2014**, 319–331. Available online: http://www.infraeco.pl/pl/art/a_17317.html (accessed on 11 August 2020).

42. Masser, I.; Crompvoets, J. *Building European Spatial Data Infrastructures*; Esri Press: Relands, CA, USA, 2015.

43. Nushi, B.; van Loenen, B.; Crompvoets, J. The STIG—A new SDI assessment method. *Int. J. Spat. Data Infrastruct. Res.* **2015**, *10*, 55–83.

44. Mijić, N.; Bartha, G. Infrastructure for Spatial Information in European Community (INSPIRE) through the Time from 2007 until 2017. In *Advanced Technologies, Systems, and Applications III*; Lecture Notes in Networks and Systems; Avdaković, S., Ed.; Springer: Cham, Switzerland, 2019; Volume 60.

45. Jarząbek, J.; Surma, E. The Draft Project Concerning Development of the Spatial Information Infrastructure in the Period 2016–2017. Available online: http://www.radaiip.gov.pl/__data/assets/pdf_file/0016/34342/Projekt-programu-budowy-IIP-w-latach-2016-2017.pdf (accessed on 15 May 2019).

46. van Oosterom, P.; Groothedde, A.; Lemmen, C.; van der Molen, P.; Uitermark, H. Land administration as a cornerstone in the global spatial information infrastructure. *Int. J. Spat. Data Infrastruct. Res.* **2009**, *4*, 298–331.

47. Enemark, S. From Cadastre to Land Governance: A Cadastre 2014 Outlook. In Proceedings of the 15th FIG International Congress, Kuala Lumpur, Malaysia, 16–21 June 2014.

 *land*

Article

Land Registration, Adjustment Experience, and Agricultural Machinery Adoption: Empirical Analysis from Rural China

Xin Deng [1,†] , Zhongcheng Yan [1], Dingde Xu [2,†] and Yanbin Qi [1,*]

1   College of Economics, Sichuan Agricultural University, Chengdu 611130, China;
    dengxin@sicau.edu.cn (X.D); 2019108007@stu.sicau.edu.cn (Z.Y.)
2   Sichuan Center for Rural Development Research, College of Management, Sichuan Agricultural University,
    Chengdu 611130, China; dingdexu@sicau.edu.cn
*   Correspondence: qybin@sicau.edu.cn
†   These authors contributed equally to this work and should be regarded as co-first authors.

Received: 22 January 2020; Accepted: 16 March 2020; Published: 17 March 2020

Abstract: Land property security and advanced factor inputs play critical roles in agricultural modernization in developing countries. However, there are unclear relationships between land property security and advanced factor inputs. This study aims to clarify these relationships from the perspective of the differentiation of the realization process of land property security. From the perspective of property rights theory and endowment effects, data from 2934 farming households in rural China are used to determine the quantitative impacts of land registration and adjustment experience on the adoption of agricultural machinery. The results are as follows: (i) Land registration does not affect the adoption of agricultural machinery. (ii) Adjustment experience has a negative impact on the adoption of agricultural machinery. (iii) The interaction of land registration and adjustment experience has a positive impact on the adoption of agricultural machinery. This study provides some policy references with which developing countries can achieve agricultural modernization and revitalize the countryside by improving property rights security.

Keywords: land property security; land registration; adjustment experience; advanced agricultural factor inputs; agricultural machinery; China

---

1. Introduction

Agricultural mechanization is an important factor in agricultural modernization in developing countries [1–3]. It matters not just because agricultural machinery helps to improve agricultural productivity [4–6], but because it is correlated with agricultural economic growth [7,8]. In developing countries, urbanization is developing rapidly and a large number of rural laborers leave home to work, seeking economic benefits [9–12]. A lack of agricultural laborers and serious aging of the remaining population have led to a desolate countryside [10]. Agricultural machinery is a labor-saving technology [13] that has gradually become the main way by which developing countries cope with agricultural labor shortages [14,15]. In addition, the adoption of agricultural machinery helps improve agricultural productivity [14,16,17]. For example, Paudel et al. [17] found that the adoption of agricultural machinery could improve rice productivity by 1110 kg/ha. Thus, agricultural mechanization is the key method for developing countries to realize agricultural modernization [18,19]. However, farmers often do not adopt it or take a long time to start adopting it [20]. Thus, it is important to explore the key drivers of the adoption of agricultural machinery.

Meanwhile, developing countries have paid special attention to the reform of their property rights systems in their modernization processes. China is the world's largest developing country and one of the world's largest agricultural countries [21,22]. China feeds 20% of the world's population with 7% of the world's cropland [23], thus, agricultural modernization is important to China [24,25]. Thus, this study shows the reform of Chinese rural land property rights system as an example. In rural China, land rights are divided into ownership, contract rights, and management rights (ownership belongs to the village collective; contract and management rights belong to farmers) [26]. Chinese government vigorously promotes land registration program since 2009. Land registration program means the contract rights and management rights of farmers are officially registered by Chinese government. And the rights of farmers are protected by the law [27,28]. More specifically, (i) in 2009, the Chinese agricultural department selected eight villages for a trial of rural contracted land registration; (ii) in 2012, the Chinese government began trialing the registration of rural contracted land across the whole county (50 pilot counties); (iii) in 2013, the Chinese government expanded the number of pilot counties for rural contracted land registration to 55; (iv) at the end of 2018, most of China's rural contracted land had been officially registered.

Land registration program can help protect farmers' interests. Land registration gains official recognition and legal protection, which means that others who want to obtain the land management rights of farmers need to obtain authorization from farmers. Thus, the impacts of land registration on farmers are undoubtedly huge. In particular, there has been much discussion in the academic community about whether land registration motivates farmers to invest in agriculture [29]. Agricultural machinery plays an important role in sustainable agriculture [15,30]. Thus, this study aims to explore whether land registration motivates farmers to adopt agricultural machinery.

Previous studies disagree about whether land registration motivates farmers to increase their agricultural investment. While some say that it does [26,31–35], others suggest that the effect is not obvious [36–40]. In reality, the Chinese government is trying to stimulate agricultural investment by stabilizing land rights. As shown in Figure 1, the scale of the land registration pilot program has gradually expanded from 8 villages in 2009 to 28 provinces in 2017. However, Figure 1 also shows that the per capita power of agricultural machinery has not increased with the scale of land registration. Thus, the case of China seems to indicate that land registration is not a clear incentive to adopt agricultural machinery.

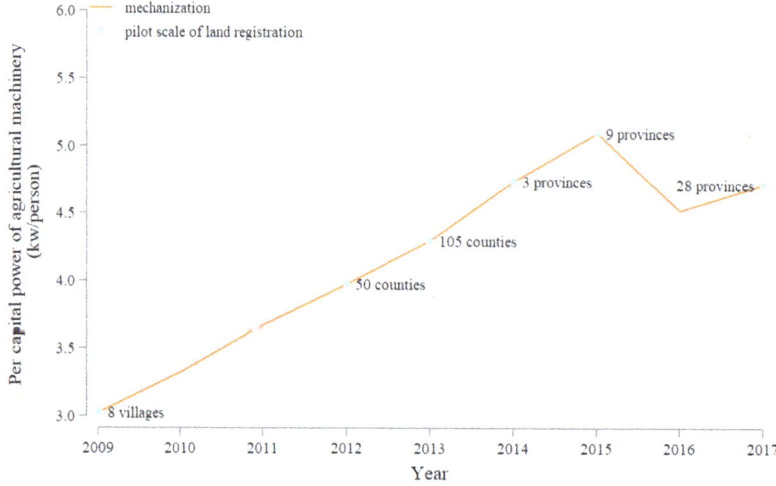

Figure 1. The relationship between land registration and agricultural machinery in China. Source: National Bureau of Statistics of China 2009–2017

Perhaps, the above dispute originates from insufficient consideration of differences in initial property rights distribution [41]. For example, under the premise of ensuring that the duration of land contracts remains unchanged, China's land management law allows appropriate adjustment of ownership of land contract rights among some farmers. Thaler [42] believed that the initial allocation of property rights plays a decisive role in the final allocation of resources. In rural China, the adjustment of the ownership of land contract rights must be approved at a villagers' meeting, and its goal is to optimize the allocation of resources. Thus, land registration may be better with appropriate adjustment of the ownership of land contract rights than without it. However, in previous studies, when discussing whether land registration stimulates agricultural investment, little consideration has been given to whether the land has been undergone appropriate adjustments before registration. Meanwhile, experience may leave long-term effects [43–45], and Ren et al. [27] and Hong et al. [41] found that farmer's experience of land adjustment may affect land investment. Thus, this study focuses on the combined impacts of land registration and adjustment experiences on the adoption of agricultural machinery.

In addition, the Chinese government has proposed a "Village Revitalization Strategy" [46–50], which aims to improve agricultural productivity and enhance rural vitality [51,52]. However, at present, the world is facing difficulties in revitalizing the countryside [10]. Thus, this study explores the combined impacts of land registration and adjustment experiences on the adoption of agricultural machinery from the perspective of Chinese farmers. The results may provide policy references for developing countries to realize agricultural modernization and revitalize the countryside.

2. Theoretical Analysis

In general, land fragmentation hinders the adoption of technologies such as agricultural machinery [53–55]. Governors hope farmers will expand the scale of land management by land registration [26,56]; this, in return, will also help to facilitate the adoption of agricultural machinery by farmers. However, differences in initial property rights may lead to different economic outcomes [57]. Empirical studies show an unclear relationship between land registration and the scale of land management [58,59]. Therefore, the impacts of land registration on the adoption of agricultural machinery require further investigation.

Differences in land registration may lead to different levels of adoption of agricultural machinery. Coase [60] believed that if the market transaction cost is zero, no matter how the initial property rights are arranged, resource allocation will automatically achieve Pareto optimality under the market mechanism. However, Thaler [42] believed that there is an "endowment effect", which does not change an individual's preferences but strengthens their motivation to maintain the status quo [61,62]. Thus, improper land registration will increase the endowment effect in farmers, which may hinder the transfer of land. As a consequence, it may be disadvantageous for farmers to adopt agricultural machinery. Hence, when we discuss the relationship between land registration and agricultural machinery adoption, we should identify the differences in land registration involved.

Differences in land registration may stem from the property rights experiences of farmers. In rural China, with the consent of two-thirds of the farmers, a village collective can adjust the land between farmers on a small scale. Land adjustment is a coherent collective action that aims to optimize land allocation. Samuelson and Zeckhauser [62] indicated that adjustment may enable individuals to form new endowment effects and make new choices. Adjustment experiences may impact the status quo and weaken endowment effects. That is, land registration with adjustment makes it possible for farmers to rationalize land valuations and investments. In return, it can help to enhance land transfer and improve the scale of land management, which may facilitate the adoption of agricultural machinery.

In summary, under the background of the reform of China's rural property rights system, and based on property rights theory and endowment effects, this study intends to provide empirical evidence for the following two issues:

1. How do the land registration and adjustment experiences affect farmers' adoption of agricultural machinery?
2. Can land registration with adjustment encourage farmers to adopt agricultural machinery?

3. Data Source, Variable Definition, and Empirical Approach

3.1. Data Source

The farmers' households play an essential role in the agricultural and rural studies [52,63–65]. According to the previous studies, this study uses the household-level data of Chinese famers belonging to the China Labor-force Dynamics Survey in 2014 (Hereinafter, CLDS2014). More specifically, the CLDS2014 was implemented by the Center for Social Science Survey at Sun Yat-sen University (Guangzhou, China) in 2014, which collected the details about the social and economic development in China, such as, rural land use, rural land registration, and agricultural production (more details can be found on the Web site http://css.sysu.edu.cn). CLDS2014 can help us to understand Chinese reality by the scientific sampling. And the sampling method employed the multistage cluster, stratified, probability-proportional-to-size (PPS) sampling to cover 29 Chinese mainland provinces (excluding Tibet and Hainan). Firstly, CLDS2014 sampled 209 counties from 29 provinces; secondly, CLDS2014 sampled 401 villages/communities from 209 counties; finally, CLDS2014 sampled 14,214 households from 401 villages/communities. In addition, the CLDS2014 is the latest open access data from the survey institutions.

This study aims to explore the relationship among land registration, adjustment experience, and agricultural machinery adoption. Thus, we clean the data of CLDS2014, and the cleaning processes are as follows: (1) the households living in urban area are not directly engaged in agriculture; thus, this study only retains the households living in rural area; and (2) this study also excludes the households living in rural areas but not engaged in agricultural production. In summary, through the above cleaning process, this study employs 2934 valid household-level questionnaires to perform empirical analysis. In addition, grain plays an important role in China with a large population, and China has a long history of planting grain. Meanwhile, CLDS2014 collected the details of planting grain. However, it did not provide the details that process farmer-adopted-agricultural machinery. Thus, the term "planting grain" used in this study is not just about planting, and may also involve cultivation and harvesting.

3.2. Variable Definition

3.2.1. Dependent Variable

At present, the Chinese government is committed to improving the level of mechanization of grain planting. Thus, this study assumes that if farmers have adopted machinery for this, they are considered to adopt agricultural machinery. Therefore, the dependent variable is binary. More specifically, 1 if a farming household adopts agricultural machinery in any planting grain processes (planting, cultivation and harvesting) or 0 otherwise.

3.2.2. Predicator Variables

Land registration is defined as whether the land contract and management rights of farmers are officially registered. Thus, it is defined as a binary variable. More specifically, 1 if the land right of the farming household has been officially registered or 0 otherwise.

Meanwhile, in rural China, with the consent of two-thirds of the farmers, a village collective can adjust land between farmers on a small scale. Hence, land adjustment is a coherent collective action that aims to optimize land allocation. In general, land adjustment occurs before land registration. Thus, an adjustment experience occurs when a farming household experiences land adjustment before the land rights are officially registered. It is defined as a binary variable: 1 if the farming household had an adjustment experience or 0 otherwise.

3.2.3. Control Variables

To improve the accuracy of empirical estimates, referencing to the studies of Ji et al. [66], Ma et al. [15], Adu-Baffour et al. [16], Belton and Filipski [14], Deng et al. [67], and Hong et al. [41], this study controls householder-level variables, household-level variables, and location-level variables. Table 1 shows the definitions and descriptive statistics of all variables for empirical model.

Table 1. The definition and data description of the variables in the model.

Variables	Definition	Mean	Standard Deviation
	Dependent variable		
Adoption	1 if farm household adopts agricultural machinery in any planting grain processes; 0 otherwise	0.59	0.49
	Predicator variables		
Registration	1 if land right of farm household has been officially registered; 0 otherwise	0.50	0.50
Adjustment	1 if farm household has experienced land adjustment before the land right officially registered; 0 otherwise	0.95	0.21
Registration × Adjustment	The interaction item of Registration and Adjustment. 1 if both Registration and Adjustment are equal to 1; 0 otherwise	0.48	0.50
	Householder-level variables		
Gender	1 if householder is male; 0 female	0.88	0.32
Age	Age of householder in years (year)	52.39	10.96
Education	1 if householder has received a high school diploma or above; 0 otherwise	0.11	0.32
Health	1 if householder has a healthy status; 0 otherwise	0.84	0.36
Job	1 if householder engages in agriculture; 0 otherwise	0.56	0.50
	Household-level variables		
Farm employment	The ratio of members engaging in agriculture to total members (%)	31.46	27.51
Off-farm employment	The ratio of off-farm members to total members (%)	27.46	26.29
Farm income	The ratio of farm income to total income (%)	50.72	39.70
Land size	The area that farm household is managing land (mu [a])	9.92	28.65
Loan	1 if farm household has borrowed the production fund; 0 otherwise	0.06	0.25
Specialty	1 if farm household is good at planting grain; 0 otherwise	0.05	0.23
Cooperation	1 if farm household belongs to cooperative organization; 0 otherwise	0.02	0.13
Subsidy	The amount of agricultural subsidy from government (RMB [b])	0.70	0.46
Internet	1 if farm household can use the Internet; 0 otherwise	0.27	0.45
	Location-level variables		
Distance	Distance between household and the nearest business center (Km)	7.25	9.22
Plain	1 if farm household belongs to plain village; 0 otherwise	0.32	0.47
Road	The share of concrete road in total road (%)	59.88	29.71

Note: [a] 1 mu is approximately equal to 667 m^2 or 0.067 ha; during the survey period, [b] 1 US dollar was approximately equal to 6.12 RMB (Chinese Yuan).

3.3. Method

This study focuses on exploring the quantitative impacts of land registration and adjustment experience on the adoption of agricultural machinery. The dependent variable for Adoption is the binary variable. Therefore, this study employs the binary Probit model for econometric regression. The basic model is set as follows Equation (1):

$$Adoption_{pci} = \beta_0 + \beta_1 Registration_{pci} + \beta_2 Adjustment_{pci} + \\ \beta_3 Registration_{pci} \times Adjustment_{pci} + \gamma X + \delta_c + \tau_p + \varepsilon_{pci} \tag{1}$$

where the subscripts of p, c, and i represent province, county, and household, respectively; Adoption is the binary variable, which value 1 means that farm household adopts agricultural machinery in planting grain and 0 means otherwise; Registration is a dummy variable, which value 1 represents that land right of farm household has been officially registered and 0 represents otherwise; Adjustment is the binary variable, which value 1 means that farm household has experienced land adjustment before the land right officially registered and 0 means otherwise; Registration × Adjustment represents the interaction item of Registration and Adjustment; X is the vector of other control variables; β_0 is the constant; β_1, β_2, and β_3 are estimated parameters; γ is the vector of estimated parameters for control variables; δ values are the county dummies; τ values are the province dummies; ε is the random error term.

4. Results

4.1. Descriptive Results

Figure 2 shows a heatmap of Pearson's correlation coefficients for the dependent and focal variables of the model. The results show that: (i) there is a positive correlation between *land registration* and the *agricultural machinery adoption*; (ii) there is a positive correlation between *adjustment experience* and the *agricultural machinery adoption*; (iii) there is a positive correlation between the interaction of *land registration*, *adjustment experience*, and *agricultural machinery adoption*.

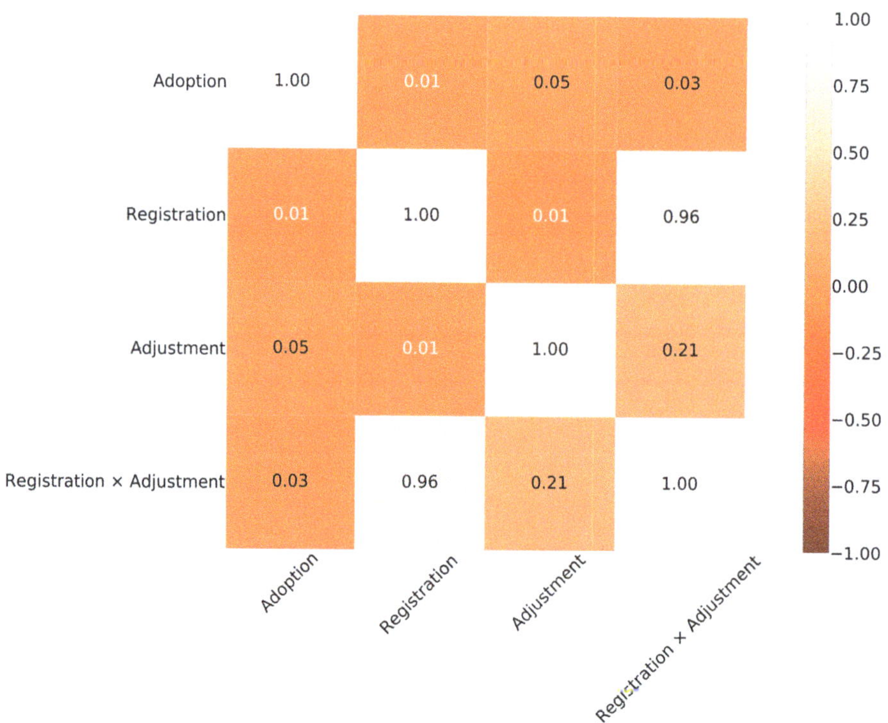

Figure 2. The heatmap of Pearson's correlation coefficients.

In addition, the mean difference can help us understand the sample structure and provide a basis for the choice of an econometric model. Figure 3 shows the mean differences in the adoption of agricultural machinery by land registration, adjustment experience, and their interaction. The results show that the groups that registered land or experienced adjustment, or both, are more inclined to adopt agricultural machinery. However, only the mean difference between groups with and groups without adjustment experience is significant ($p < 0.05$).

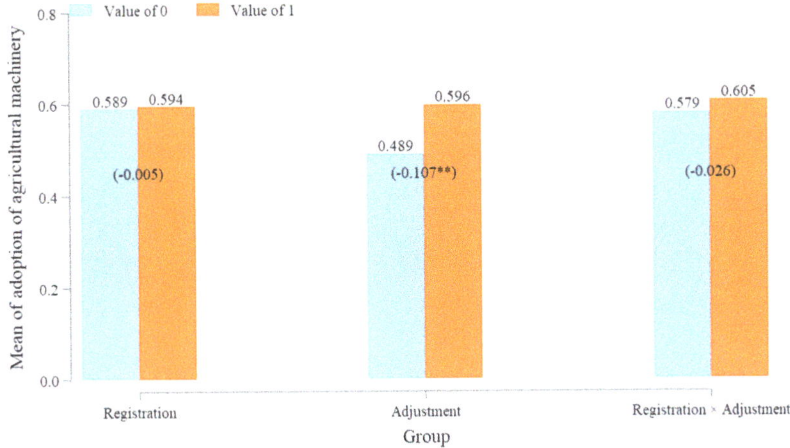

Note: Difference in parentheses,** p < 0.05;
Value of 1 means that farm household belongs to the groups of Registration,
Adjustment or Registration × Adjustment, and value of 0 for otherwise

Figure 3. Mean difference of adoption of agricultural machinery by groups.

In summary, both the Pearson's correlations and mean differences help us understand data structure. Although the statistical results show that land adjustment experience may play an important role in the adoption of agricultural machinery, it is still necessary to discuss the relationship by econometric models. However, previous studies have paid little attention to this relationship. Thus, this study uses an econometric model to discuss the quantitative impacts of land registration, adjustment experience, and their interactions on the adoption of agricultural machinery.

4.2. Empirical Results

4.2.1. Impacts of Registration and Adjustment on Agricultural Machinery Adoption

Table 2 presents the empirical estimates. In Table 2, the dependent variables for all models are binary discrete variables (whether or not farmers adopt agricultural machinery). Meanwhile, this study used a causal identification strategy that gradually adds explanatory variables. More specifically, in Models (1) to (5), a stepwise process was used to add the focal variables, county and province dummy variables, householder variables, household variables, and location variables. For all models, the value of Wald χ^2 was significant at a level of 1%, and the R^2 values gradually increase, indicating that the identification strategy was suitable. Additionally, since the Probit model was non-linear, a marginal effect (i.e., Model (6)) was calculated on the basis of Model (5) to quantify the relationship.

As shown in Models (1) to (5) in Table 2, the coefficient of Registration was not significant except in Model (1), which indicates that the impact of land registration on the adoption of agricultural machinery may be uncertain. The coefficient of Adjustment was significantly negative ($p < 0.01$) except in Model (1), which indicates that the impact of adjustment experience on the adoption of agricultural machinery may be negative. The coefficient of Registration × Adjustment was significantly positive ($p < 0.10$), which indicates that the combined impact of land registration and adjustment experience on the adoption of agricultural machinery was positive. As shown in the marginal effects estimates (Model (6) of Table 2), compared with other farmers, those who have experienced land adjustment before land registration are 14.2% more likely to adopt agricultural machinery. In addition, in Model (5) of Table 2, the variables Off-farm employment, Subsidy, and Internet can also increase farmers' enthusiasm for adopting agricultural machinery.

Table 2. The impact of registration and adjustment on the adoption of agricultural machinery.

	Model (1)	Model (2)	Model (3)	Model (4)	Model (5)	Model (6)
Registration	−0.645 ***	−0.227	−0.213	−0.354	−0.357	−0.080
	(0.217)	(0.268)	(0.268)	(0.279)	(0.279)	(0.063)
Adjustment	−0.061	−0.724 ***	−0.729 ***	−0.885 ***	−0.905 ***	−0.203 ***
	(0.154)	(0.232)	(0.231)	(0.236)	(0.237)	(0.053)
Registration × Adjustment	0.691 ***	0.502 *	0.489 *	0.623 **	0.635 **	0.142 **
	(0.222)	(0.278)	(0.278)	(0.290)	(0.290)	(0.065)
Gender			0.149	0.144	0.141	0.032
			(0.094)	(0.096)	(0.096)	(0.022)
Age			−0.004	−0.003	−0.003	−0.001
			(0.003)	(0.003)	(0.003)	(0.001)
Education			0.193 *	0.134	0.136	0.030
			(0.103)	(0.104)	(0.103)	(0.023)
Health			0.121	0.072	0.081	0.018
			(0.087)	(0.089)	(0.089)	(0.020)
Job			0.035	0.149 *	0.138	0.031
			(0.066)	(0.085)	(0.086)	(0.019)
Farm employment				−0.001	−0.001	−0.000
				(0.002)	(0.002)	(0.000)
Off-farm employment				0.005 ***	0.005 ***	0.001 ***
				(0.002)	(0.002)	(0.000)
Farm income				−0.001	−0.001	−0.000
				(0.001)	(0.001)	(0.000)
Land size				0.003	0.003	0.001
				(0.003)	(0.003)	(0.001)
Loan				−0.010	−0.012	−0.003
				(0.134)	(0.135)	(0.030)
Specialty				0.236	0.158	0.035
				(0.185)	(0.185)	(0.041)
Cooperation				0.007	0.006	0.001
				(0.259)	(0.260)	(0.058)
Subsidy				0.420 ***	0.424 ***	0.095 ***
				(0.077)	(0.077)	(0.017)
Internet				0.243 ***	0.222 ***	0.050 ***
				(0.074)	(0.075)	(0.017)
Distance					−0.025 ***	−0.006 ***
					(0.006)	(0.001)
Plain					0.488 ***	0.109 ***
					(0.156)	(0.035)
Rode					−0.002	−0.001
					(0.003)	(0.001)
Constant	0.282 *	0.935 **	0.845 *	0.674	1.036 **	
	(0.150)	(0.384)	(0.438)	(0.455)	(0.471)	
County dummies	No	Yes	Yes	Yes	Yes	Yes
Province dummies	No	Yes	Yes	Yes	Yes	Yes
Wald χ^2	15.651 ***	825.349 ***	833.258 ***	875.000 ***	882.002 ***	882.002 ***
R^2	0.004	0.366	0.369	0.386	0.396	0.396
Obs.	2934	2934	2934	2934	2934	2934

Note: Robust standard errors in parentheses; * $p < 0.1$, ** $p < 0.05$, *** $p < 0.01$

4.2.2. Estimated Results of Robustness Tests

To ensure that the estimates in Table 2 are reliable, robustness tests were used, with the results shown in Table 3. In Table 3, Model (1) represents the sub-sample regression (farmers without land transfer), while Model (2) changes the regression method to a logit model.

As shown in Table 3, we also controlled for householder-level variables, household-level variables, location-level variables, and county and province dummy variables. The estimates in Table 3 are similar to those in Table 2. More specifically, the coefficient of Registration was not significant, the coefficient of Adjustment was negative ($p < 0.01$), and the coefficient of Registration × Adjustment was positive ($p < 0.10$). Thus, the results of Table 3 indicate that the results of Table 2 are robust.

Table 3. The estimated results of robustness tests.

	Model (1)	Model (2)
Registration	−0.223	−0.570
	(0.290)	(0.460)
Adjustment	−0.737 ***	−1.608 ***
	(0.261)	(0.401)
Registration × Adjustment	0.512 *	1.080 **
	(0.301)	(0.483)
Gender	0.199 **	0.250
	(0.101)	(0.173)
Age	−0.003	−0.006
	(0.003)	(0.006)
Education	0.197 *	0.211
	(0.116)	(0.187)
Health	0.068	0.118
	(0.098)	(0.159)
Job	0.070	0.223
	(0.093)	(0.152)
Farm employment	−0.000	−0.001
	(0.002)	(0.003)
Off-farm employment	0.004 **	0.008 ***
	(0.002)	(0.003)
Farm income	−0.001	−0.001
	(0.001)	(0.002)
Land size	−0.001	0.006
	(0.003)	(0.006)
Loan	0.033	−0.037
	(0.156)	(0.244)
Specialty	−0.050	0.255
	(0.192)	(0.351)
Cooperation	−0.151	−0.109
	(0.295)	(0.488)
Subsidy	0.417 ***	0.748 ***
	(0.084)	(0.136)
Internet	0.203 **	0.376 ***
	(0.083)	(0.134)
Distance	−0.026 ***	−0.044 ***
	(0.006)	(0.010)
Plain	0.453 ***	0.928 ***
	(0.169)	(0.297)
Rode	−0.003	−0.006
	(0.003)	(0.005)
Constant	1.004 **	1.860 **
	(0.496)	(0.804)
County dummies	Yes	Yes
Province dummies	Yes	Yes
Wald χ^2	753.363 ***	656.835 ***
R^2	0.380	0.398
Obs.	2215	2934

Note: Robust standard errors in parentheses; * $p < 0.1$, ** $p < 0.05$, *** $p < 0.01$; Model (1)–(3) means the models of sub-sample data, the Logit model, and the instrumental regression, respectively.

5. Discussion

Based on data from 2934 farming households in rural China, this study focuses on the quantitative impacts of land registration, adjustment experience, and their interactions on the adoption of agricultural machinery. The contributions of this study are as follows: (i) under the guidance of property rights theory and endowment effects, this study focuses on the quantitative impact of heterogeneous land registration on agricultural inputs; (ii) it further enriches the understanding of property rights theory and endowment effects. China is the world's largest developing country and empirical evidence from there may provide a reference for land property reform in other developing countries. This study may also provide some policy references for developing countries to realize agricultural modernization and revitalize the countryside.

The results of this study have some similarities and differences from previous studies. First, we found no significant impact of land registration on the adoption of agricultural machinery. This is consistent with Brasselle et al. [40], Beekman and Bulte [37], Domeher and Abdulai [38], Lovo [36], and Goldstein et al. [39], who report that property rights security may not obviously affect agricultural input. Second, there was a negative impact of adjustment experience on the adoption of agricultural machinery. Finally, there was a positive impact of the interaction of land registration and adjustment experience on the adoption of agricultural machinery. These findings differ from those of Hong et al. [41], who reported that land registration positively affects the investment incentive of farmers without land adjustment experience.

The findings of this study are interesting because property rights are important [60]. However, due to the endowment effect [42], the registration process of property rights is also very important [57]. The endowment effect does not change individuals' preferences, but strengthens their motivation to maintain the status quo [61,62]. Thus, when the land rights of a farming household have been officially registered without land adjustment, famers may be less willing to transfer land due to the endowment effect. This may be a barrier to solving the problem of land fragmentation. In return, there was no impact of land registration without adjustment experience on the adoption of agricultural machinery. Therefore, when land has been adjusted without land registration, farmers' property rights may be insecure, which may decrease their willingness to invest in agriculture [26,31–35]. Additionally, there was a negative impact of adjustment experience without land registration on the adoption of agricultural machinery. When the land rights of a farming household have been officially registered after land adjustment, the adjustment helps optimize land resource allocation [9], while registration helps improve property security [68]; in return, there is a positive impact of the interaction of land registration and adjustment experience on the adoption of agricultural machinery. In summary, to explore the relationship between the security of property rights and agricultural inputs, we should not only pay attention to the results of property rights registration, but also to the process of property rights registration.

In addition, this study has several deficiencies, which can be addressed in future studies. Specific among them are as follows: (i) This study focused on the quantitative impacts of land registration, adjustment experience, and their interactions on the adoption of agricultural machinery. Future studies could further explore the driving mechanisms behind these quantitative relationships. (ii) Agricultural machinery is only one important agricultural input. Future studies could further discuss whether the findings of this study are applicable to other important agricultural inputs (e.g., soil improvement, irrigation facilities, etc.). (iii) The data of this study is set such that land registration and land adjustment were prior to agricultural machinery adoption, which may partly solve the problem of mutual causality. Future studies could further test the findings of this study by instrumental variable method. (iv) China has a special land ownership institution; namely, ownership belongs to the village collective, while contract and management rights belong to individual farmers. Future studies could further explore whether the findings of this study are applicable to developing countries where rural land ownership is private.

6. Conclusions and Implications

From the perspective of property rights theory and endowment effects, data from 2934 farming households in rural China are used to determine the quantitative impacts of land registration and adjustment experience on the adoption of agricultural machinery. The results are as follows:

1. Land registration does not affect the adoption of agricultural machinery.

2. Adjustment experience has a negative impact on the adoption of agricultural machinery.
3. The interaction of land registration and adjustment experience has a positive impact on the adoption of agricultural machinery.

Based on the above findings, we can also derive some policy implications. Although the security of land property rights is important for agricultural investment, we should also pay attention to the process of making land property rights secure. That is, when the government promotes land registration to ensure the security of land property rights, the first thing that the government should do is respect farmers' willingness to optimize the allocation of land resources via land adjustment. In addition, this study finds that using the Internet can improve the adoption of agricultural machinery. The internet can help farmers obtain information on agricultural technology, which may increase their likelihood of adopting agricultural technology. This suggests that the government increase internet access in rural areas.

Author Contributions: Conceptualization, X.D., D.X. and Y.Q.; formal analysis, X.D.; funding acquisition, Y.Q.; methodology, X.D.; visualization, X.D.; writing – original draft, X.D., Z.Y., D.X. and Y.Q.; writing—review and editing, X.D., Z.Y., D.X. and Y.Q. All authors have read and agreed to the published version of the manuscript.

Funding: The National Social Science Foundation of China (Grant No. 14XGL003) funded this study.

Acknowledgments: All authors gratefully acknowledge the support from the National Social Science Foundation of China (Grant No. 14XGL003). We also extend great gratitude to the anonymous reviewers and editors for their helpful review and critical comments. Additionally, all authors are very grateful to the Center for Social Science Survey at Sun Yat-sen University who provided the data.

Conflicts of Interest: All authors declare no conflict of interest.

References

1. Pingali, P. Agricultural Mechanization: Adoption Patterns and Economic Impact. *Handbook Agric. Econ.* **2007**, *3*, 2779–2805.
2. Sims, B.; Kienzle, J. Sustainable Agricultural Mechanization for Smallholders: What Is It and How Can We Implement It? *Agriculture* **2017**, *7*, 50. [CrossRef]
3. Nguyen, H.Q.; Warr, P. Land Consolidation as Technical Change: Economic Impacts in Rural Vietnam. *World Dev.* **2020**, *127*, 1047. [CrossRef]
4. Mottaleb, K.A.; Rahut, D.B.; Ali, A.; Gérard, B.; Erenstein, O. Enhancing Smallholder Access to Agricultural Machinery Services: Lessons from Bangladesh. *J. Dev. Stud.* **2017**, *53*, 1502–1517. [CrossRef]
5. Zhang, M.; Duan, F.; Mao, Z. Empirical Study on the Sustainability of China's Grain Quality Improvement: The Role of Transportation, Labor, and Agricultural Machinery. *Int. J. Environ. Res. Public Health* **2018**, *15*, 271. [CrossRef]
6. Yi, Q.; Chen, M.; Sheng, Y.; Huang, J. Mechanization Services, Farm Productivity and Institutional Innovation in China. *China Agric. Econ. Rev.* **2019**. [CrossRef]
7. Zhang, X.; Yang, J.; Thomas, R. Mechanization Outsourcing Clusters and Division of Labor in Chinese Agriculture. *China Econ. Rev.* **2017**, *43*, 184–195. [CrossRef]
8. Devkota, R.; Pant, L.P.; Gartaula, H.N.; Patel, K.; Gauchan, D.; Hambly-Odame, H.; Thapa, B.; Raizada, M.N. Responsible Agricultural Mechanization Innovation for the Sustainable Development of Nepal's Hillside Farming System. *Sustainability* **2020**, *12*, 374. [CrossRef]
9. Deng, X.; Xu, D.-D.; Zeng, M.; Qi, Y.-B. Does Labor Off-Farm Employment Inevitably Lead to Land Rent Out? Evidence from China. *J. Mt. Sci.* **2019**, *16*, 689–700. [CrossRef]
10. Liu, Y.; Li, Y. Revitalize the World's Countryside. *Nature* **2017**, *548*, 275–277. [CrossRef]
11. Huang, K.; Deng, X.; Liu, Y.; Yong, Z.; Xu, D. Does Off-Farm Migration of Female Laborers Inhibit Land Transfer? Evidence from Sichuan Province, China. *Land* **2020**, *9*, 14. [CrossRef]
12. Xu, D.; Yong, Z.; Deng, X.; Zhuang, L.; Qing, C. Rural-Urban Migration and Its Effect on Land Transfer in Rural China. *Land* **2020**, *9*, 81. [CrossRef]
13. Lin, J.Y. Prohibition of Factor Market Exchanges and Technological Choice in Chinese Agriculture. *J. Dev. Stud.* **1991**, *27*, 1–15. [CrossRef]

14. Belton, B.; Filipski, M. Rural Transformation in Central Myanmar: By How Much, and for Whom? *J. Rural Stud.* **2019**, *67*, 166–176. [CrossRef]

15. Ma, W.; Renwick, A.; Grafton, Q. Farm Machinery Use, Off-Farm Employment and Farm Performance in China. *Aust. J. Agric. Resour. Econ.* **2018**, *62*, 279–298. [CrossRef]

16. Adu-Baffour, F.; Daum, T.; Birner, R. Can Small Farms Benefit from Big Companies' Initiatives to Promote Mechanization in Africa? A Case Study from Zambia. *Food Policy* **2019**, *84*, 133–145. [CrossRef]

17. Paudel, G.P.; KC, D.B.; Justice, S.E.; McDonald, A.J. Scale-Appropriate Mechanization Impacts on Productivity among Smallholders: Evidence from Rice Systems in the Mid-Hills of Nepal. *Land Use Policy* **2019**, *85*, 104–113. [CrossRef]

18. Li, W.; Wei, X.; Zhu, R.; Guo, K. Study on Factors Affecting the Agricultural Mechanization Level in China Based on Structural Equation Modeling. *Sustainability* **2019**, *11*, 51. [CrossRef]

19. Kansanga, M.; Andersen, P.; Kpienbaareh, D.; Mason-Renton, S.; Atuoye, K.; Sano, Y.; Antabe, R.; Luginaah, I. Traditional Agriculture in Transition: Examining the Impacts of Agricultural Modernization on Smallholder Farming in Ghana under the New Green Revolution. *Int. J. Sustain. Dev. World Ecol.* **2019**, *26*, 11–24. [CrossRef]

20. Mottaleb, K.A. Perception and Adoption of a New Agricultural Technology: Evidence from a Developing Country. *Technol. Soc.* **2018**, *55*, 126–135. [CrossRef]

21. Deng, X.; Xu, D.; Zeng, M.; Qi, Y. Landslides and Cropland Abandonment in China's Mountainous Areas: Spatial Distribution, Empirical Analysis and Policy Implications. *Sustainability* **2018**, *10*, 3909. [CrossRef]

22. Deng, X.; Xu, D.; Qi, Y.; Zeng, M. Labor Off-Farm Employment and Cropland Abandonment in Rural China: Spatial Distribution and Empirical Analysis. *Int. J. Environ. Res. Public Health* **2018**, *15*, 1808. [CrossRef] [PubMed]

23. Zhang, J. China's Success in Increasing Per Capita Food Production. *J. Exp. Bot.* **2011**, *62*, 3707–3711. [CrossRef] [PubMed]

24. Cui, Z.; Zhang, H.; Chen, X.; Zhang, C.; Ma, W.; Huang, C.; Zhang, W.; Mi, G.; Miao, Y.; Li, X. Pursuing Sustainable Productivity with Millions of Smallholder Farmers. *Nature* **2018**, *555*, 363–366. [CrossRef]

25. Ma, L.; Long, H.; Zhang, Y.; Tu, S.; Ge, D.; Tu, X. Agricultural Labor Changes and Agricultural Economic Development in China and Their Implications for Rural Vitalization. *J. Geogr. Sci.* **2019**, *29*, 163–179. [CrossRef]

26. Cheng, W.; Xu, Y.; Zhou, N.; He, Z.; Zhang, L. How Did Land Titling Affect China's Rural Land Rental Market? Size, Composition and Efficiency. *Land Use Policy* **2019**, *82*, 609–619. [CrossRef]

27. Ren, G.; Zhu, X.; Heerink, N.; Feng, S.; van Ierland, E. Perceptions of Land Tenure Security in Rural China: The Impact of Land Reallocations and Certification. *Soc. Nat. Resour.* **2019**, *32*, 1399–1415. [CrossRef]

28. Zhou, Y.; Li, X.; Liu, Y. Rural Land System Reforms in China: History, Issues, Measures and Prospects. *Land Use Policy* **2020**, *91*, 104330. [CrossRef]

29. Lemel, H. Land Titling: Conceptual, Empirical and Policy Issues. *Land Use Policy* **1988**, *5*, 273–290. [CrossRef]

30. Mottaleb, K.A.; Krupnik, T.J.; Erenstein, O. Factors Associated with Small-Scale Agricultural Machinery Adoption in Bangladesh: Census Findings. *J. Rural Stud.* **2016**, *46*, 155–168. [CrossRef]

31. Alston, L.J.; Libecap, G.D.; Schneider, R. The Determinants and Impact of Property Rights: Land Titles on the Brazilian Frontier. *J. Law Econ. Organ.* **1996**, *12*, 25–61. [CrossRef]

32. Bambio, Y.; Agha, S.B. Land Tenure Security and Investment: Does Strength of Land Right Really Matter in Rural Burkina Faso? *World Dev.* **2018**, *111*, 130–147. [CrossRef]

33. Higgins, D.; Balint, T.; Liversage, H.; Winters, P. Investigating the Impacts of Increased Rural Land Tenure Security: A Systematic Review of the Evidence. *J. Rural Stud.* **2018**, *61*, 34–62. [CrossRef]

34. Goldstein, M.; Udry, C. The Profits of Power: Land Rights and Agricultural Investment in Ghana. *J. Polit. Econ.* **2008**, *116*, 981–1022. [CrossRef]

35. Ma, X.; Heerink, N.; van Ierland, E.; van den Berg, M.; Shi, X. Land Tenure Security and Land Investments in Northwest China. *China Agric. Econ. Rev.* **2013**, *5*, 281–307. [CrossRef]

36. Lovo, S. Tenure Insecurity and Investment in Soil Conservation. Evidence from Malawi. *World Dev.* **2016**, *78*, 219–229. [CrossRef]

37. Beekman, G.; Bulte, E.H. Social Norms, Tenure Security and Soil Conservation: Evidence from Burundi. *Agric. Syst.* **2012**, *108*, 50–63. [CrossRef]

38. Domeher, D.; Abdulai, R. Land Registration, Credit and Agricultural Investment in Africa. *Agric. Financ. Rev.* **2012**, *72*, 87–103. [CrossRef]

39. Goldstein, M.; Houngbedji, K.; Kondylis, F.; O'Sullivan, M.; Selod, H. Formalization without Certification? Experimental Evidence on Property Rights and Investment. *J. Dev. Econ.* **2018**, *132*, 57–74. [CrossRef]

40. Brasselle, A.-S.; Gaspart, F.; Platteau, J.-P. Land Tenure Security and Investment Incentives: Puzzling Evidence from Burkina Faso. *J. Dev. Econ.* **2002**, *67*, 373–418. [CrossRef]

41. Hong, W.; Luo, B.; Hu, X. Land Titling, Land Reallocation Experience, and Investment Incentives: Evidence from Rural China. *Land Use Policy* **2020**, *90*, 104271. [CrossRef]

42. Thaler, R. Toward a Positive Theory of Consumer Choice. *J. Econ. Behav. Organ.* **1980**, *1*, 39–60. [CrossRef]

43. Cassar, A.; Healy, A.; Von Kessler, C. Trust, Risk, and Time Preferences after a Natural Disaster: Experimental Evidence from Thailand. *World Dev.* **2017**, *94*, 90–105. [CrossRef]

44. Thayer, Z.; Barbosa-Leiker, C.; McDonell, M.; Nelson, L.; Buchwald, D.; Manson, S. Early Life Trauma, Post-Traumatic Stress Disorder, and Allostatic Load in a Sample of American Indian Adults. *Am. J. Hum. Biol.* **2017**, *29*, e22943. [CrossRef] [PubMed]

45. Deng, X.; Xu, D.; Zeng, M.; Qi, Y. Does Early-Life Famine Experience Impact Rural Land Transfer? Evidence from China. *Land Use Policy* **2019**, *81*, 58–67. [CrossRef]

46. Xu, D.; Liu, E.; Wang, X.; Tang, H.; Liu, S. Rural Households' Livelihood Capital, Risk Perception, and Willingness to Purchase Earthquake Disaster Insurance: Evidence from Southwestern China. *Int. J. Environ. Res. Public Health* **2018**, *15*, 1319. [CrossRef]

47. Xu, D.; Peng, L.; Liu, S.; Wang, X. Influences of Risk Perception and Sense of Place on Landslide Disaster Preparedness in Southwestern China. *Int. J. Disaster Risk Sci.* **2018**, *9*, 167–180. [CrossRef]

48. Xu, D.; Deng, X.; Guo, S.; Liu, S. Labor Migration and Farmland Abandonment in Rural China: Empirical Results and Policy Implications. *J. Environ. Manag.* **2019**, *232*, 738–750. [CrossRef]

49. Du, J.; Zeng, M.; Xie, Z.; Wang, S. Power of Agricultural Credit in Farmland Abandonment: Evidence from Rural China. *Land* **2019**, *8*, 184. [CrossRef]

50. Xu, D.; Ma, Z.; Deng, X.; Liu, Y.; Huang, K.; Zhou, W.; Yong, Z. Relationships between Land Management Scale and Livelihood Strategy Selection of Rural Households in China from the Perspective of Family Life Cycle. *Land* **2020**, *9*, 11. [CrossRef]

51. Deng, X.; Xu, D.; Zeng, M.; Qi, Y. Does Outsourcing Affect Agricultural Productivity of Farmer Households? *Evidence from China China Agric. Econ. Rev.* **2020**. [CrossRef]

52. Deng, X.; Xu, D.; Zeng, M.; Qi, Y. Does Internet Use Help Reduce Rural Cropland Abandonment? Evidence from China. *Land Use Policy* **2019**, *89*, 104243. [CrossRef]

53. Chen, Z.; Huffman, W.E.; Rozelle, S. Farm Technology and Technical Efficiency: Evidence from Four Regions in China. *China Econ. Rev.* **2009**, *20*, 153–161. [CrossRef]

54. Niroula, G.S.; Thapa, G.B. Impacts and Causes of Land Fragmentation, and Lessons Learned from Land Consolidation in South Asia. *Land Use Policy* **2005**, *22*, 358–372. [CrossRef]

55. Zeller, M.; Diagne, A.; Mataya, C. Market Access by Smallholder Farmers in Malawi: Implications for Technology Adoption, Agricultural Productivity and Crop Income. *Agric. Econ.* **1998**, *19*, 219–229. [CrossRef]

56. Min, S.; Waibel, H.; Huang, J. Smallholder Participation in the Land Rental Market in a Mountainous Region of Southern China: Impact of Population Aging, Land Tenure Security and Ethnicity. *Land Use Policy* **2017**, *68*, 625–637. [CrossRef]

57. Gould, K.A. Land Regularization on Agricultural Frontiers: The Case of Northwestern Petén, Guatemala. *Land Use Policy* **2006**, *23*, 395–407. [CrossRef]

58. Jacoby, H.; Minten, B. *Land Titles, Investment, and Agricultural Productivity in Madagascar: A Poverty and Social Impact Analysis*; World Bank: Washington, DC, USA, 2006.

59. Deininger, K.; Jin, S. The Potential of Land Rental Markets in the Process of Economic Development: Evidence from China. *J. Dev. Econ.* **2005**, *78*, 241–270. [CrossRef]

60. Coase, R.H. The Problem of Social Cost. *J. Law Econ.* **1960**, *3*, 87–137. [CrossRef]

61. Kahneman, D.; Knetsch, J.L.; Thaler, R.H. The Endowment Effect, Loss Aversion, and Status Quo Bias: Anomalies. *J. Econ. Perspect.* **1991**, *5*, 193–206. [CrossRef]

62. Samuelson, W.; Zeckhauser, R. Status Quo Bias in Decision Making. *J. Risk Uncertain.* **1988**, *1*, 7–59. [CrossRef]

63. Xu, D.; Zhang, J.; Rasul, G.; Liu, S.; Xie, F.; Cao, M.; Liu, E. Household Livelihood Strategies and Dependence on Agriculture in the Mountainous Settlements in the Three Gorges Reservoir Area, China. *Sustainability* **2015**, *7*, 4850–4869. [CrossRef]

64. Xu, D.; Peng, L.; Liu, S.; Su, C.; Wang, X.; Chen, T. Influences of Migrant Work Income on the Poverty Vulnerability Disaster Threatened Area: A Case Study of the Three Gorges Reservoir Area, China. *Int. J. Disaster Risk Reduct.* **2017**, *22*, 62–70. [CrossRef]

65. Xu, D.; Deng, X.; Guo, S.; Liu, S. Sensitivity of Livelihood Strategy to Livelihood Capital: An Empirical Investigation Using Nationally Representative Survey Data from Rural China. *Soc. Indic. Res.* **2019**, *144*, 113–131. [CrossRef]

66. Ji, Y.; Yu, X.; Zhong, F. Machinery Investment Decision and Off-Farm Employment in Rural China. *China Econ. Rev.* **2012**, *23*, 71–80. [CrossRef]

67. Deng, X.; Zeng, M.; Xu, D.; Wei, F.; Qi, Y. Household Health and Cropland Abandonment in Rural China: Theoretical Mechanism and Empirical Evidence. *Int. J. Environ. Res. Public Health* **2019**, *16*, 3588. [CrossRef]

68. Xu, W.; Li, M.; Bell, A.R. Water Security and Irrigation Investment: Evidence from a Field Experiment in Rural Pakistan. *Appl. Econ.* **2019**, *51*, 711–721. [CrossRef]

 *land*

Project Report

Participatory Land Administration in Indonesia: Quality and Usability Assessment

Trias Aditya [1,*], Eva Maria-Unger [2], Christelle vd Berg [2], Rohan Bennett [2,3], Paul Saers [2], Han Lukman Syahid [4], Doni Erwan [4], Tjeerd Wits [5], Nurrohmat Widjajanti [1], Purnama Budi Santosa [1], Dedi Atunggal [1], Imam Hanafi [6] and Dewi Sutejo [6]

[1] Department of Geodetic Engineering, Faculty of Engineering, Universitas Gadjah Mada, Jalan Grafika no. 2 Yogyakarta 55284, Indonesia; nwidjadjanti@ugm.ac.id (N.W.); purnamabs@ugm.ac.id (P.B.S.); dediatunggal@ugm.ac.id (D.A.)

[2] Kadaster International, Kadaster, Hofstraat 110, 7311 KZ Apeldoorn, The Netherlands; eva.unger@kadaster.nl (E.M.-U.); christelle.vandenberg@kadaster.nl (C.v.B.); rohan.bennett@kadaster.nl (R.B.); Paul.Saers@kadaster.nl (P.S.)

[3] Swinburne Business School, Swinburne University, John Street, Hawthorn, Victoria 3122, Australia

[4] Center of Cadastral Survey and Mapping, Directorate-General of Agrarian Infrastructure, Ministry of Agrarian Affairs & Spatial Planning Affairs/National Land Agency (BPN), Jl Kuningan Barat 1 No. 1, Jakarta 12710, Indonesia; hanhan.syahid@atrbpn.go.id (H.L.S.); dony.erwan@atrbpn.go.id (D.E.)

[5] Meridia, Landmapp B.V., Mauritskade 63, 1092 AD Amsterdam; The Netherlands; tjeerd.wits@meridia.land

[6] Jaringan Kerja Pemetaan Partisipatif (JKPP), Jl. Cimanuk Blok B7 No. 6 Komp. Bogor Baru, Bogor 16152, Indonesia; imamh@jkpp.org (I.H.); dewisutejo@jkpp.org (D.S.)

* Correspondence: triasaditya@ugm.ac.id; Tel.: +62-274-520226

Received: 10 January 2020; Accepted: 5 March 2020; Published: 9 March 2020

Abstract: This paper presents the results from a quality and usability analysis of participatory land registration (PaLaR) in Indonesia's rural areas, focusing on data quality, cost, and time. PaLaR was designed as a systematic community-centered land titling project collecting requisite spatial and legal data. PaLaR was piloted in two communities situated in Tanggamus and Grobogan districts in Indonesia. The research compared spatial data accuracy between two approaches, PaLaR and the normal systematic land registration approach (PTSL) with respect to point accuracy and polygon area. Supplementary observations and interviews were undertaken in order to evaluate the effectiveness of the spatial and legal data collection, as well as logical consistency of the data collected by the community committee, using a mobile application. Although the two pilots showed a lower spatial accuracy than the normal method (PTSL), PaLaR better suited local circumstances and still delivered complete spatial and legal data in a more effective means. The accuracy and efficiency of spatial data collection could be improved through the use of more accurate GNSS antennas and a seamless connection to the national land databases. The PaLaR method is dependent on, amongst other aspects, inclusive and flexible community awareness programs, as well as the committed participation of the community and local offices.

Keywords: quality; usability; boundary data collection; legal data collection; first titling; land administration

---

1. Introduction

In Indonesia, like other contexts, spatial and legal data collection for systematic land titling projects are often considered challenging tasks, especially for local land offices. It is not easy for local land offices to collect and verify the required documents completely, especially considering the pluralism inherent to the underlying land tenure structure. The issue makes formal land registration in Indonesia challenging [1]. Under the current legal and institutional framework, Indonesian systematic land titling activities are procedurally demanding and rigid, requiring active participation from communities, villages, and government officers, owing to uncoordinated and sporadic registration activities in the past [2].

Fit for purposes land administration (FFP LA) principles aim to accelerate land registration activities utilizing spatial, institutional, and legal framework and also call for incremental improvement [3]. FFP LA has been tested, if not implemented, worldwide [4,5]. Although there are comprehensive FFP LA implementing guidelines available [6], see also [7], finding the best-fit land registration and spatial data collection method suitable for the country context remains a significant task in itself: there are no one size fits all approaches. Further, managing the financial, political, legal, and administrative aspects regarding large-scale registration campaigns remains challenging, even when FFP LA approaches are used.

Indonesia's current progress on land registration provides an example to examine how quality, cost, and speed can be leveraged to reach the Indonesian government's goal of registering all unregistered land parcels by 2025. The central government launched PTSL (*Pendaftaran Tanah Sistematik Lengkap*—a complete systematic land registration for all land parcels using fixed boundary approaches with terrestrial and photogrammetry surveys) as mandated by the President through President Instruction No. 2/2018. Before PTSL was launched in 2017, the capacity for land mapping and certification was around one and a half million land parcels per year. Since 2017, the land registration campaign has resulted in a massively increased workload for ATR/BPN. In 2017, PTSL covered five million land parcels, and in 2018 the number of parcels increased to seven million. The target is to complete nine million land parcels in 2019[1]. The remaining, more than 50 million land parcels, are aimed to be registered completely by 2025. PTSL was designed to map all land parcels and to certify unregistered land parcels nationwide covering each village.

Two years post-implementation of PTSL, completeness is still seen as a big challenge as land offices frequently focus only on unregistered parcels, leaving parcels with conflicts, floating titles, and unregistered parcels, decidedly unmapped. From PTSL results in 2018, it was shown that of the 7.7 million land parcels covered, 62.1% in total could be followed up with formal registration, whilst 24.6% could not be certified due to uncertain landowners' legal status. Meanwhile, 13.2% from the total were unmapped land titles and about 2,200 cases were either conflicting or in the court. On the issue of uncertain landowners, constituting almost 25%, this is caused by several factors: (i) local land offices lack access to formal documents regarding the underlying rights; (ii) unknown and in-absentia landowners; and (iii) unsettled family disputes due to disagreement over land inheritances. Moreover, the central government has in place stringent standards for land offices to produce land titles. Low-level problems are rooted in institutional arrangements and contribute more than technical problems, these being related to the land office's capacity to survey and map land parcels. This leaves the mapping of many land parcels in villages incomplete. Incompleteness leads to uncertainties of rights of registered land parcels in villages: previous studies have therefore correctly questioned the links between formal land titling, tenured security and livelihood improvement [8–11].

In order to have a systematically registered and complete land administration system, a country-specific approach, which is fast but reliable is required. Community-driven, participatory, and crowdsourced approaches promote an efficient and complete land boundary inventory that can be used for many purposes, including environmental protection and land certification [12–17]. The body of knowledge for community-driven, participatory, crowdsourced, and volunteered data collection in the research domain of land administration has evolved in recent years [4,5,18–21]. Participatory mapping practices for customary and indigenous land rights and land protection, especially for forested and rural areas, including in Indonesia, have evolved in decades [22]. However, participatory mapping

---

[1] https://www.thejakartapost.com/news/2018/10/19/agrarian-ministry-distributes-6-2m-land-certificates.html.

that produces "citizen cadastre" [23] that followed up with formal recognition of rights and land titles in collaboration with national land offices is just a few or just emerging. FFP LA and systematic land titling emphasizes working with communities to fill gaps, for example, in accelerating adjudication processes [24], or for eliminating the social constraints [25] related to land governance. However, measuring the acceptance and consequences of such a community-based land registration has not been conducted for the land administration domain, especially in Indonesia.

In response, this paper aims to provide an analysis, based on a comparative framework, between the existing systematic land registration approach (i.e., PTSL) on the one hand, and the participatory approach, (i.e., participatory land registration—PaLaR), on the other. The analysis focuses specifically on the areas of cost, time, usability, and quality of the methods. Specific attention is given to the spatial quality of the methods used in PaLaR. According to the FFP LA guidelines, policy reforms in the spatial, legal, and institutional framework are required to accelerate land administration completeness. In the Indonesian context, the country has surveyed and mapped more than 20 million land parcels for three years using PTSL. Nevertheless, as mentioned above, about 24% out of mapped land parcels are not ready for first-titling due to unclear ownership, or due to low participation from the community. Thus, the challenges lie more in increasing community participation to provide legal and administrative data, whilst preserving the quality of spatial data collection when using PTSL. Taking this into account, in order to achieve the aim, the research applied two participatory land registration pilots to facilitate the spatial and legal data collection led by community representatives in Tanggamus and Grobogan. The results of spatial and legal data collection in the pilots were compared with the results using the PTSL approach. This paper aims to provide a collaborative analysis regarding spatial quality, effectiveness, and efficiency of participatory approach and tools, which are lacking in the current FFP literature. This paper will also then identify lessons learned with regard to the institutional and legal framework.

This paper is structured as follows. Section 2 presents the materials and methods including the background and the location of the pilots and the methods of the comparative assessments. Further, it discusses the challenge and impact of increasing participation during the registration process and the modernization of the data collection methods [7,26] using usability perspectives. Two pilots, presented here, applied some FFP LA principles, but harmonized, as much as possible, with the current underlying spatial and legal frameworks. Section 3 presents the results of the comparative assessments. Section 4 discusses the lessons learned and required improvements and arrangements to increase the speed and usability degree of participatory land administration, in order to scale from pilots to national policy.

2. Materials and Methods

2.1. Methods in the Field: PTSL vs. PaLaR

PTSL is a government program dealing with systematic land title registration of all unregistered land parcels in rural areas in Indonesia. Land registration activities cover survey, mapping, registration, and certification of all land parcels in a village. As in one village, there can be land parcels that have been certified previously, the mapping should also deal with boundaries of registered land parcels, in order to create an up-to-date and complete representation of land ownership boundaries in the village. PTSL is conducted as a top-down approach program starting with the determination of the village as a PTSL location.

Indonesia applies mandatory boundary demarcation in land titling projects based upon Government Regulation (No 24/1997). For surveying and mapping land parcels, a special task force consisting of government surveyors or licensed surveyors are mandated to collect spatial data of land boundaries. In parallel to that team, a juridical team is deployed to collect and verify the legal data concerning the landowner identity and underlying ownership data. Juridical teams are land office employees assigned by the head of the land office. The budget to run PTSL was allocated from either the national budget, or local budget, or Corporate Social Responsibility Funds, or funds by the local community. Standards and procedures for the technical implementation and budget have been officially assigned through the Ministry of ATR/BPN. Steps of the PTSL approach include (a) determination of PTSL location; (b) spatial data collection; (c) legal data collection; and (d) data processing that includes legal data verification, validation, and titling. As suggested in previous research [27], mandatory boundary demarcation still faces challenges in terms of social and non-technical issues, leading to low participation.

This paper also assesses an approach, called PaLaR (participatory land registration), a community-centered data collection and facilitation for village-level land registration campaigns. In contrast to PTSL, PaLaR tends towards a bottom-up approach rather than a top-down: spatial and legal data collection is done completely through the community representatives with the guidance of government officers.

The PaLaR approach was implemented in two villages residing in Central Java and Lampung, through a cooperation between the Ministry of ATR/BPN from Indonesia and Kadaster from the Netherlands. Community representatives are referred here as the community-based land registration committees (CLRC, or interchangeably called "the committee" herein), who performed the field activities (spatial and legal data collection). PaLaR explores the use of community-centered data collection and production, to ensure that tenure security for all can be achieved successfully. It is hypothesized that by trading-off between efficiency and quality, in implementing fit for purpose land registration for Indonesia, the method can provide a design for massive scale and fast land titling.

As PaLaR is intended to produce the same official titles as PTSL, without changing the administrative processes, a comparison between PTSL (government-led land titling activities) and PaLaR (government-facilitated and community-based land titling activities) should be done using the same activities of both PTSL and PaLaR (i.e., socialization, spatial data collection, legal data collection, data processing, and certification).

- Socialization refers to field activities disseminating information about the systematic land registration campaigns.
- Spatial data collection relates to field activities done by surveyors to survey and store information regarding boundary points, adjacency and ownerships of land parcels for first-titling purposes.
- Legal data collection refers to field and office activities to collect underlying formal documents specifying the owner and the ownership status of land parcels. The required data include signed application forms, official personal identity verified by the civil registry, an underlying proof of land ownership (e.g., a statement letter from the village office) and the newest land tax bill from the Municipality Office.
- Data processing refers to office activities to validate the completeness and the validity of the submitted documents.

The differences between PTSL and PaLaR methods on implemented four plans are presented in the Table 1.

Data processing and certification using GeoKKP software for all land parcels in two village pilots were done by the corresponding Land Office's staff. For the GeoKKP entry, the local staff must validate the digital data against the paperwork. Figure 1 provides a graphical overview of the described workflow designed for the PaLaR pilots.

Table 1. Differences between the Pendaftaran Tanah Sistematik Lengkap (PTSL) and the participatory land registration (PaLaR) implementation.

Stages	PTSL	PaLaR
Socialization	• A hierarchical top-down activity in disseminating information regarding PTSL from government officers to village leaders and neighborhood leaders. • An information session usually takes two hours and consists of briefings from leaders and is followed by questions and answers (Q/A) on the procedure and prerequisite documents needed to participate in the PTSL program.	• The government involves a non-governmental organization (NGO) called JKPP (the network for participatory mapping actions) that have been advocating counter-mapping for villages and indigenous rights. • The socialization includes a group discussion related to land ownership and conflict resolution. It also covers technical training to community representatives (CLRC) that took place four days of classes and practices in the village office, attended by village leaders and community representatives.
Boundary data collection	• Government surveyors or licensed surveyors facilitate the recordation process and land surveying of unregistered land parcels using modern surveying tools (i.e., Total Station/RTK-GNSS).	• The community representatives (CLRC) acted as facilitators to conduct awareness-raising campaigns. CLRC also facilitated mapping and social/legal data collection activities. They (CLRC) also did individual parcel boundaries' measurements, legal data verification, and submission. The data was collected digitally using a tablet with the Meridia Collect App, connected to a GNSS Antenna.
Legal Data Collection	• A specific/task force team was assigned to collect and verify the application data from landowners. The legal data collection is done manually (paper-based) as current procedures are demanding paper works to be in place.	• All required data was collected in PaLaR during the interview sessions and captured as digital data using mobile applications operated by the committee. The major difference with the PTSL approach: legal data were collected by the community itself, through the community committee.

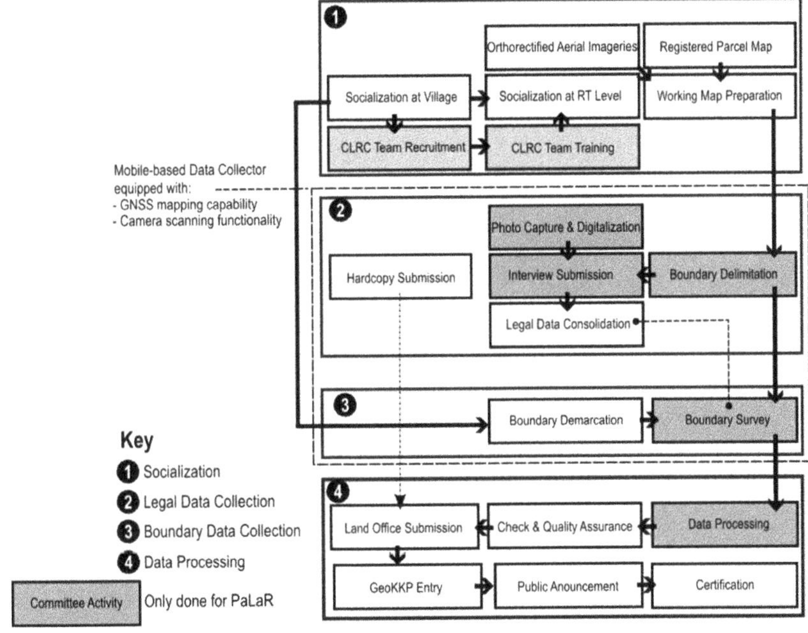

Figure 1. Workflow of PTSL and PaLaR activities (grey boxes were applied for PaLaR only, not for PTSL).

2.2. Materials in the Field

Narrowing to the work underpinning this work, subsequent to the above procedures, a spatial data quality assessment was performed to compare geometric quality against the reference data for both the test data and the resulting data from survey and mapping activities done by the CLRC team, using GNSS tools during the field survey of boundary points. Reference data was acquired by the research team using RTK GNSS/GPS mapping using L1/L2 geodetic devices for parcels that had been previously surveyed by the CLRC. CLRC accompanied the research team to show the boundary markers and in communicating with landowners during the collection of reference data.

The study utilized an app 'Meridia Collect', to collect the spatial and legal data. It ran on a tablet that was connected to a GNSS device, which is an L1 low-cost GNSS, known as Emlid Reach RS. On the backend, Meridia Collect was supported with Podio to support the online data management, and to cover data quality checking. Data integration with the national system, i.e., GeoKKP, was completed manually by local land offices after receiving the registration data from the community team.

Core functionalities in Meridia Collect were the Carta and Terra functionality. Carta was used to delineate parcels, including the use of snapping tools. Terra was used to map agreed-upon point boundaries in the field, using the GNSS mapping tool. Legal data collection frequently used Register to enable data entry (interview/typing), document digitalization, photos, and fingerprints. In addition, the dashboard application was provided to do spatial data quality checking including topology and snapping tools, and legal and spatial data integration.

2.2.1. Reference Data

Field measurements of the reference data were completed using the RTK method with L1/L2 GNSS geodetic devices. The RTK method was undertaken using the available CORS (continuously operating reference system). Here, the correction parameters were transferred using a standard NTRIP (network transport of RTCM via the Internet protocol). The base station was not a fixed CORS station, but, rather, a GNSS receiver that was positioned freely in the field adjusting to possible field obstructions.

Field measurements were conducted using GNSS geodetic devices (L1/L2) with an achievable accuracy of 10 cm. During the field measurements, boundary points that were demarcated in the field through markers were collected/measured by the CLRC members, under the supervision of the landowners. The results of the measurements were represented as shapefiles that could be directly compared with the other data sample. For the field measurements, the following GNSS receiver was used: (Brand: South Type: Galaxy G1), produced in 2017, with the following specifications (gathered from the handbook of the device):

Channel	220 Channels
Signal Tracking	GPS, GLONASS, Galileo, Beidou, QZSS, WAAS, MSAS, EGNOS, GAGAN, E5A, E5B
RTK Survey	Horizontal 8 mm + 1 ppm; Vertical 15 mm + 1 ppm
RTK initialization time	2 m, 8 s
Data Format (differential)	CMR+, CMRx, RTCM 2.x, RTCM 3.x
Data format (GPS output)	NMEA 0183, PJK, binary code
Network support	VRS, FKP, MAC, NTRIP protocol; NFC module, USB, Radio, Bluetooth
Market price in 2018	IDR 250.000.000, 00 (2 pcs set as Base-Rover)

2.2.2. Evaluation Data

The evaluated datasets were the data collected in PaLaR using a local coordinate/reference system, captures through a Base (ground control point) and Rover configuration, and a tablet, which was connected to a low-cost GNSS receiver, Emlid Reach RS+, which had the following specifications:

Channel	72 Channels
Signal Tracking	GPS/QZSS L1, GLONASS G1, BeiDou B1, Galileo, E1, SBAS
RTK Survey	Horizontal 7 mm + 1 ppm; Vertical 14 mm + 2 ppm
RTK initialization time	2 m, 8 s
Data Format (differential)	RTCM2, RTCM3
Data format (GPS output)	NMEA, ERB, plain text
Network support	USB, RS232, PPS, Bluetooth, Radio, NTRIP
Market price in 2018	IDR 26.500.000, 00 (2 pcs set as Base-Rover)

In addition to land boundaries, the app also captures registration documents, including ID cards, application forms, tax receipts, letters of ownership from the village office, required for first-titling. Data collection done by CLRC was stored in Meridia Collect. After data cleaning was completed by a Meridia data officer, the data was submitted to Tanggamus and Grobogan local land offices. Data collected were classified as parcels with registration data (ready for land certification), parcels without registration data (considered as K4/already registered or K3/not ready to be registered), and interviews only (without spatial data collection).

2.3. Methods in the Office

2.3.1. Quality Assessments

This study was looking at the data quality and the procedure quality that resulted from participatory land registration. Here, 'quality' refers to the "totality of characteristics of a product that bear on its ability to satisfy stated and implied needs" (ISO 2002, originally in ISO standard 8402). The data quality that was seen as essential here were spatial data quality, logical consistency, and completeness. Meanwhile, the procedure quality that was considered essential for land registration for this study included time, cost, and usability of the application used by the community. The set of measures for these two quality themes and their corresponding attributes is given as follows.

From the literature, spatial data quality can be broken down into elements of quality, characterizing the fitness of the product against specific standards (also known as producer accuracy). Some known international standards define spatial data quality: i.e., ISO 1957:2013 on Spatial Data Quality, and ASPRS (American Society for Photogrammetry and Remote Sensing) on ASPRS Positional Accuracy Standards [28]. From a different perspective, quality can also be seen as fitness for use, where quality is seen to meet the specific purposes for the use, either by an expert or common users following a possible list of fitness- for-use criteria (see, e.g., [29,30]).

According to geospatial information literature, five to eleven elements can be used to distinguish spatial data quality [31]. For assessing the spatial data collected, at least six elements are commonly used to describe quality: lineage, positional accuracy, attribute accuracy, logical consistency, completeness, and semantic accuracy. In these two pilots, the geospatial data quality assessment was focused only on positional accuracy, logical consistency, and completeness.

- Positional accuracy can be seen as the accuracy of the coordinate values. This can be calculated as relative or absolute positional accuracy. Absolute positional accuracy is defined as the accuracy of test coordinate values against the reference coordinate values. As the land parcels form polygonal areas, the accuracy here is assessed per the differences in point position and area between the test dataset (dataset collected by CLRC) and reference dataset (dataset collected by the research team).
- Logical consistency deals with the contradictions that violate compliance between the schema and structure of the spatial dataset and the values represented in the collected data. Logical consistency includes topological consistency and validity of attributes compared to the conceptual schema [32]. It may also include format consistency [33].
- Completeness refers to "the relationship between database objects and the abstract universe of all such objects", [34] or measures related to accessing data and missing data [33].

Spatial data quality was assessed through the comparison of the evaluation dataset against the reference dataset. The reference dataset represented an ideal PTSL spatial data collection using the RTK GNSS method, following the normal land titling/PTSL technical guidelines. The surveyed points of the PALAR's land parcels were assessed using a point accuracy test. For assessing the quality of polygon areas of PALAR's land parcels, this study implemented the polygonal area assessment.

Point accuracy shows the closeness of the evaluation data to the reference data and is assumed to have a better result and be closer to the standard of point measurement. It yields a displacement line of length between the evaluation and reference point datasets. Point accuracy was often used to evaluate OpenStreetMap (OSM) data quality (see [35,36]).

In addition to point and area evaluation, an assessment of the completeness and logical consistency of the data collection was also done. Completeness herein refers to the completion of the spatial data collection for one village. The national (PTSL) standard demands that results of the cadastral mapping of a village are categorized as existing certified land parcels (K4), undisputed land parcels not ready for certification (K3), disputed land parcels (K2), and undisputed land parcels ready for certification (K1). A score between 0% and 100% will determine the completeness of the mapping. Completeness can also be seen as the completion of legal data collection. Legal data collection will result through the entry of the legal data regarding the legal information of the landowners' identity (e.g., family card and identity card) and land ownership underlying status (e.g., a letter from the village, transaction deed, tax receipt, and so forth). Legal data completeness is specified as "Yes" or "No".

Logical consistency refers to the fidelity of the relationships in the dataset collected in PaLaR. For land registration purposes, the topology of land parcels is a logical data consistency measure that is considered important. This ensures that the shape of the individual land parcel is closed properly and that adjacent land parcels are topologically correct. The topological and validity cleansing of collected data in Tanggamus and Grobogan was assessed.

2.3.2. Time Assessment

Time in this assessment refers to the number of hours and days to complete all activities, excluding the certification process, which is excluded as the process is within the government's control. Some delays and breaks occurred in both pilots. For the assessment, use of time for spatial and legal data collection, idleness and break sessions were not counted. Time was seen as a stop-watch measurement for completing the spatial and legal data collection in the field.

The use of time to complete the land registration campaign during the PaLaR project was limited to the socialization and data collection time only. Other stages of the campaign, including data processing and certification, were not assessed. The actual data collection time was measured by the system embedded in the Meridia Collect app.

2.3.3. Cost Assessment

Significant cost in this assessment was money allocated from the government budget to finish the activities. As the budget of PaLaR only covered socialization and data collection, and partly data processing but no certification, the focus of cost assessment is on socialization and data collection activities only. The results of the cost assessment will be compared with international guidelines and can be used by stakeholders to plan the participatory land registration budget.

For this study, the land office and the committees in Tanggamus and Grobogan were interviewed regarding the costs for data collection[2]. Costs that were not yet considered in the results of the project

---

2   Costs that are used to facilitate the PaLaR activities are comprised of:

- Cost to mobilize and provide grants (seen as fees) for CLRC members;
- Cost required by village leaders to provide boundary monuments and administrative materials (e.g., paper copy and stamp expenses);

include a group discussion facilitated by JKPP, field training, hardware and software investments (e.g., tablets, GPS and document scanner, and software developments).

2.3.4. Usability Assessments

The term "usability", as it is used in this paper, refers to the usability of the geospatial information interface used in the pilots and not to the geospatial data itself, as used in other works. Usability assessment was done to evaluate the effectiveness, efficiency, and usefulness of the mobile application to gather spatial and legal data from the community. Previous works have applied usability measures of mobile applications for participatory data collection [37,38], but no work has been done in support of land titling projects. The research measures were focused on effectiveness, efficiency, understandability of the interface, and how error-tolerant and engaging it was as recommended in the '5E usability measures' [39].

Usability assessments were completed by conducting interviews with users and observations during the use of the app to collect the field data. Here, "users" refer to CLRC members in Tanggamus and Grobogan. The app refers to the Meridia Collect app, the data collection tool used by CLRC.

3. Results

3.1. Statistics of the Overall Data

In total, 1000 points of land parcels in Tanggamus and 1697 points on land parcels in Grobogan were collected. The status of the data completion was captured at the end of the pilot project and is shown in Table 2.

Table 2. PaLaR land parcels of Tanggamus and Grobogan pilots.

No	Parcels	Tanggamus	Grobogan	Total	Percentage
1	Parcels with legal data	465	237	702	26%
2	Parcels without legal data	535	1460	1995	74%
3	Exported parcels data	956	291	1247	46.20%
1 + 2	Total of parcels	1000	1697	2697	100%

As of September 2018, final updates were provided by the two local land offices. The land parcel status in the Village of Kuripan, Tanggamus, and Wandan kemiri, Grobogan can be seen in Table 3. In Tanggamus, PaLaR land parcels that were ready for land titling (K1) were 532, meanwhile in the Village of Wandan Kemiri, Grobogan, land parcels that were ready for land titling, were 686. It was about 50% in total, out of PaLaR land parcels that can be followed up with land titling. Other cases were K3 and K4. A typical example for K3 is landowners did not want to continue to participate in land registration programs for unconfirmed reasons (see its explanation in the Discussion). K4 means that land offices must take actions to improve the quality of land records as previously the land titles were either not mapped correctly or with no spatial information (known as floating titles).

- Cost to cover logistics for socialization and during the field data collection.

Table 3. PaLaR land parcels of Tanggamus and Grobogan pilots.

No	Parcels	Tanggamus	Grobogan	Total	Percentage
1	Parcels to be registered (K1)	532	686	1218	50%
2	Parcels without legal data (K3)	304	703	1006	42%
3	Parcels registered previously (K4)	100	94	194	8%
	Total of parcels	1036	1483	2418	100%

3.2. Results of Spatial Data Quality Evaluation in Tanggamus

3.2.1. Results of Evaluation for Point Accuracy

Evaluation for positional accuracy was completed first by selecting land parcels identified in the referenced and evaluation dataset through the same number of points. The authors limited the evaluation of positional accuracy to paired boundary points found in both the referenced and evaluation dataset. In Tanggamus 15 parcels out of 33 had a different number of points.

The research team that recorded the point boundaries of the reference data was accompanied by CLRC during the survey. During the field survey to individual parcels, most landowners attended the field check. CLRC checked the points with the previously surveyed ones, by using the map provided by the tablet used in PaLaR or by consulting with the landowners. Most land parcels were demarcated in the field, hence point identification was easier.

Next, parcels in both the reference and evaluation dataset that had the same number of points per parcel were compared to determine the difference of the coordinates. This comparison produced a length of line (distance) between the reference and evaluation of boundary points. The average and standard deviation of the shiftings were calculated. The differences of the point coordinates between the reference and evaluation dataset for Tanggamus can be seen in Figure 2. The average difference was 0.521 meters while the standard deviation was 0.570 meters. Poor results (>1.5 m) resulted from either poor data correction or points surveyed with GNSS floated solutions.

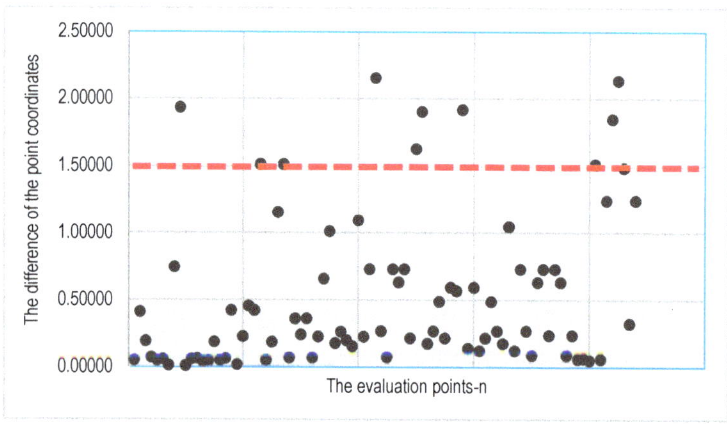

Figure 2. Positional accuracy (in meters) plotted as shifting of points found between reference and evaluation dataset, shifting > 1.5 m is indicated through the red dashed line.

3.2.2. Results of Evaluation for Polygonal Features

The polygonal quality of land parcels was assessed using the area comparison method. The comparison method was used to compare parcels of the same unique ID from the evaluation and reference dataset. There were 33 samples of land parcels. From the polygonal area, an evaluation was done in Tanggamus, the results of the classification of the spatial data quality of area comparisons can be seen in Table 4. Equal interval grouping was used to show the distribution of values of the results for area comparison.

Table 4. Classification of area comparison in Tanggamus using the equal interval.

Classification of Area Comparison using the Equal Interval Method			
Classes		Quality	Frequency
0.000	41.667	Very good	29
41.667	83.333	Good	2
83.333	125.000	Quite Good	0
125.000	166.667	Not Good	1
166.667	208.333	Bad	0
208.333	250.000	Very Bad	1

Graphical views of spatial data quality derived from all methods can be seen in Figure 3.

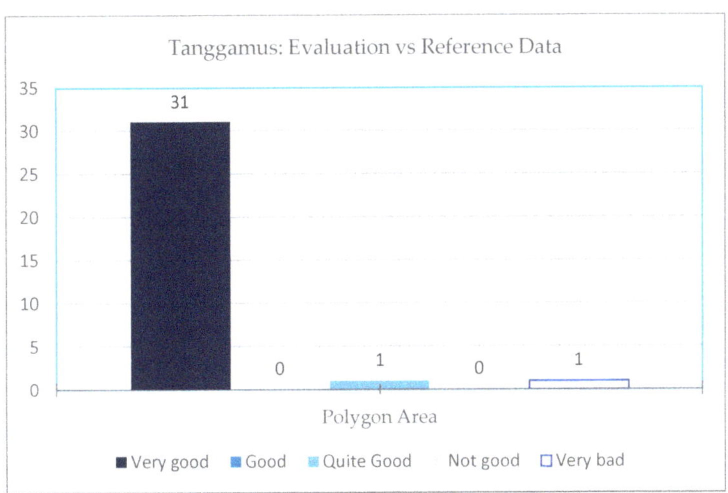

Figure 3. Analysis of the evaluation and reference dataset in Tanggamus.

The visualization of polygon data for the evaluation and referenced land parcels can be seen in Figure 4. It shows the differences of lengths of parcel boundaries between the reference and evaluation data.

Differences were caused by at least two factors: (1) location of points measured by CLRC were not exactly the same points as measured by the research team. In some cases, although the research team was accompanied by the CLRC during the measurements, there was uncertainty within the CLRC team regarding the location of the point because the field demarcation was not permanently installed or the landowner was not present and (2) effect of data cleaning in the post-processing by the Meridia data officers that included snapping or adding points, which produced different parcel boundaries. As seen in the parcel with the red circle, 22 different points existed between the reference and the evaluation datasets. This could have been caused by having no boundary demarcation in the field. A similar case also occurred with the land parcel indicated through the yellow circle.

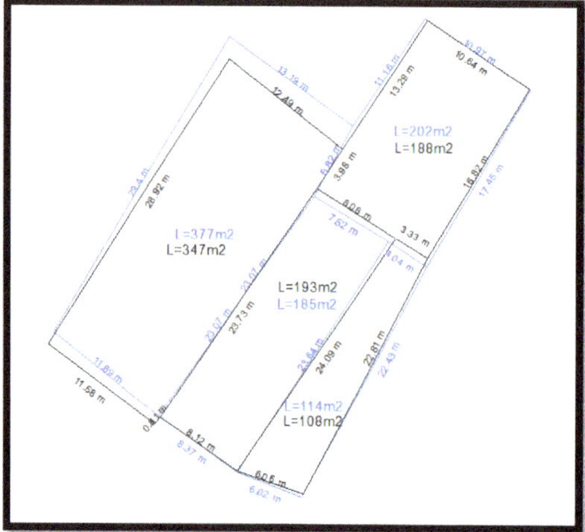

Figure 4. Visualization of differences in the results of evaluation (blue) and reference (black) polygonal data in the study area of Tanggamus pilot.

Figure 5 (top) shows the evaluation versus the reference dataset, although the evaluation dataset here was not cleaned (before the geometric data cleaning process). In some cases, Figure 5 (below), the differences may have resulted from wrong snapping during the data cleaning process. The differences between the reference and the evaluation dataset were less before than those after the data cleaning.

Figure 5. *Cont.*

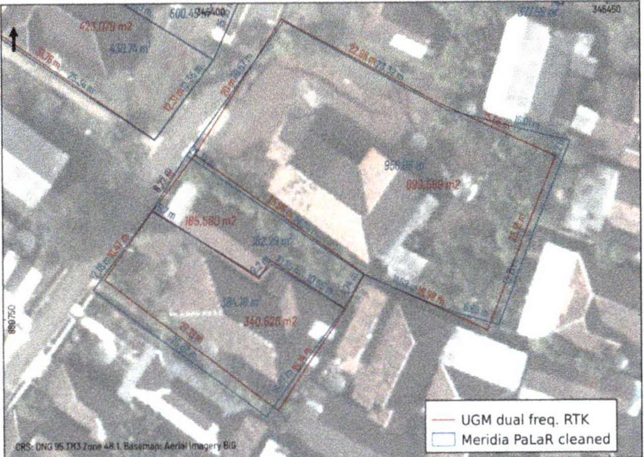

Figure 5. Differences in results in the area and length of the evaluation (uncleaned) and reference (cleaned) data.

Furthermore, the difference in quality was visualized for both polygonal feature assessments. Darker colors represented lower quality, hence, for areal comparison, darker green meant lower quality. This meant a greater difference between the reference and evaluation datasets. Maps of polygonal quality of collected land parcels can be seen in Figure 6. Darker means bigger differences or larger gaps.

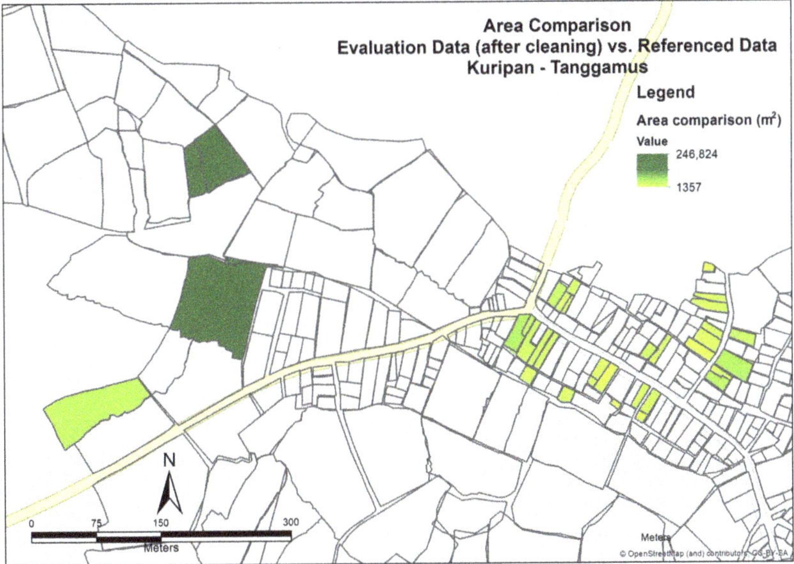

Figure 6. Map of area comparisons between the evaluation and reference dataset in Tanggamus.

3.2.3. Results of Logical Consistency

All parcel data collected from Tanggamus were checked for logical consistency. From checking activity, it can be concluded that topological consistency and attributes' validity were good.

3.2.4. Results of Completeness

Completeness here refers to the completion of spatial data collection for one village. The final map produced should give clear information as to which land was categorized as existing certified land parcels (K4), undisputed land parcels not ready for certification (K3), disputed land parcels (K2), and undisputed land parcels ready for certification (K1). Completeness can also be seen as the completion of identification of all parcels and their ownership status. It can be concluded that the completeness was well achieved with the PaLaR method, as 100% of all land parcels were mapped.

In terms of legal data collection, as of August 2018, 465 out of 535 parcels were without legal data. This could have been because incomplete parcels (red) either were not registered or not ready to be registered (Figure 7). Landowners who were not registered were generally less motivated to participate during the land registration.

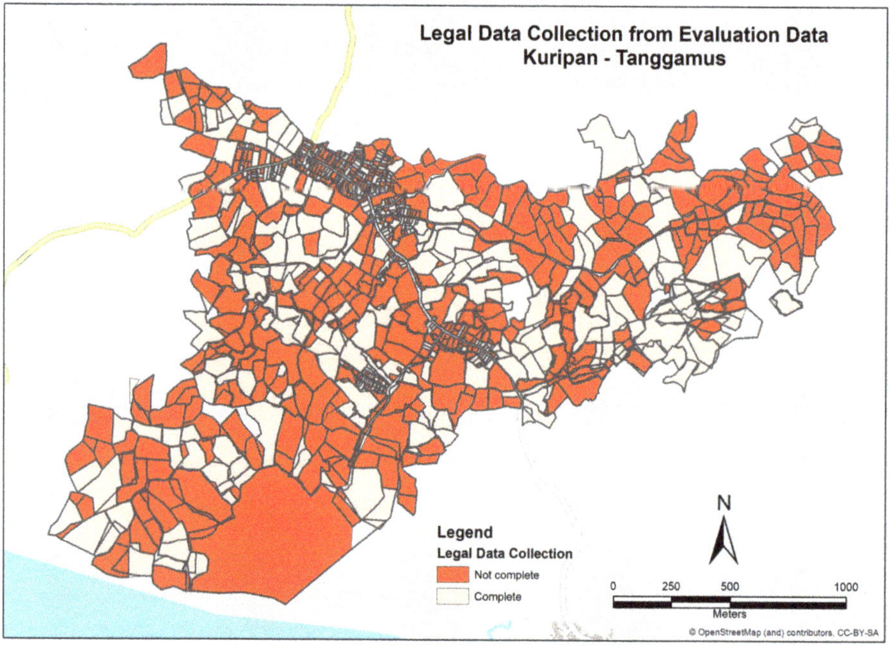

Figure 7. Completeness of legal collection in Tanggamus.

For clarity, the results of the spatial accuracy assessment of the Grobogan pilot are presented separately in Appendix A.

3.3. *Legal Data Quality*

Results of legal data quality were compiled based on interviews with Grobogan and Tanggamus land officers on the 24th of September 2018. Both local land offices believed the errors found in PaLaR data collection were acceptable, as similar errors were also recognized in PTSL. Examples of errors included: typographical errors, wrong addresses, invalidated or missing identity numbers, and family card numbers, and underlying ownership letters from villages. In terms of legal data collection, there were no serious issues in the data recorded through the data collector app. It can be said that the legal data quality from PaLaR data collection was acceptable, or at least comparable to PTSL.

3.4. Results of Cost Assessment

The costs of the PaLaR project were covered by the land office using the same unit price per parcel as targeted in the PTSL activity. Hence, it can be said that the costs allocated to the PaLaR activities, particularly in dealing with CLRC mobilization and field data collection, was well-covered using the PTSL budget.

In Tanggamus:

1. Legal and spatial data collection for K1 (ready to be registered) was paid at IDR 85.000/parcel. This was done by the CLRC. At the time of the report, there were 532 land parcels ready to be certified.
2. Legal data verification and data processing were done by the local land office using the PTSL budget.

The total expense was estimated to be IDR 65.2 million, comprising IDR 45.2 million for data collection (fees to CLRC) and about 20 million for preparation and socialization. For 532 land parcels that were certified, this meant that the cost for land registration was about IDR 122.600 /parcel with PaLaR (before legal data processing and certification by the land office), i.e., the land registration cost was 8.4 USD per parcel before legal data verification and certification activities of the Local Land Office.

In Grobogan:

1. Legal and spatial data collection for K1 (ready to be registered) was IDR 50.000/parcel. The activity was done by the CLRC. In total, there were 686 parcels in K1 status (ready to be registered);
2. Preparation, logistics, and legal data verification support from the village office was paid by the local land office with the amount of IDR 7.500/parcel;
3. Legal data verification and data processing was done by local land office using the standard PTSL budget; cost: 90.000/parcel;
4. The certification process would be 10.000/parcel using the standard PTSL budget.

The total expense was estimated to be IDR 59.5 million, comprising IDR 39.5 million for data collection (fees to CLRC) and 20 million for preparation and socialization. For the 686 land parcels that were certified, this meant the cost for land registration was IDR 86.650 /parcel for socialization and data collection activities, i.e., the cost was 5.9 USD/parcel before legal data processing and certification by the Land Office.

PTSL implementation in Tanggamus (field survey cost for Area III) was set to IDR 214.980 (by government surveyor), or IDR 330.240 (by licensed surveyor). Meanwhile, field cost for Area V, e.g., Grobogan, was IDR 114.340 (by government surveyor) or IDR 170.000 (by licensed surveyor). The cost specified in the PTSL project was all-inclusive of hardware and software. For example, costs for utilizing GNSS receivers (L1/L2 types) and CAD/GIS software were the responsibility of consultant/private surveyors. The summary of the cost comparison between PaLaR and PTSL is given in Table 5.

Table 5. Cost simulation and comparison between PTSL and PaLaR methods.

Activity	Cost per Land Parcel			
	Tanggamus PTSL	Tanggamus PaLaR	Grobogan PTSL	Grobogan PaLaR
Socialization	7500	7500	7500	7500
Data collection	330,240	50,000	170,000	85,000
Data processing	90,000	90,000	90,000	90,000
Certification	10,000	10,000	10,000	10,000
Cost per parcel (IDR)	437,740	157,500	277,500	192,500
Cost per parcel (USD)	31 USD	11.25 USD	19.82 USD	13.75 USD

The calculation of the cost only considered land parcels with K1 and K3 status, as currently the government only considers K1 and K3 land parcels for project costing. As seen in Table 6, the PaLaR method offered a lower cost than PTSL. However, it is important to note that this measure was not considered the following cost used in the pilots: the group discussions facilitated by participatory facilitators outside community and government organization and training, training resources, and hardware and software investments. The PaLaR method implemented in this study is comparable to adjudication and survey activities done by the community specified in the cost and financing guideline [40]. Still, the output of PaLaR is comparable to the type of fix boundary demarcation of systematic registration activity. The use of technology and the collaboration between government, community facilitator and community representatives produced a value for the money solution for participatory land registration.

Table 6. The actual and hypothetical time required in PaLaR pilots.

Pilot Areas	Registered Parcels	Activity	Actual Time Units/Parcel	Hypothetical Time (vol × Time) (minutes)	Total Time Done by 4 Teams/Village with 6 Working hours/day
Tanggamus	532	Boundary	32 min	532 × 32	11.8 day
	532	Legal	11.8 min	532 × 11.8	4.3 day
Grobogan	686	Boundary	25 min	686 × 25	11.9 day
	686	Legal	9.1 min	686 × 9.1	4.3 day

Notes: 1. Legal data collection used the maximum length of activity times: 26.4 days or 6240 min; 2. For spatial data collection in Kuripan, Tanggamus district, the average time required to survey boundary points of each parcel in residential areas and rice paddies was 32 min.

3.5. Results of Time Assessment

Using the data provided by the application system, the duration of spatial data collection for both Tanggamus and Grobogan can be summarized as follows:

Mapping Completion Time

1. Kuripan: Rice paddy 39.67 min, Farming land 37.21 min, Residential 20.39 min;
2. Wandankemiri: Rice paddy 30 min, Farming land 18.11 min, Residential 24.48 min.

One should consider the difference in field terrain between the two pilots. Tanggamus has a steep topography with some rolling areas, especially towards the forested mountain and farming areas. Meanwhile, Wandan Kemiri, Grobogan, has a plain topography. This may lead to differences in field survey times.

Meanwhile, the total duration of the legal data collection can be extracted from the data capture completion as follows:

1. Interview submissions: 16.3 days
2. Mapping submissions: 25.0 days
3. Drawing submissions: 26.4 days

Based on the data given, the time spent for the completion of data collection in the Tanggamus and Grobogan cases can be seen in Table 6.

It could be concluded that the actual time spent for legal and spatial data collection in PaLaR, both in Tanggamus and Grobogan, was consistent at 11.8–11.9 days for spatial data collection and 4.3 for legal data collection. This clearly indicates that the PaLaR method could accelerate the spatial and legal data collection. According to the PTSL procedure, the data collection time should not exceed 60 days, in total, for each village.

3.6. Usability Issues of the Data Collector Application

With the data collection process having been divided into two main activities, i.e., spatial and legal data collection, the results of the usability assessments are presented in Table 7. The focus of the usability assessment was limited to the apps used to support CLRC. The feedback on usability parameters was mainly taken from structured interviews with CLRC members in Kuripan (Tanggamus) and Wandan Kemiri (Grobogan).

Table 7. Assessment of usability indicators during the field data collection (as-is).

	Legal Data Collection	Spatial Data Collection	Overall Evaluation
Effective	Yes, when data requirements were complete, typing and data capturing can be done effectively	Yes, it is well connected with legal data	Acceptable with minor improvement
Efficient	Yes, when all relevant documents were complete, data input can be done optimally (e.g., max 45–60 applicants/day)	Yes, it is well connected with legal data. Synchronization is necessary to be done during delineation and determination of land parcels, but this feature relies heavily on internet connection and device reliability. More efficient synchronization would be better	Acceptable with minor improvement
Easy to Use	Yes, but complains from CLRC members because of the overheating of the tablets. This influenced the reliability (force closed of app), e.g., most devices in Wandan Kemiri need to be restarted 4 times to collect one parcel	Yes, but complains from CLRC members for its: - Overheating and force closed issues - Lack of "snapping to line" feature for delineation on the screen or survey in the field (note: this functionality was implemented in the Meridia Collect app after the two pilots)	Acceptable with major improvement
Error tolerant	Yes, the CLRC members could trackback and revise the error	Yes, but many issues were about the unique IDs of the parcel that were defined as long and difficult to type (note: Automatic identification using barcode or fingerprint, for example, were proposed by some CLRC members)	Acceptable with minor improvement
Engaging	Yes, digitalization approach receives positive response	Yes, but the typical slow response from the device and overheating of the devices created frustration for the CLRC members. The typical "float" solution and sometimes difficult connection between the GNSS device and the tablet also decreased the trust and enthusiasm of CLRC members	Acceptable with major improvement

In terms of collaboration mechanics in participatory mapping activities [41,42], the role of CLRC members as facilitators and the tablet as the map interface (enabler) was assessed based on the communication and coordination criteria.

3.6.1. Communication

Some notes that were considered important were:

1. Synchronized map (linked legal and spatial data) positively speed up the process of collecting legal data and land parcels' claims.
2. Village leaders and CLRC skills/competency to facilitate discussion and communication, as seen in Tanggamus, accelerated the data collection progress.
3. Maps were used as media to facilitate discussions of village boundaries and distribution of land parcels in RT.

3.6.2. Coordination

In terms of coordination matters, some notes can be addressed to PaLaR pilots, including:

1. Synchronized maps on tablets were used to effectively schedule and plan daily targets;
2. Wrong identification of land parcels was minimized when maps were used and local knowledge from village leaders was well-applied in the process;
3. Efficient field surveys and legal data collection was coordinated visually among CLRC members;

4.   Validation of legal and spatial data by the Land Office staff was based on the submitted data.

Collaboration mechanics, in terms of coordination and communication across stakeholders in community-based data collection, were not deliberately prepared. As a result, data collection and validation processes, as well as stakeholders' interactions, were not clearly understood by all village leaders and CLRC members. Regarding mobile and GPS technology use, software usability and utility for supporting data collection were still not optimal. Lessons learned from the two pilots regarding effectiveness, efficiency, ease of learning, error tolerance, and usability of the interface suggested software improvements to optimize the potential benefits of community-centered land titling projects.

4. Discussion

Whilst the comparative analysis was organized around the themes of data quality and procedure quality, and the derivative attributes of positional accuracy, logical consistency, completeness, time, cost, and usability—here, the discussion uses the three overarching FFP LA themes of spatial framework, legal framework, and institutional framework, to organize the discussion.

Lessons learned regarding the FFP LA spatial framework are as follows. First, the use of very high-resolution drone/aerial imagery for boundary delimitation and legal data collection was found to increase the clarity of the data collection. Second, great efficiency was found in ensuing legal data collection was well integrated with spatial data collection. This improves the digital capture of documents, photographs, and personal identity that eases data integration and entry to the KKP. However, more efficient results could be derived by the local land office if the connection between KKP and the mobile app was made earlier during the mapping process to the data checks. This integration would ease CLRC members in detecting parcels that were already mapped and in avoiding overlapping ownership claims. Additionally, regarding the overall efficiency of two pilots, a faster field data processing and cheaper cost can be obtained by the community when a seamless synchronization between field data collection and the KKP is in place.

In comparison to the normal PTSL approach, the PaLaR spatial data collection was proven to be cheaper but the results were less accurate in comparison to the typical PTSL spatial parcel data. This assessment was an important finding from the pilots, given the normal fix boundary approach that PTSL did (using field and photogrammetry surveys) has produced approximately 20 million land parcels within three years. The parcel data has grown from about 1 million land parcels before PTSL to seven million parcels a year during PTSL project. Nevertheless, the completeness of spatial and legal data collection, as the two PaLaR pilots produced, seem to be more promising than the typical PTSL method produced. The spatial accuracy for PaLaR method can be improved once a higher precision dual GNSS antenna was used as the survey sensor for the data collector app. Thus, the PaLaR method would increase the level of spatial completeness while reducing the cost of registration per parcel significantly.

This paper argues that the detail assessments on the quality of the products and the usability of methods as well as cost and time efficiency resulted from the study could benefit professionals working in land administration in designing the standards and workflow for community-based systematic land registration activities. Rwanda project costs 7 USD/parcel [43] and Laos pilot costs 15.1 USD/parcel [44] applied the general boundary. Tanzania Pilot costs 14–48 USD/parcel [45] and this Indonesia PaLaR Pilot costs 12.5 USD/parcel applied a mix solution that tends towards the fix boundary. The collaborative work of community committee and government and NGO supported with a low-cost digital survey and mapping platform in Indonesia pilot promises completeness and low-cost solution for a country-context FFP implementation. According to costing and financing land administration system (COFLAS) guideline [40], the similar output to PaLaR (fix boundary with little to big investments of GNSS network) would cost from 10–50 USD/parcel.

The outcome of this research is relevant to justify that FFP LA implementation is more than just switching from a field survey to aerial imageries in determining boundary points. As demonstrated in the pilots, the spatial framework can be solved through the participatory approach with the help of mobile technology. However, the biggest challenge is a lack of guidelines to relax institutional arrangements and legal incentives to attract landowners [46,47] to participate in the process entirely until their land certificates are delivered. Although the CLRC mapped all land parcels eligible for first titling in two villages, many landowners were still not submitting the registration requirements until the pilots were over. These failures to complete land registration were caused by several reasons, which have not been confirmed by the landowners. However, the CLRC told authors that some landowners voiced concerns regarding future yearly tax and informal payments after their lands are certified. Some others informed the CLRC about their internal family disputes regarding property inheritance, which makes land certification impossible to take place. After the pilot finished, we also learned that few landowners had less trust for village leaders. The real reasons why eligible landowners avoid participating in the first titling campaign are essential issues to explore which are not the focus of this paper, see, e.g., [48]. However, this information strengthens the research topic questioning the relationship between the formal certification and the community perceived security to their property [8,9], or land registration effectiveness [49], which are very diverging and heterogenic in developing countries [50].

Four parties collaborate during the pilots, i.e., government, village leaders, community, and community facilitator. FFP LA literature has mainly focused on data collection processes [4,7,51], especially on the use of aerial photos/satellite imageries to solve boundary determination issues. The literature that evaluates the community as the central actor in the participatory adjudication process for a systematic land registration project is very few if not available [23]. The key for successful participation during the two pilots includes information delivery and clarity regarding the objectives and significance of the project as well as its benefits to the community. This information transformation took place during the socialization and the training. The government surveyor and administrator could then focus on the quality check of the collected data. We also learned that trust to the village leader as well as signs of a trustworthy process are essential for systematic land registration, see [5,48]. Although the two pilots successfully collected spatial and legal data, follow-up actions of individual landowners to process further the formal land registration, which includes submitting signed forms and making registration payments have proven problematic. This research has not delved into the investigation of the real reasons behind landowners' decision to decline from the formal registration.

In terms of the legal framework, the current statutory requirement was fulfilled by the PaLaR method and it was compliant with the PTSL land registration method, in a way that fixed the boundary survey for residential areas. Moreover, visible boundary determinations using photogrammetric methods for paddy fields and farmland were done effectively within the community (see Appendix A—Appendix A.1). In terms of legal mandate, community members acted as data collectors, while local land offices acted as data validators. One issue still remains: contradictory boundary delimitation requires landowners and neighbors to agree and be in place during data collection. In the PaLaR pilots, this was still challenging in cases where existing landowners were living in other cities. From the pilots, it was shown that better participatory aspects in the PaLaR process, than those in PTSL approach, does not automatically mean a higher willingness for the community to register their land. Until the pilots were finalized, many land parcels were still in K3 status, as owners and their families did not show up to submit their signed registration forms and to show their original documents to the land officers. Legal instruments and incentives to ease their participation could increase their trust in the government's land registration projects and thus reduce K3 problems. It is also thought that if boundary mapping could be a compulsory activity in each village, this will be promising for field validation, which is necessary to reduce unmapped certificates or K4. In this regard, all landowners, or their trusted representatives, who are receptive to the committee, will participate; otherwise, specific disincentives could emanate from the government. Government-facilitated participatory data collection with the PaLaR method needs legal backup (i.e., ministry regulation) to run smoothly.

Regarding institutional arrangements, the CLRC, or the committee, was set up by community members themselves and endorsed by the village and higher rank government officers. The composition of the CLRC should represent trusted community members or village leaders (seen as a trusted intermediary) from both genders, in order to make the data collection process trusted and followed by all landowners. Village leaders, who usually have strong knowledge of the history of the village and its land, were important in providing explanations to landowners and in mediating any land boundary disputes. Young residents were usually familiar with digital technology and were also expected to be active in assisting senior leaders in data communication to help mediate disagreements and conflicts. From the pilots, it was clear that transparency in managing the project budget at the community level was required from the beginning. This included transparency in financial allocation for CLRC members; salaries needed to be agreed upon local land offices, village offices, and the PaLaR team before the project could start. Further adoption of FFP LA principles in order to meet the country's specific context, as suggested by Barry (2018) was made. For example, the strategy to set up a village committee as a formal team to conduct survey and mapping activities to all applied land boundaries rather than the boundary delineation only was made.

Coordination and activity arrangements facilitated by local offices should be done among all stakeholders. Budget specifications should be specified at the beginning of the project, including contracts and agreements with CLRC. As argued in Section 4, PaLaR offers an accelerated time and cost-effectiveness for governments. Still, a legal framework for enabling the PaLaR budget model (with some improvements), to facilitate PaLaR implementation needs to be developed.

5. Conclusions

This work sought to explore whether an even more participatory approach to land registration in Indonesia, underpinned by FFP LA principles, could compare and even improve on the existing approach with respect to data quality and procedural effectiveness. In this regard, the developed PaLaR methodology was used to successfully create two complete village parcel maps in Kuripan, Tanggamus, and Wandan Kemiri, Grobogan. Although PaLaR lacked point accuracy compared to the reference data, their polygonal area differences were considered sufficiently small, at 5%, to comply with the current system requirements. The differences between the textual and graphical area led to an aerial validation of the cadastral map that was assumed to be relevant for this work. In this regard, according to FFP LA principles [26], as discussed earlier, PaLaR suits local circumstances and can represent a complete title recording process with legal data quality and logical consistency. Eventually, the less accurate spatial boundaries (in comparison to the reference data) can be improved by replacing low-cost GNSS devices, as used in PaLaR, with more accurate, but more expensive, double frequency antenna GNSS receivers. Though in remote areas, where the canopy is very dense and topography varies heavily, and in areas where radio or internet connections did not exist, as seen in the hilly part of Kuripan Tanggamus, neither low nor high-cost GNSS will be useful.

The PaLaR pilot offered good prospects in regard to time efficiency that could be useful in accelerating land parcel mapping and registration. Using the PaLaR method, spatial data collection was completed within 11.8 days for each campaign in Tanggamus and Grobogan. The actual time to complete the legal data collection was 4.3 days. This is faster than the data collection using PTSL. However, the overall time to organize and complete participatory data collection still depends, among other factors, especially on community awareness as well as CLRC and local office preparedness. The method can be made more efficient when the connection to national land databases can be established. In this way, existing land boundaries and their corresponding data can be accessed by CLRC through the mobile app so overlapping surveys and duplication can be prevented.

The PaLaR pilot provides a good showcase for possibly cheaper (using only 36%–69% PTSL budget allocation) data collection for a possible countrywide FFP LA application for rural areas in Indonesia. The total money requested to collect spatial and legal data, as well as to create a complete village parcel map, was 8.4 USD/parcel before legal data processing and titling by land offices in

Kuripan, Tanggamus and 5.9 USD/parcel before legal data processing and certification by the land office in Wandan Kemiri, Grobogan. However, it should be noted that the costs mentioned here were only related to data collection involving CLRC (before legal data processing and certification), while hardware and software investments, training, supervision, and validation costs were not considered and depend on the chosen tools and technologies. Nevertheless, the budget simulation from PaLaR pilots provides outstanding value for money that competitive to other similar projects [43–45]. Additionally, a budgeting standard to support these community projects needs to be well-formulated. The authors suggest an extra-budgetary allocation to facilitate community participation enabling second-round field validation and information completion through the committee. This would ensure both quantitative and qualitative validity of mapped land information.

Many aspects related to technology used in the new approach, as demonstrated in the PaLaR project, should be reconsidered. More flexible but clear legal and institutional frameworks to enable CLRC with the support of NGOs (e.g., JKPP facilitators), working as para-surveyors in the process of data collection and verification, should be in place. Registrars and government surveyors in local offices should step back and take more a role as facilitators and validators of the results. This shift paradigm from "data collector" to "data validator and quality assessor" for government surveyors deserves attention and consideration. The expected changes should also include improvements in the policy for national surveys and improvements of budgeting standards when the PaLaR method is implemented as a mix of government-facilitated and community-centered land registration projects.

Author Contributions: Conceptualization, T.A., T.W., and C.v.B.; methodology, T.A., P.S.; software, T.W.; validation, T.A., E.M.-U. and R.B.; formal analysis, T.A., N.W.; investigation, D.A., P.B.S.; resources, H.L.S., I.H., D.S., C.v.B. and D.E.; data curation, T.A. and T.W.; writing—original draft preparation, T.A.; writing—review and editing, R.B., E.M.-U., T.A.; visualization, T.A.; supervision, C.v.B., D.E.; project administration, H.L.S. and C.v.B.; funding acquisition, T.A. All authors have read and agreed to the published version of the manuscript.

Funding: This research was funded by Kadaster International, project number 9001948.

Acknowledgments: Authors from Universitas Gadjah Mada would like to thank Febriyan F. Susanta for data processing as well as Kusmiarto, Fahmi C, Mustofa, Reza and Fauzi for collecting reference data during the research project. Authors would also like to thank Simon Ulvund and Johannes Eberenz from Meridia for their support.

Conflicts of Interest: The authors declare no conflict of interest.

Appendix A. Results of Spatial Data Quality Evaluation in Grobogan

Appendix A.1. Results of Evaluation for Point Accuracy

Evaluation of positional accuracy was done first by selecting land parcels found in the referenced and evaluation data with the same point numbers. We limited the evaluation of positional accuracy to paired boundary points found in the reference and evaluation data. Overall, 29 of 37 land parcels in the Grobogan evaluation data had the same number of points in one polygon as land parcels in the reference data.

Next, parcels in both the reference and evaluation data having the same number of points were compared to determine the differences in point coordinates between the two data sets. This comparison produced length of line (distance) between reference and evaluation boundary points. Averages and standard deviations of the shifting were calculated. Positional accuracy can be seen in Figure A1 (paddy field) and Figure A2 (residential areas). The average of the differences in land parcels in the residential areas was 0.71 meters while the standard deviation was 0.548 meters. The average of differences in land parcels located in the paddy field was 0.612 meters while the standard deviation was 0.562 meters. Poor results (>1.5 m) resulted from either poor data correction or floated solutions; as a result, absolute positioning could be used in place of RTK positioning.

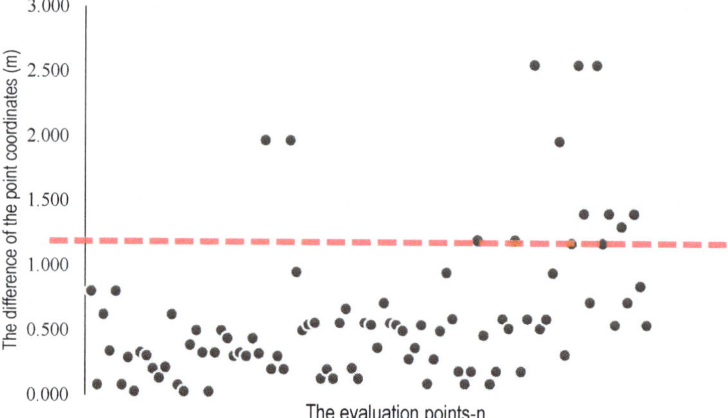

Figure A1. Positional accuracy (in meters) plotted as a shifting of points between the reference and evaluation data for paddy fields in Grobogan.

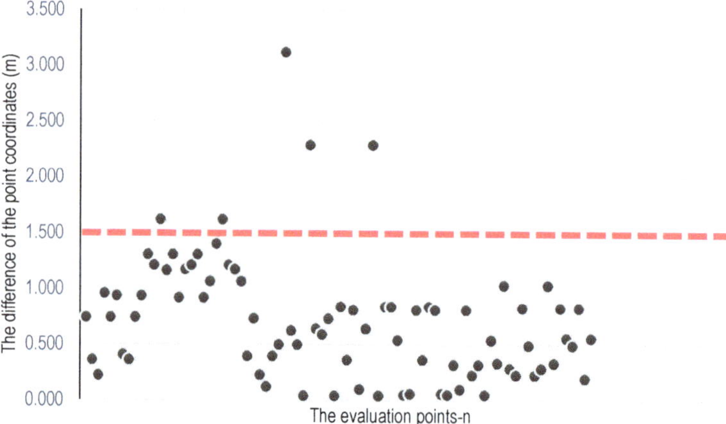

Figure A2. Positional accuracy (in meters) plotted as a shifting of points between the reference and evaluation data of parcels in residential areas in Grobogan.

Appendix A.2. Results of Evaluation of Polygonal Features

Polygonal quality of land parcels was assessed using area comparison. The method was used to compare parcels with the same identification from the evaluation and referenced data. There were 33 sample parcels. Results are summarized in Table A1 and depicted in Figure A3. As done in Tanggamus, classification of the results, the equal interval grouping method was chosen to represent the values.

Table A1. Classification of area comparisons in Grobogan using the equal interval method between evaluation and reference datasets.

Classification of Area Comparison Using the Equal Interval Method			
Classes		Quality	Frequency
0.000	41.667	Very good	33
41.667	83.333	Good	1
83.333	125.000	Quite Good	1
125.000	166.667	Not Good	2
166.667	208.333	Bad	0
208.333	250.000	Very Bad	0

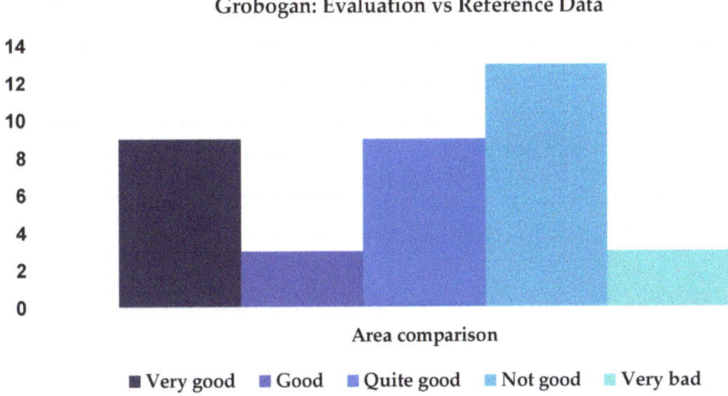

Figure A3. Evaluation and reference dataset analysis in Grobogan.

Visualization of the polygonal data for the evaluation and reference land parcels can be seen in Figure A4.

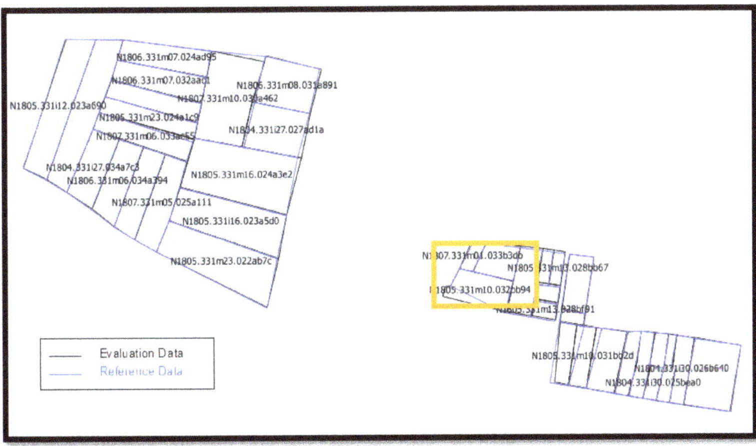

Figure A4. Visualization of land parcels used as evaluation and reference data in Grobogan using AutoCAD.

In Figure A4, differences between reference and evaluation parcels as indicated by yellow circles resulted after data cleaning performed by the data officer. The left bottom corner point was attached to the road.

Quality visualization was done for the polygonal feature assessment. Darker colors represented lower quality. For area comparison, darker green meant lower quality, i.e., the differences between reference and evaluation areas in the corresponding polygons were greater than those in the polygons with the brighter green color. For near distance, the greater the distance of the centroid, the lower the quality. Maps of polygon quality for collected land parcels can be seen Figure A5. In Figure A5, the greatest differences were found in two darkest green polygons. The data officer snapped the left corner point from one of the parcels to the road during the data processing. Also, during the field evaluation, the owner seemed to show some boundary markers to the research team that were in different locations from previously surveyed by the CLRC.

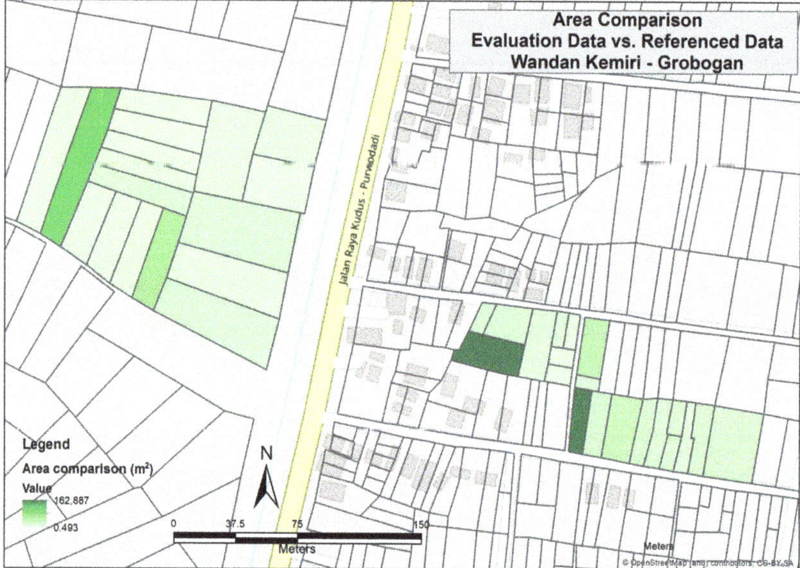

Figure A5. Map of area comparisons between the evaluation and reference data in Grobogan.

Appendix A.3. Results of Logical Consistency

All parcel data collected from Grobogan were checked for logical consistency. From checking activity, it could be concluded that topological consistency and attributes' validity were perfect.

Appendix A.4. Results of Completeness

The completeness achieved with the PaLaR method was 100% as all land parcels were mapped. In terms of legal data collection, as of August 2018, 1460 out of 1900 parcels were completely collected but not yet validated. This meant that this dataset required validation by the Local Land Office before the publication of land titles. Red parcels in Figure 6 indicates that the status of these parcels either was registered or not ready for first registration (K3). In these cases, the adjudication team was required to go to the field to validate the legal data kept by the landowners, otherwise, landowners with K3 labels still represented more than 500 parcels in the village (see Figure A6).

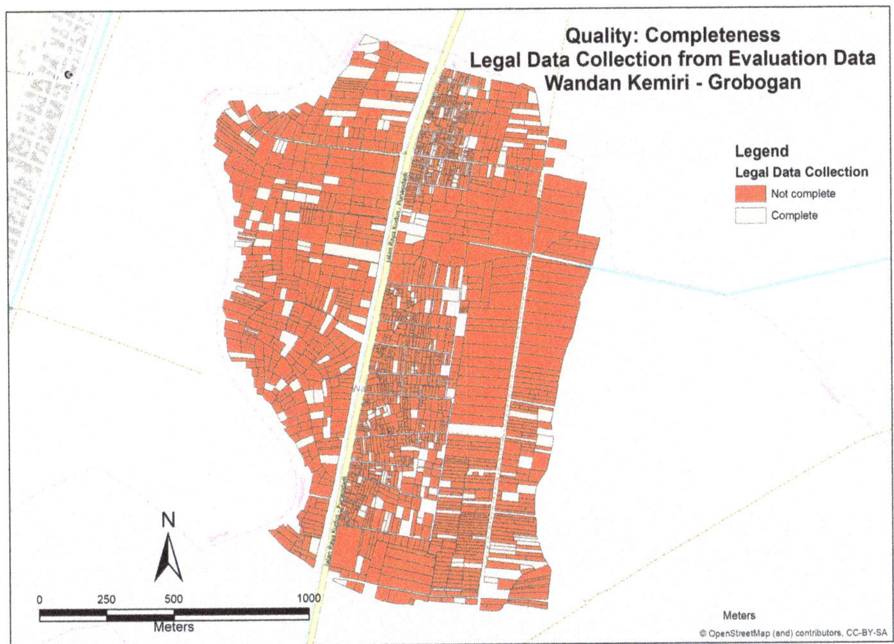

Figure A6. Completeness of the legal data collection (those that were checked by the land office and ready for first land titling) from the evaluation data in Grobogan.

References

1. Van Der Eng, P. 10 After 200 years, why is Indonesia's cadastral system still incomplete. In *Land and Development in Indonesia*; ISEAS–Yusof Ishak Institute Singapore: Singapore, 2016; pp. 227–244.
2. Zevenbergen, J.A. *Systems of Land Registration: Aspects and Effects*; Netherlands Geodetic Commission (NCG): Delft, The Netherlands, 2002.
3. Enemark, S.; Bell, K.C.; Lemmen, C.; McLaren, R. *Fit-For-Purpose Land Administration*; Enemark, S., Bell, K.C., Lemmen, C., McLaren, R., Eds.; FIG (International Federation of Surveyor) & The World Bank: Copenhagen, Denmark, 2014.
4. Bennett, R.; Alemie, B.K. Fit-for-purpose land administration: Lessons from urban and rural Ethiopia. *Surv. Rev.* **2016**, *48*, 11–20. [CrossRef]
5. Moreri, K.; Fairbairn, D.; James, P. Issues in developing a fit for purpose system for incorporating VGI in land administration in Botswana. *Land Use Policy* **2018**, *77*, 402–411. [CrossRef]
6. Enemark, S.; McLaren, R.; Lemmen, C. *Fit-For-Purpose Land Administration Guiding Principles for Country Implementation*; UN-HABITAT-Kadaster-GLTN: Nairobi, Kenya, 2016.
7. Rahmatizadeh, S.; Rajabifard, A.; Kalantari, M.; Ho, S. A framework for selecting a fit-for-purpose data collection method in land administration. *Land Use Policy* **2018**, *70*, 162–171. [CrossRef]
8. Bromley, D.W. Formalising property relations in the developing world: The wrong prescription for the wrong malady. *Land Use Policy* **2009**, *26*, 20–27. [CrossRef]
9. Durand-Lasserve, A.; Selod, H. The Formalization of Urban Land Tenure in Developing Countries. In *Urban Land Markets: Improving Land Management for Successful Urbanization*; Lall, S.V., Freire, M., Yuen, B., Rajack, R., Helluin, J.-J., Eds.; Springer: Dordrecht, The Netherlands; Berlin/Heidelberg, Germany, 2009; pp. 101–132.
10. Payne, G. Land tenure and property rights: An introduction. *Habitat Int.* **2004**, *28*, 167–179. [CrossRef]
11. Uwayezu, E.; De Vries, W. Indicators for Measuring Spatial Justice and Land Tenure Security for Poor and Low Income Urban Dwellers. *Land* **2018**, *7*, 84. [CrossRef]

12. Beaudoin, G.; Rafanoharana, S.; Boissiere, M.; Wijaya, A.; Wardhana, W. Completing the Picture: Importance of Considering Participatory Mapping for REDD+ Measurement, Reporting and Verification (MRV). *PLoS ONE* **2016**, *11*, e0166592. [CrossRef]

13. Chapin, M.; Lamb, Z.; Threlkeld, B. Mapping indigenous lands. *Annu. Rev. Anthr.* **2005**, *34*, 619–638. [CrossRef]

14. French, M.; Popal, A.; Rahimi, H.; Popuri, S.; Turkstra, J. Institutionalizing participatory slum upgrading: A case study of urban co-production from. *Environ. Urban.* **2018**, *4*, 1–22.

15. Gallo, D.S.; Cardonha, C.; Avegliano, P.; Carvalho, T.C.M.D.B. Taxonomy of Citizen Sensing for Intelligent Urban Infrastructures. *IEEE Sensors J.* **2014**, *14*, 4154–4164. [CrossRef]

16. Meredith, T.; Macdonald, M. Community-supported slum-upgrading: Innovations from Kibera. *Habitat Int.* **2017**, *60*, 1–9. [CrossRef]

17. Mohamed, M.A.; Ventura, S.J. Use of Geomatics for Mapping and Documenting Indigenous Tenure Systems. *Soc. Nat. Resour.* **2000**, *13*, 223–236. [CrossRef]

18. Basiouka, S.; Potsiou, C. The volunteered geographic information in cadastre: Perspectives and citizens' motivations over potential participation in mapping. *GeoJournal* **2013**, *79*, 343–355. [CrossRef]

19. Rahmatizadeh, S.; Rajabifard, A.; Kalantari, M. A conceptual framework for utilising VGI in land administration. *Land Use Policy* **2016**, *56*, 81–89. [CrossRef]

20. Siriba, D.N.; Dalyot, S. Adoption of volunteered geographic information into the formal land administration system in Kenya. *Land Use Policy* **2017**, *63*, 279–287. [CrossRef]

21. Asiama, K.; Bennett, R.; Zevenbergen, J. Participatory Land Administration on Customary Lands: A Practical VGI Experiment in Nanton, Ghana. *ISPRS Int. J. Geo-Information* **2017**, *6*, 186. [CrossRef]

22. Peluso, N.L. Whose woods are these? counter-mapping forest territories in kalimantan, indonesia. *Antipod.* **1995**, *27*, 383–406. [CrossRef]

23. Di Gessa, S.; Poole, P.; Bending, T. *Participatory Mapping as a Tool for Empowerment: Experiences and Lessons Learned from the ILC Network*; ILC/IFAD: Rome, Italy, 2008.

24. Mustofa, F.C.; Aditya, T.; Sutanta, H. Evaluation of participatory mapping to develop parcel-based maps for village-based land registration purpose. *Int. J. Geoinformatics* **2018**, *14*, 2.

25. Hendriks, B.; Zevenbergen, J.; Bennett, R.; Antonio, D. Pro-poor land administration: Towards practical, coordinated, and scalable recording systems for all. *Land Use Policy* **2019**, *81*, 21–38. [CrossRef]

26. Enemark, S.; McLaren, R.; Lemmen, C. *Fit-For-Purpose Land Administration–Guiding Principles*; Global Land Tool Network (GLTN): Copenhagen, Denmark, 2015.

27. Arruñada, B. Evolving practice in land demarcation. *Land Use Policy* **2018**, *77*, 661–675. [CrossRef]

28. ASPRS. New Asprs Positional Accuracy Standards for Digital Geospatial Data Release. *Photogramm. Eng. Remote Sens.* **2015**, *81*, 277.

29. Wentz, E.; Shimizu, M. Measuring Spatial Data Fitness-for-Use through Multiple Criteria Decision Making. *Ann. Am. Assoc. Geogr.* **2018**, *108*, 1150–1167. [CrossRef]

30. Devillers, R.; Bedard, Y.; Jeansoulin, R.; Moulin, B. Towards spatial data quality information analysis tools for experts assessing the fitness for use of spatial data. *Int. J. Geogr. Inf. Sci.* **2007**, *21*, 261–282. [CrossRef]

31. van Oort, P.A.J. Spatial Data Quality: From Description to Application. PhD Thesis, Wageningen Universiteit, Wageningen, The Netherlands, 2006.

32. ISO. ISO/FDIS 19157: 2013. In *Geographic Information: Data Quality*; International Organization for Standardization: Geneva, Switzerland, 2013.

33. Leibovici, D.G.; Pourabdollah, A.; Jackson, M.J. Which spatial data quality can be meta-propagated? *J. Spat. Sci.* **2013**, *58*, 3–14. [CrossRef]

34. Veregin, H. Data quality parameters. *Geogr. Inf. Syst.* **1999**, *1*, 177–189.

35. Haklay, M. How good is volunteered geographical information? A comparative study of OpenStreetMap and Ordnance Survey datasets. *Environ. Plan. B Plan. Des.* **2010**, *37*, 682–703. [CrossRef]

36. Arsanjani, J.J.; Mooney, P.; Zipf, A.; Schauss, A. Quality Assessment of the Contributed Land Use Information from OpenStreetMap Versus Authoritative Datasets. In *Lecture Notes in Geoinformation and Cartography*; Springer Science and Business Media LLC: Berlin/Heidelberg, Germany, 2015; pp. 37–58.

37. Moumane, K.; Idri, A.; Abran, A. Usability evaluation of mobile applications using ISO 9241 and ISO 25062 standards. *SpringerPlus* **2016**, *5*, 548. [CrossRef]

38. Mugisha, A.; Nankabirwa, V.; Tylleskär, T.; Babic, A. A usability design checklist for Mobile electronic data capturing forms: The validation process. *BMC Med. Informatics Decis. Mak.* **2019**, *19*, 4. [CrossRef]

39. Quesenbery, W. Balancing the 5Es: Usability set your mind at Es. *Cut. IT* **2004**, *17*, 4–11.

40. Burns, T.; Fairlie, K. *Framework for Costing and Financing Land Administration Services (CoFLAS)*; UN-HABITAT-FIG-GLTN: Nairobi, Kenya, 2018; Available online: https://mirror.gltn.net/index.php/component/jdownloads/download/2-gltn-documents/2196-framework-for-costing-and-financing-land-administration-services-coflas (accessed on 25 February 2020).

41. Aditya, T. Usability Issues in Applying Participatory Mapping for Neighborhood Infrastructure Planning. *Trans. GIS* **2010**, *14*, 119–147. [CrossRef]

42. Gutwin, C.; Greenberg, S. The mechanics of collaboration: Developing low cost usability evaluation methods for shared workspaces. In Proceedings of the IEEE 9th International Workshops on Enabling Technologies: Infrastructure for Collaborative Enterprises (WET ICE 2000), Gaithersburg, MD, USA, 14–16 June 2000.

43. Ngoga, T.H. *A Quick, Cost-Effective Approach to Land Tenure Regularisation: The Case of Rwanda*; International Growth Centre: London, UK, 2019.

44. Florian, J.R.; Sisoulath, V.; Metzger, C.; Chanhtangeun, S.; Phayalath, X.; Derbidge. *Systematic Land Registration in Rural Areas of Lao PDR: Concept document for Countrywide Application*; GIZ: Bonn and Eschborn, Germany, 2015.

45. Lauren Persha and Gwynne Zodrow. *Cost and Time Effectiveness Study—Final Report: Mobile Application to Secure Tenure—Pilot in Tanzania*; NORC and MSI for USAID: Chicago, IL, USA, 2017.

46. Barry, M. Fit-for-purpose land administration—Administration that suits local circumstances or management bumper sticker? *Surv. Rev. TA TT* **2018**, *50*, 383–385. [CrossRef]

47. Murtazashvili, I.; Murtazashvili, J. Can community-based land adjudication and registration improve household land tenure security? Evidence from Afghanistan. *Land Use Policy* **2016**, *55*, 230–239. [CrossRef]

48. Törhönen, M.-P. Sustainable land tenure and land registration in developing countries, including a historical comparison with an industrialised country. *Comput. Environ. Urban Syst.* **2004**, *28*, 545–586. [CrossRef]

49. Barry, M.; Whittal, J. Land registration effectiveness in a state-subsidised housing project in Mbekweni, South Africa. *Land Use Policy* **2016**, *56*, 197–208. [CrossRef]

50. Ma, X.; Heerink, N.; Van Ierland, E.; Berg, M.V.D.; Shi, X. Land tenure security and land investments in Northwest China. *China Agric. Econ. Rev.* **2013**, *5*, 281–307. [CrossRef]

51. Ramadhani, S.A.; Bennett, R.M.; Nex, F.C. Exploring UAV in Indonesian cadastral boundary data acquisition. *Earth Sci. Informatics* **2017**, *11*, 129–146. [CrossRef]

Article

Improving Infrastructure Installation Planning Processes using Procedural Modeling

Nae-Young Choei [1], Hyungkyoo Kim [1] and Seonghun Kim [2,*]

1 Department of Urban Design and Planning, Hongik University, Seoul 04066, Korea; nychoei@hongik.ac.kr (N.-Y.C.); hkkim@hongik.ac.kr (H.K.)
2 Research Institute of Science and Technology, Hongik University, Seoul 04066, Korea
* Correspondence: monotaxism@gmail.com; Tel.: +82-2-320-1635

Received: 6 January 2020; Accepted: 6 February 2020; Published: 10 February 2020

Abstract: Time and costs are often the most critical constraints in implementing a development impact fee (DIF) for local infrastructure installation planning in South Korea. For this reason, drafting quality plan alternatives and calculating precise DIFs for improvement remain challenging. This study proposes an application of a procedural modeling method using CityEngine as an alternative to traditional methods, which rely on AutoCAD. A virtual low-density suburban development project in Jeju, South Korea was used to compare the workability of the two methods. The findings suggest that procedural modeling outperforms the other approach by significantly reducing the number of steps and commands required in the planning process. This paper also argues that procedural modeling provides real-time 2- and 3-dimensional modeling and design evaluation and allows for a more efficient assessment of plan quality and calculation of DIF. We also argue for the need to diffuse procedural modeling to better support local planning practices.

Keywords: infrastructure installation planning; procedural modeling; development impact fees; Jeju

1. Introduction

The growth of cities and regions requires planning practices to address the increasing demand for infrastructure. Supplying infrastructure is a service usually provided by central and local governments [1]. One frequently observed issue in infrastructure supply is that the cost of installation is a burden on the public sector, while the benefits are seized by a handful of individuals [2–5]. This generates a problem of equity. In response, many local governments in the United States combine development permit issues with development impact fees (DIF) as a betterment levy. DIF is a policy instrument used to control urban growth [6,7] and is recognized by many critics as a reasonable action of police power to promote equity [8,9] by imposing an installation cost of the infrastructure for new developments, theoretically equivalent to the social marginal cost, on the developers and householders. Today, DIF has triggered lively debates on housing prices, housing supply, and regional economic growth [10–20], as well as on its environmental contributions, such as restricting vehicle use and reducing air pollution levels [14,21–23].

Unsustainablity is found in many suburban developments [24–28]. For this reason, infrastructure installation planning in South Korea mandates the imposition of DIF. Following a four-stage process (Figure 1), local governments calculate DIF-based infrastructure installation plans by relating capital development plans and infrastructure cost allocation plans or by applying coefficients like the land conversion factor in limited circumstances. These plans attempt to balance development permits with the provision of infrastructure and to levy a part of the costs of installation to developers [29,30]. DIF in South Korea faces several issues that need to be resolved, as argued by local critics. Institutional and legal settings are criticized for lacking consistency [31]. The arbitrary designation of infrastructure supply zones, for example, is questioned [32–36]. Other concerns include double taxation issues and bubble effects [31].

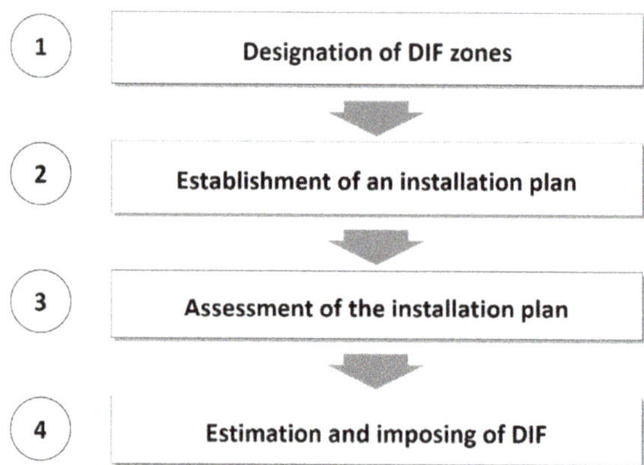

Figure 1. The four stages of infrastructure installation planning in South Korea.

One clearly overlooked aspect in the literature, especially in the South Korean context, is the applicability of DIF in local planning practice. Time and cost are often the most critical concerns in implementing DIF in infrastructure installation planning, though it is difficult to find empirical studies that investigate this. Typically, zoning requirements are automatically cleared if a plan is not established within a year after going through public hearings and planning commission meetings. We found, from a series of interviews with local planners and practitioners, that their work efficiency and consistency are often challenged when drafting plan alternatives and calculating precise DIFs; these problems require major improvements. For these reasons, with a specific focus on the infrastructure installation planning process, we propose the application of procedural modeling using CityEngine in the planning process as an alternative to the existing method for calculating DIF, which relies on AutoCAD, to enhance workability. We do so by applying the traditional method and procedural modeling to a virtual single-family housing development project in Jeju, South Korea. We compare the work processes of the two methods and identify the benefits of the latter by investigating the detailed steps and commands required for each method. The findings of the research may inform planners and policymakers at the local levels.

2. Procedural Modeling

Technologies heavily used in planning, like computer-aided design (CAD) and the geographic information system (GIS), have significantly enhanced efficiency in practice but retain limitations. In practice, these two tools require the extensive drawing of objects in the plan. These drawings are produced manually using a computer mouse of tablet pencil. On the other hand, procedural modeling provides distinct advantages. It automatically generates and updates 3-dimensional models of urban environments by incorporating information on roads, blocks, buildings, and other physical elements in real-time.

In general, procedural modeling refers to a computer graphics technology that automatically implements a completion model from spatial and attribute information with a set of rule files, similar to the use of grammar in linguistics. This methodology has already been used in many 3-dimensional modeling programs, including 3ds Max, Blender, Cinema 4D, and OpenSCAD. CityEngine is one of the many applications that apply procedural modeling [37] based on shape grammar [38]. It is widely accepted as an effective tool in designing and implementing street patterns and subdivision plans for a given site during the initial states of urban development.

Shape grammar, introduced by Stiny and Gips, is a type of grammatical system that generates geometric figures and spaces [39]. It focuses on interpreting orders through which a complex shape can be delineated with a group of simple shapes, just as sentences, paragraphs, and texts are ultimately produced by combining individual words using grammar. The composition of a shape grammar is built upon an initial form to be transformed; a group of shape rules, which define the initial form transformation; and a generation engine for selecting and executing the shape rules. Efforts to apply shape grammar to realistic modeling works have long been carried out in a wide range of industrial fields like architecture and urban planning [40–44], coffee makers [45], and Harley-Davidson motorcycles [46].

Procedural modeling is based on the theoretical basis of the L-System, derived from biology [37] and first introduced by Lindenmayer [47]. This system is capable of simultaneously generating new forms from each initial form [47] and could be useful for modeling an organism such as a tree, because the tree's genesis and growth in all directions are accurately and simultaneously explained by this theory. Procedural modeling perceives the city as an organism, in which elements such as buildings, roads, and parcels are intertwined. Recent studies of procedural modeling (mostly by computer graphics engineers) have focused on various elements of the city, like topography [48,49], vegetation [50–52], water systems [53,54], roads, buildings [37,55], building interiors, and the overall urban systems [56].

3. A Virtual Development Project

Jeju is geographically the largest island in South Korea located in the south of the Korean Peninsula, as presented in Figure 2, and is a special self-governing administrative unit. Although the island is globally recognized as an area with high environmental values [57,58], its coastal waters and soil are being severely damaged by pollution [59,60] because of the growing amount of discharged wastewater that often exceeds the island's treatment capacity [61]. The indiscriminate allowance of new real estate developments with little concern for existing capacities has been criticized as one of the key reasons behind these issues [62–64]. Raising general utility rates due to new suburban residential developments is triggering serious equity debates among the residents [65].

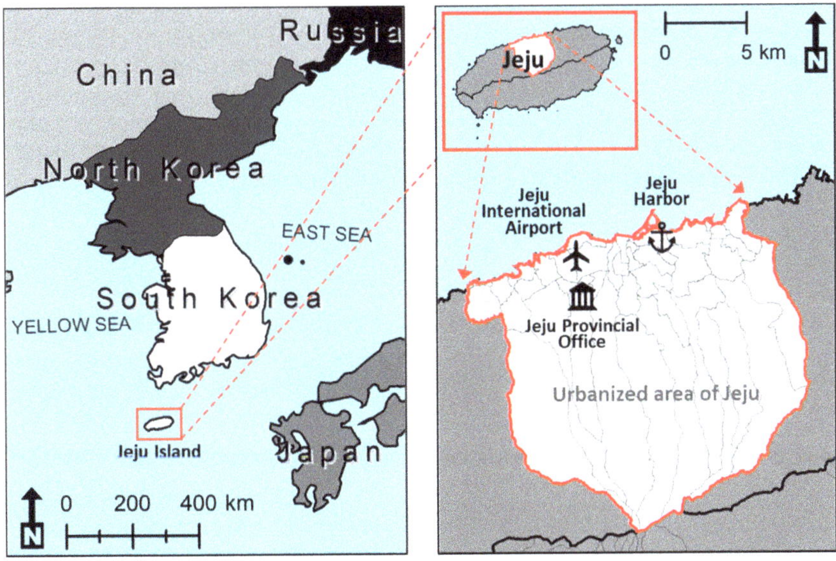

Figure 2. The location of Jeju.

Figure 3 presents the site in Jeju selected for our virtual development project. In recent years, a number of development permits have been issued in the site, and new arterial roads have been constructed along its outskirts. Designating the site as a DIF zone may generate revenue and thereby secure a relevant supply of infrastructure. For research purposes and to simulate the outcomes of various plans, we assume that the site remains undeveloped.

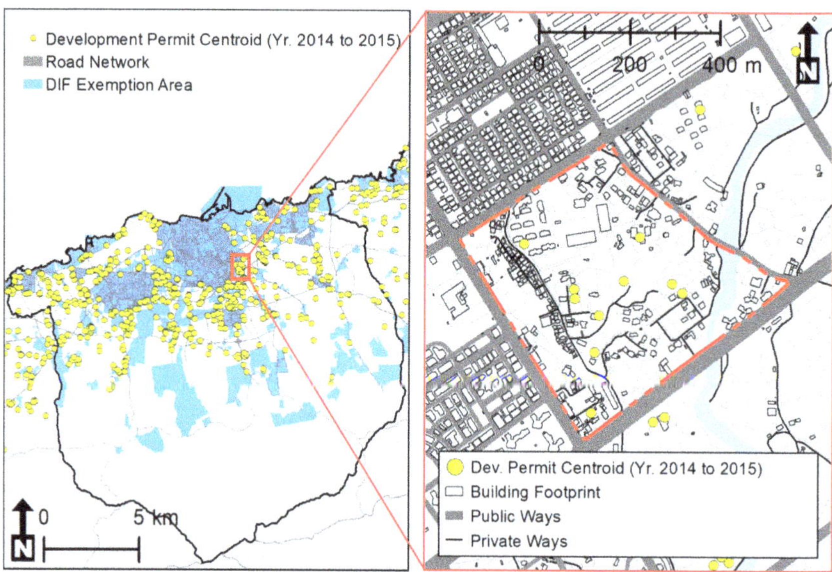

Figure 3. Location of the site for the virtual development project.

We utilize a wide range of information on topography, buildings, roads, parcels, and land prices, which are regularly gathered and updated by government agencies. The information on topography includes digital elevation maps and orthophotos. Information on buildings and roads was retrieved from geospatial vector data. Data for the parcels and their prices are from 2013.

A set of conditions is assumed for installing the infrastructure for this virtual development. Given the recent urban development trends in Jeju, we assume a low-density residential development that accommodates around 450 single-family detached houses at a density level of a 2.0 floor area ratio (FAR) and a 0.6 building coverage ratio (BCR), following local zoning requirements. The proportion of land allocated for roads is confined to between 15% and 30%, which is also based on the zoning requirements.

Six development typologies that employ the findings of Southworth and Owens [66] were established for the project, as shown in Figure 4. The first three are (1) grid, (2) loop, and (3) cul-de-sac; and the other three are combinations of the first three, which are (4) grid and loop, (5) loop and cul-de-sac, and (6) cul-de-sac and grid. As the site is located on a plain, we do not take into account local features (such as topography, conservation areas, or flood prone areas) in this process, although these features may be incorporated in procedural modeling practice when necessary. Table 1 outlines the site characteristics of the six typologies.

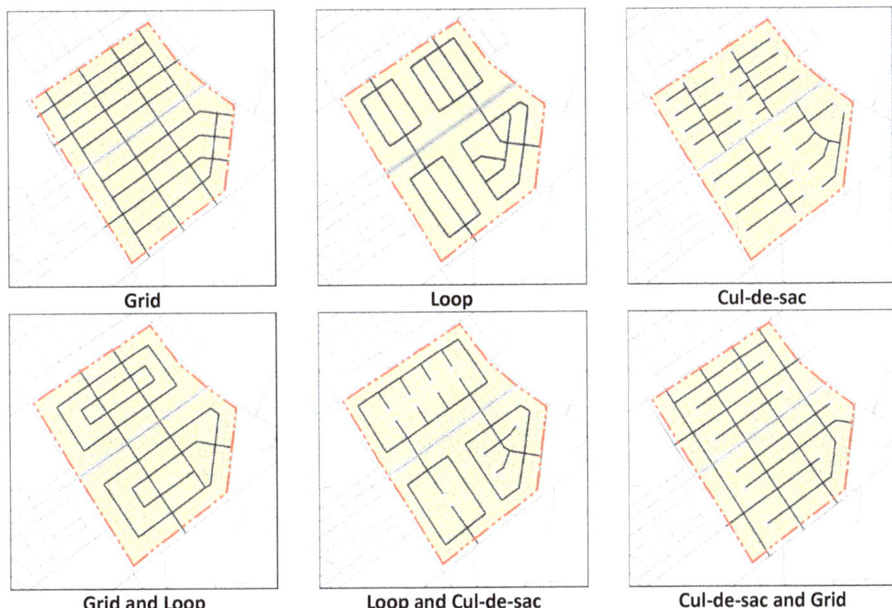

Figure 4. The six development typologies adopted in the virtual development project.

Table 1. Site characteristics of the six typologies.

Site Characteristics	Grid	Loop	Cul-de-sac	Grid and Loop	Loop and Cul-de-sac	Cul-de-sac and Grid
Site area (m^2)	208.332	208.332	208.332	208.332	208.332	208.332
Total area of parcels (m^2)	164,823	172,615	171,321	171,764	172,787	166,895
Total number of parcels	457	477	487	463	434	460
Total area of roads (m^2)	43,510	35,717	37,011	36,569	35,546	41,438
Total length of roads (m)	5414	4299	3770	4425	3972	4941
% area of roads	20.9	17.1	17.8	17.6	17.1	19.9
Total length of water and sewage (m)	5414	4299	3770	4425	3972	4941

Table 2 presents the rules and definitions of the six metrics in the two categories we adopted for a comparative evaluation of the six typologies. First, we apply three metrics to measure the quality of living for each typology: (1) isovist to reflect safety [67]; (2) integration to assess the legibility of the residential environment [68]; and (3) betweenness to judge comfort [69]. Second, we adopt measures of costs for road construction, water and sewage supply, and the compensation of land to install infrastructure. Their formulae follow local codes.

Table 3 presents calculation results of the quality of development for the six typologies. The cul-de-sac and grid typologies showed the highest isovist values, followed by grid, grid and loop, loop, loop and cul-de-sac, and cul-de-sac. For integration, grid shows the highest value, followed by cul-de-sac and grid, grid and loop, loop, cul-de-sac, and loop and cul-de-sac. Cul-de-sac and grid again present the highest betweenness, followed by grid, grid and loop, cul-de-sac, loop and cul-de-sac, and loop. Overall, the cul-de-sac and grid typology yields the best quality among the six.

Table 4 shows the calculated results of the infrastructure installation costs and expected DIF for each development typology. The total installation cost is topped by the grid, followed by loop and cul-de-sac, cul-de-sac and grid, grid and loop, cul-de-sac, and loop. The expected total DIF equals 30% of this total cost based on the local infrastructure supply codes. In the end, the largest expected total

DIF per household is found for the loop, followed by cul-de-sac, grid and loop, cul-de-sac and grid, loop and cul-de-sac, and grid.

Table 2. Metrics and their rules of definitions for measurements.

Metrics		Rules or Definitions
Quality of the development typology	Isovist [67]	Mean amount of areas visible from a specific location
	Integration [68]	$INT_i = \frac{D_N}{RA_i}$ where INT_i is integreation of street i; D_N is a normalizing factor depending on N; and RA_i is relative asymmetry
	Betweenness [69]	$B(v) = \sum_w \sum_u \frac{g_{uw}(v)}{g_{uw}}$ where $B(v)$ is betweenness at node v; u is trip origin; w is trip destination; g_{uw} is the number of paths between u and w; and $g_{uw}(v)$ is the number of paths between u and w that contain node v.
Installation Costs	Road [a]	(total road length) x (cost per meter)/(total area of residential parcels)
	Water and sewage [a]	(total water and sewage length) x (cost per meter)/(total area of residential parcels)
	Land compensation [a]	(total price of parcels acquired for road construction)/(total area of residential parcels)

Note: [a]. Follows local codes.

Table 3. Calculation of the quality of development for each typology and their *min–max normalized values*.

Metrics		Grid	Loop	Cul-de-sac	Grid and Loop	Loop and Cul-de-sac	Cul-de-sac and Grid
Quality of the development typology	Isovist	7381 *0.97*	7041 *0.27*	6601 *0.00*	7283 *0.89*	6964 *0.11*	7473 *1.00*
	Integration	5943 *1.00*	5124 *0.06*	5139 *0.07*	5617 *0.06*	4818 *0.00*	5927 *1.00*
	Betweenness	941,300 *0.99*	175,119 *0.00*	465,663 *0.17*	827,554 *0.94*	254,803 *0.01*	958,839 *1.00*
Mean normalized value		*0.99*	*0.11*	*0.08*	*0.63*	*0.04*	*1.00*
Rank		2	4	5	3	6	1

Table 4. Calculation of installation costs and expected development impact fee (DIF) for each typology.

Metrics		Grid	Loop	Cul-de-sac	Grid and Loop	Loop and Cul-de-sac	Cul-de-sac and Grid
Installation costs (million KRW)	Road	5050	4146	4296	4245	4126	4810
	Water and sewage	1560	1239	1087	1275	1145	1424
	Land compensation	9603	7491	7962	7948	9106	7693
Total installation cost (million KRW)		16,214	12,876	13,345	13,468	14,377	13,926
Expected Total DIF (million KRW)		4864	3863	4003	4040	4313	4178
Expected DIF per household (million KRW)		10.6	8.1	8.2	8.7	9.9	9.1
Rank		6	1	2	3	5	4

4. Comparing the Two Methods of Infrastructure Installation Planning

4.1. The Traditional Method and Procedural Modeling

Two methods of infrastructure installation planning are possible to implement in the virtual project, as Figure 5 illustrates. One is the traditional method that uses AutoCAD to model the project, depthmapX to assess the development quality, and ArcGIS to calculate the DIF. The other is based on procedural modeling using CityEngine as a substitute for AutoCAD. depthmapX and ArcGIS are used for the same purposes.

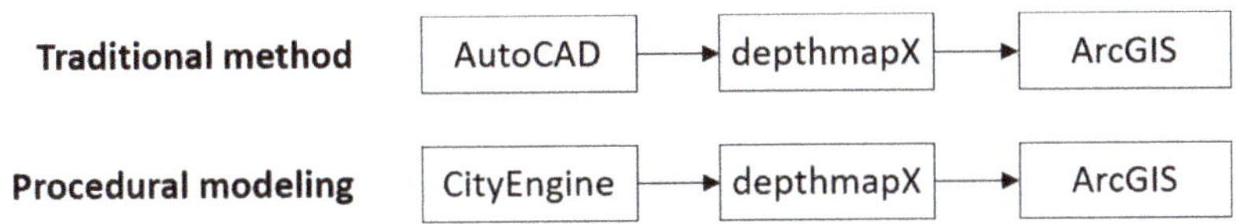

Figure 5. Two methods of infrastructure installation planning.

Figure 6 illustrates in detail the two methods for infrastructure installation planning. The traditional method follows an eight-step process. The first step, data acquisition, involves collecting and cleaning the data, typically for the built environment, zoning, topography, and land prices. The next step is to draft a schematic plan that illustrates a rough image of the development using the data. The third step models the 2-dimensional features of the plan, such as the road networks, blocks, building footprints, and infrastructures, using AutoCAD. The fourth step models the 3-dimensional features, including the topography, buildings, vegetation, and elevations of various infrastructures. This is also done by AutoCAD and provides a glimpse of the pedestrian (or a bird's eye view) of the plan. The fifth decides whether the plan is acceptable based on the planner's expertise and experience. If the plan is considered unacceptable, a new schematic plan is drafted from step two. The sixth step involves pre-processing the model by AutoCAD prior to the spatial analysis. The amount of work in this step depends on the compatibility of the model and the availability of the attribute data. The seventh step carries out a spatial analysis of the plan using depthmapX, a spatial syntax tool, and ArcGIS. The former is applied to measure plan quality and the latter to calculate the installation costs and DIF. The eighth is carried out by an expert planner by determining the plan's feasibility through an assessment of the plan's quality, installation costs, and DIF. A candidate plan is completed after these eight steps. If the plan is perceived to be unfeasible, a new schematic plan is drafted from step two.

Procedural modeling using CityEngine goes through six steps, which is two steps shorter than the traditional method. The first step simultaneously prepares a rule file and spatial analysis tool. The rule file plays a critical role by creating models and providing their appearances in detail. A substantial amount of time and costs may be required to develop coding syntax to generate models and to build assets, such as buildings and road textures. The amount can be significantly reduced when the rule file is reused or imported from external sources. The spatial analysis tool is a set of codes that are customized for procedural modeling; they are created by ModelBuilder in ArcGIS and are interoperable with the rule file. The second step, data acquisition, and the third, schematic plan, are identical to those from the traditional method. The fourth step involves 2- and 3-dimensional modeling done simultaneously in CityEngine. This is combined with a real-time evaluation of plans by an expert planner. The fifth step, spatial analysis using depthmapX and ArcGIS, and the sixth, which determines the project's feasibility by assessing plan quality, installation costs, and DIF, are again the same as those involved in the previous method. The candidate plan is finalized after these six steps.

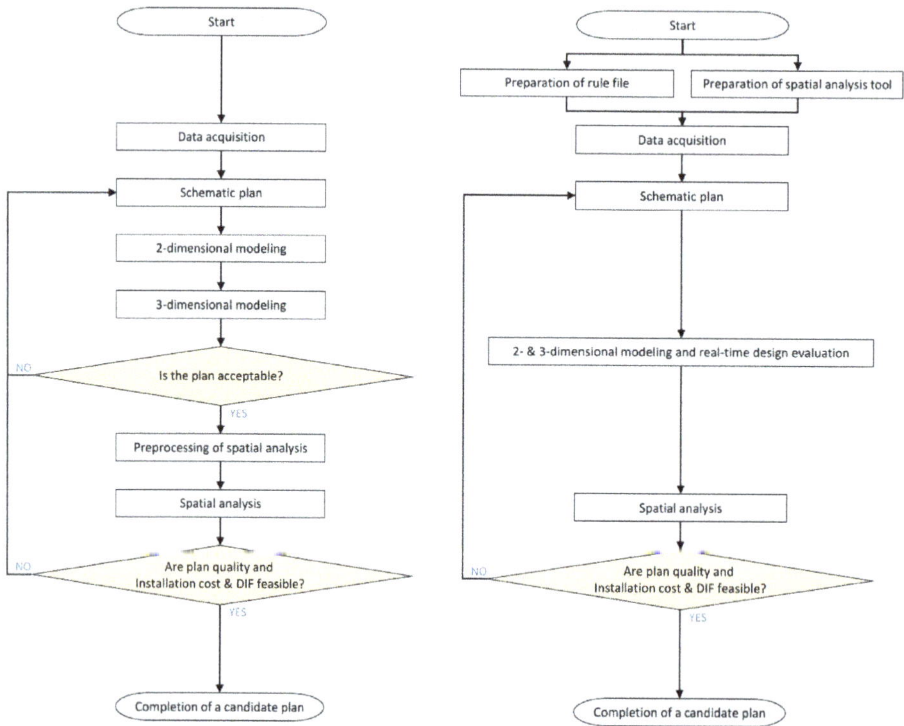

Figure 6. The two methods of infrastructure installation planning in detail: (left) traditional method; and (right) procedural modeling.

4.2. Benefits of Procedural Modeling Using CityEngine

Besides the number of steps required for the two methods of infrastructure installation planning, there are a number of differences that distinguish the two from each other. First, the new method initially requires an additional step for preparation. Second, the modeling process of the two is significantly different; 2- and 3-dimensional modeling, and their initial assessment, in the traditional method are combined into a single step in the new method. Third, the preprocessing of spatial analysis in the traditional method is omitted in the new method, as it is already taken care of in the initial preparation step.

The disadvantages of using procedural modeling can be summarized as follows. First, the time and cost involved in the initial preparation step can be unavoidable, especially when creating a rule file and spatial analysis tool for the first time. Second, advanced and specialized skills may be required to carry out this step. However, the time and cost will gradually diminish when the step is repeated multiple times. Further, the skills can be adopted without a steep learning curve.

There are clear advantages in procedural modeling. Simultaneous modeling of 2- and 3-dimensions enables real-time assessment of the draft plan, thereby considerably reducing the time and cost in the planning process to calculate the DIF. Second, the interoperability between CityEngine, depthmapX, and ArcGIS is significantly enhanced to remove the preprocessing step and avoid data loss during the spatial analysis.

One example of the benefits of procedural modeling is the automated editing of features. As shown in Figure 7, the traditional procedure usually goes through the following eight steps and ultimately requires 307 separate commands when constructing a road that cuts through an existing residential block:

1. Draw a center line to define the origin, destination, orientation, and shape of the new road feature;
2. Secure some buffer space to edit the feature;
3. Develop a rough draft of the feature in the secured space;
4. Complete design of feature details, such as nodes (i.e., intersections) and street corners;
5. Convert the remaining parts of the buffer space to residential use;
6. Dissolve the parts into adjacent residential blocks;
7. Subdivide the updated residential blocks;
8. Recalculate and assign attribute values to road and parcel features.

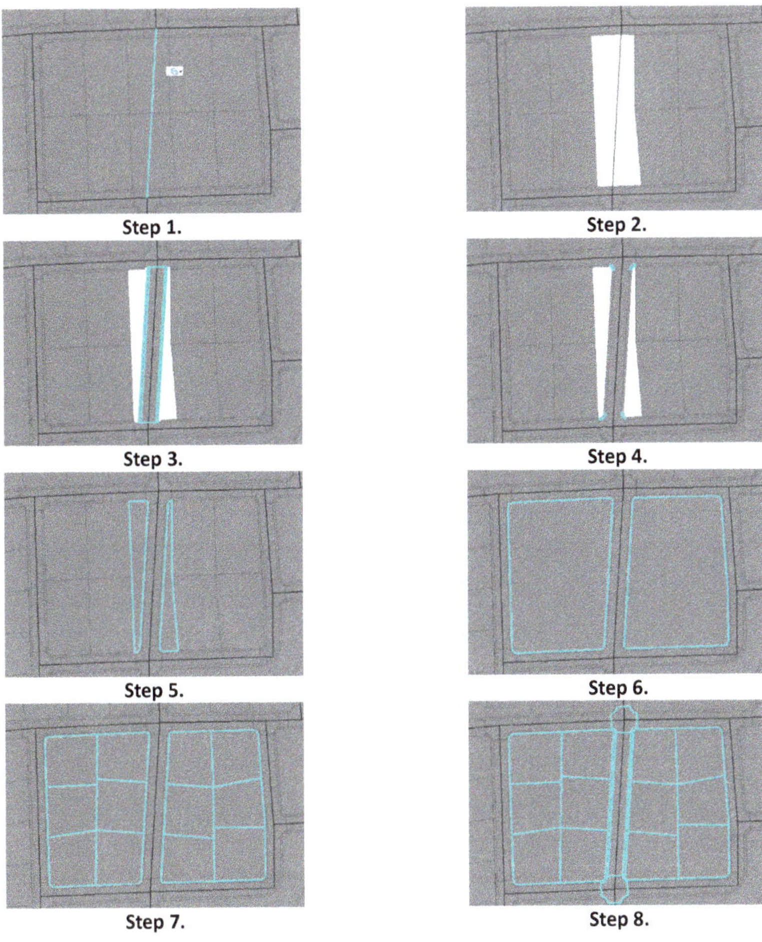

Figure 7. The existing procedure for installing a new road.

CityEngine, on the other hand, requires only the first step of the traditional procedure, which is composed of only two commands. The details and attribute values of the road feature are processed by automatically assuming that the environmental variables are input into the model beforehand. This significantly shortens the modeling process by reducing the number of steps from eight, as previously illustrated, to one, as shown in Figure 8.

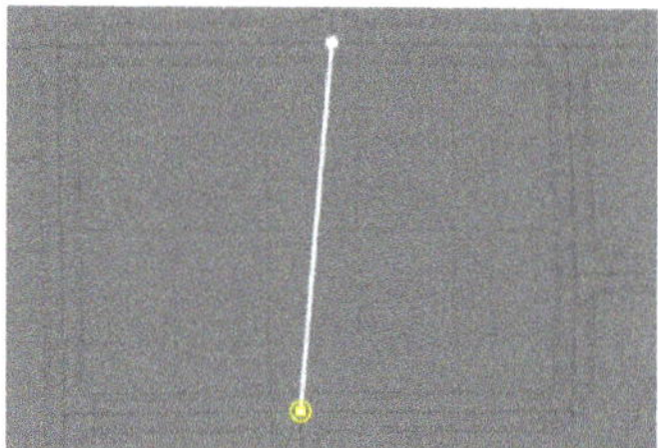 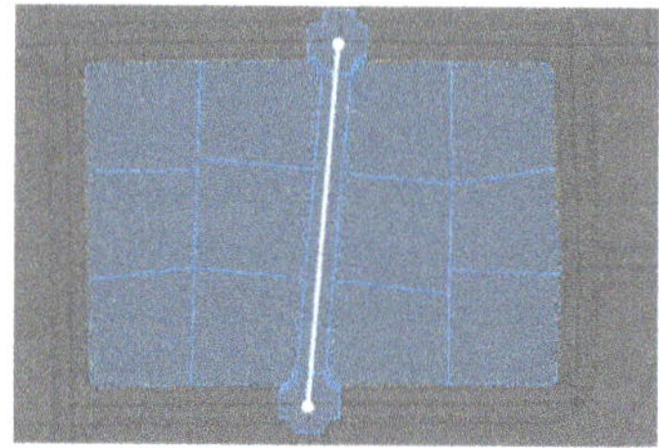

Figure 8. An alternative procedure for installing a new road using CityEngine.

The benefits of using procedural modeling do not stop here. As Figure 9 illustrates, CityEngine creates real-time 3-dimensional models using the default values embedded in a rule file or feature attributes that exist in a 2-dimensional model, when the rule file is applied. Figure 8 itself completes a modification of the 3-dimensional model. This significantly shortens the modeling process, as well as the time and cost required. It also provides a modification of the plan alternatives and offers chances to visually review the results at the same time. On the other hand, the traditional method results in a significant increase in time and cost as it does not allow 2-dimensional modeling, 3-dimensional modeling, and plan reviews take place simultaneously.

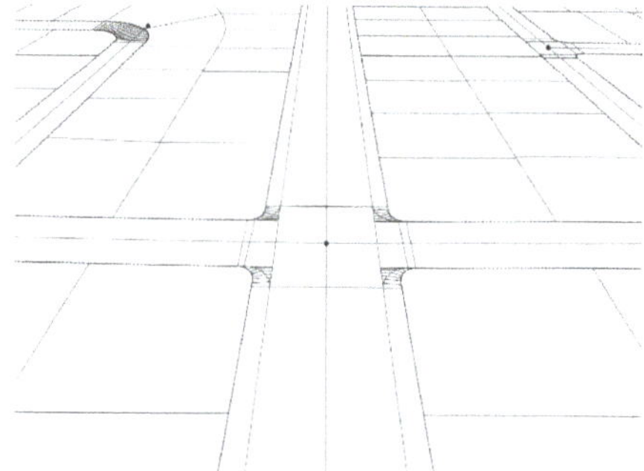 

Figure 9. An example of the automated installation of roads using a rule file in procedural modeling.

The preparation of rule files and spatial analysis tools may make procedural modeling more time and cost heavy than the traditional method. However, the time and cost spent on preparation are usually no more than a one-time expense. Rule files and spatial analysis tools, either created or acquired, can be re-applied to other projects without any additional work. Figure 10 presents this process.

From a long-term perspective, procedural modeling presents more efficient performance than the traditional method, as Figure 11 illustrates. The traditional method may reduce time and cost in the initial stages of the project. However, as more plan alternatives are developed and reviewed, procedural modeling increasingly outperforms the other. The benefits of the traditional method are likely to diminish for the following reasons. First, procedural modeling is capable of drafting 2- and 3-dimensional models and carrying out reviews simultaneously. Second, it does not require the preprocessing of spatial analysis. Third, the time and cost required for preparing rule files and spatial analysis tools do not occur repeatedly.

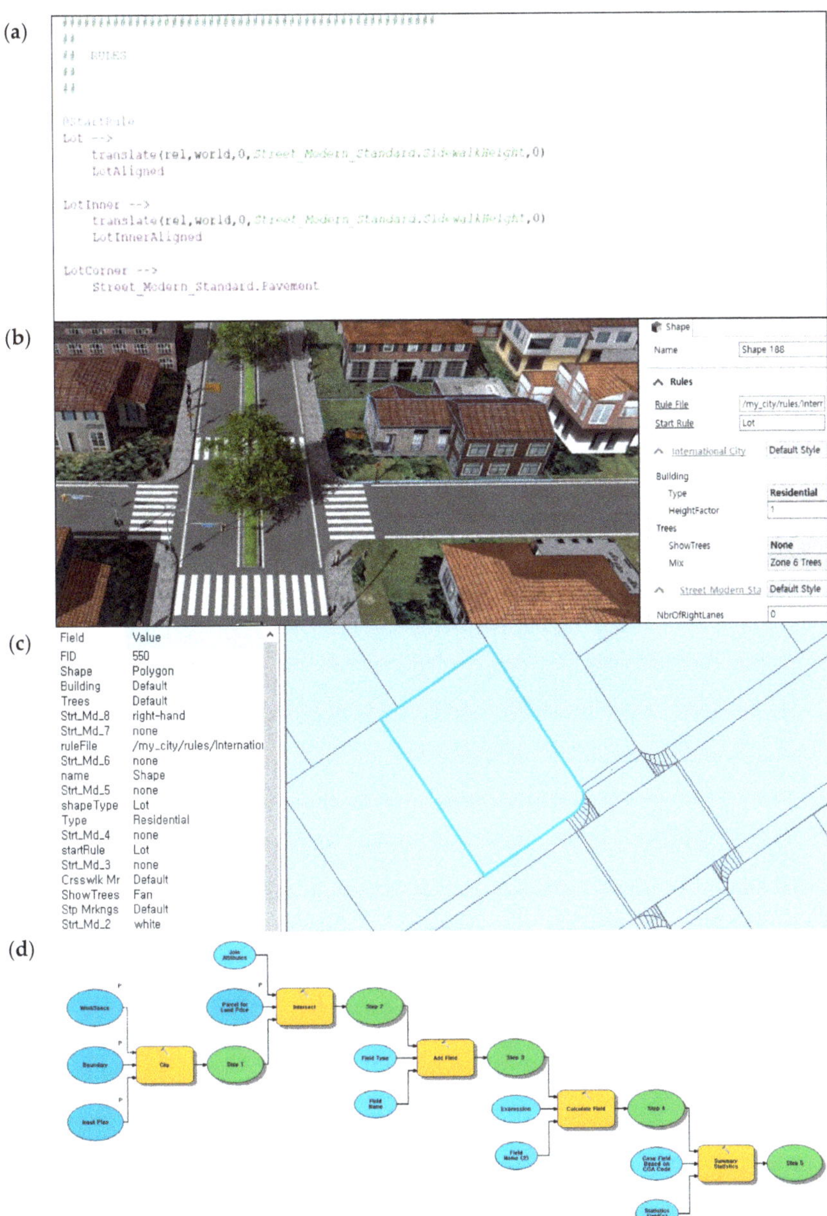

Figure 10. The process of enhancing interoperability using a rule file and spatial analysis tools: (**a**) prepare a variable in the rule file in CityEngine; (**b**) generate a model using the rule file; (**c**) Export the model to ArcGIS; and (**d**) Analyze the cost using a spatial analysis tools in ModelBuilder customized for the variable.

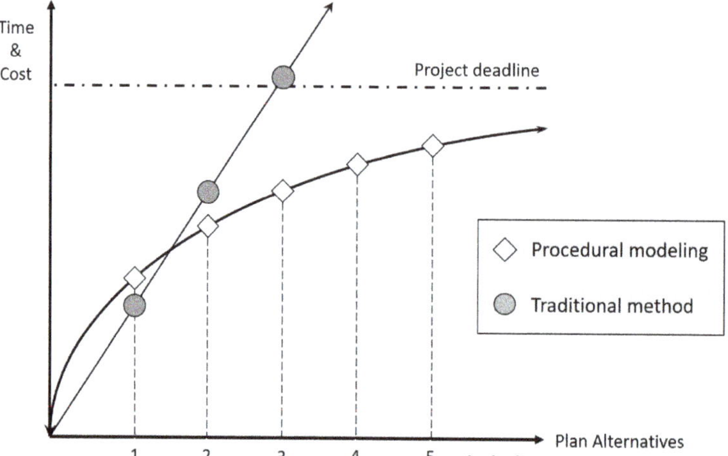

Figure 11. A comparison of the expected time and cost between the traditional method and procedural modeling.

5. Conclusions

The key focus of this study has been to overcome the time and cost constraints in infrastructure installation planning that hamper the efficacy of DIF in South Korea. The traditional method complicates the planning process and makes it difficult to evaluate plans. We propose procedural modeling as an alternative and have demonstrated its relative advantages, especially in enhancing workability by considerably reducing the number of steps and commands for planning practice to achieve the same outcome.

This study is not without its shortcomings. First, it compared two methods for infrastructure installation planning by reviewing each process in a relatively descriptive manner and did not incorporate a pre-defined set of quantifiable parameters, like the amount of time and cost spent on the actual work. Second, the rule file provided by the ESRI of CityEngine, which was developed mostly for a US context, was adopted in this study, instead of a customized rule file that may better account for the circumstances of Jeju. Third, the two methods were applied to a single low-density subdivision case in suburban Jeju and may require further applications in other density settings with varying housing typologies in different regions to produce more reliable outcomes.

However, this study provides several key contributions to the literature. First, this study fills existing research gaps by focusing on the applicability of the DIF in local planning practice and providing tangible solutions. Second, the research findings may be extended to the competitive local public goods equilibrium [70]. Improvements in the planning process for supplying public infrastructure may remove some sources of inefficiency.

Several policy suggestions have emerged from this study. First, financial assistance for the initial costs for preparing rule files and spatial analysis tools may help the diffusion of procedural modeling in local planning practice. Second, government-led training programs and technical support for planners will promote faster adoption of procedural modeling and lighten their burden. Third, procedural modeling may foster more efficient creation of local environmental policies on air pollution and natural open spaces to overcome the side effects of sprawl. Fourth, the findings of this study may inform the installation process of other types of infrastructure that come with new residential development, thereby reducing time and cost. Lastly, the development of manuals that ensure the interoperability of procedural modeling for the calculation of DIFs may significantly increase applicability in local planning practice.

Author Contributions: Conceptualization, N.-Y.C.; methodology, S.K.; formal analysis, S.K.; writing—original draft preparation, N.-Y.C.; writing—review and editing, H.K.; visualization, S.K.; project administration, N.-Y.C.; funding acquisition, H.K. All authors have read and agreed to the published version of the manuscript.

Funding: This work was supported by the Ministry of Education of the Republic of Korea and the National Research Foundation of Korea (NRF-2018S1A3A2075332).

Conflicts of Interest: The authors declare no conflict of interest.

References

1. Nahrin, K. Urban development policies for the provision of utility infrastructure: A case study of Dhaka, Bangladesh. *Util. Policy* **2018**, *54*, 107–114. [CrossRef]
2. Cui, Y.; Sun, Y. Social benefit of urban infrastructure: An empirical analysis of four Chinese autonomous municipalities. *Util. Policy* **2019**, *58*, 16–26. [CrossRef]
3. Levine, J.C. Equity in Infrastructure Finance: When Are Impact Fees Justified? *Land Econ.* **1994**, *70*, 210–222. [CrossRef]
4. Manaugh, K.; Badami, M.G.; El-Geneidy, A.M. Integrating social equity into urban transportation planning: A critical evaluation of equity objectives and measures in transportation plans in North America. *Transp. Policy* **2015**, *37*, 167–176. [CrossRef]
5. Page, S.N.; Ankner, D.W.; Jones, C.; Fetterman, R. The Risks and Rewards of Private Equity in Infrastructure. *Public Works Manag. Policy* **2008**, *13*, 100–113. [CrossRef]
6. Mullen, C. State Impact Fee Enabling Acts. In *Impact Fees: Principles and Practice of Proportionate-Share Development Fees*; Nelson, A.C., Nicholas, J.C., Juergensmeyer, J.C., Eds.; Routledge: New York, NY, USA, 2017.
7. Porter, D.R. *Managing Growth in America's Communities: Second Edition*, 2nd ed.; Island Press: Washington, DC, USA, 2012; ISBN 978-1-59726-610-9.
8. Nelson, A.C. Development impact fees. *J. Am. Plan. Assoc.* **1988**, *54*, 3–6. [CrossRef]
9. Stroud, N. Legal Considerations of Development Impact Fees. *J. Am. Plan. Assoc.* **1988**, *54*, 29–37. [CrossRef]
10. Altshuler, A.A.; Gomez-Ibanez, J.A. *Regulation for Revenue: The Political Economy of Land Use Exactions*; Brookings Institution Press: Washington, DC, USA, 1993; ISBN 978-0-8157-9127-0.
11. Bae, S.-S.; Kwon, S.-W.; Coutts, C.; Park, S.-C.; Feiock, R.C. Development Impact Fees: A Vehicle or Restraint for Land Development? *Lex Localis* **2015**, *13*, 1047–1065. [CrossRef]
12. Blanco, A.G.; Steiner, R.L.; Kim, J.; Chung, H. Effects of Impact Fees on Urban Form and Congestion in Florida. *Transp. Res. Rec.* **2012**, *2297*, 38–46. [CrossRef]
13. Burge, G.S.; Nelson, A.C.; Matthews, J. Effects of proportionate-share impact fees. *Hous. Policy Debate* **2007**, *18*, 679–710. [CrossRef]
14. Burge, G.S.; Ihlanfeldt, K.R. Promoting Sustainable Land Development Patterns through Impact Fee Programs. *Cityscape* **2013**, *15*, 83–105.
15. Delaney, C.J.; Smith, M.T. Pricing Implications of Development Exactions on Existing Housing Stock. *Growth Chang.* **1989**, *20*, 1–12. [CrossRef]
16. Delaney, C.J.; Smith, M.T. Impact Fees and the Price of New Housing: An Empirical Study. *Real Estate Econ.* **1989**, *17*, 41–54. [CrossRef]
17. Downing, P.B.; McCaleb, T.S. *The Economics of Development Exactions*; Planners Press: Chicago, IL, USA, 1987.
18. Huffman, F.E.; Nelson, A.C.; Smith, M.T.; Stegman, M.A. Who Bears the Burden of Development Impact Fees? *J. Am. Plan. Assoc.* **1988**, *54*, 49–55. [CrossRef]
19. Singell, L.D.; Lillydahl, J.H. An Empirical Examination of the Effect of Impact Fees on the Housing Market. *Land Econ.* **1990**, *66*, 82–92. [CrossRef]
20. Snyder, T.; Stegman, M.A.; Moreau, D.H. *Paying for Growth: Using Development Fees to Finance Infrastructure*, 3rd ed.; Urban Land Institute: Washington, DC, USA, 1989; ISBN 978-0-87420-663-0.
21. Burge, G.S.; Trosper, T.L.; Nelson, A.C.; Juergensmeyer, J.C.; Nicholas, J.C. Can Development Impact Fees Help Mitigate Urban Sprawl? *J. Am. Plan. Assoc.* **2013**, *79*, 235–248. [CrossRef]
22. Jepson, E.J. Could Impact Fees Be Used for CO_2 Mitigation? *J. Urban Plan. Dev.* **2011**, *137*, 204–206. [CrossRef]
23. Nicholas, J.C.; Juergensmeyer, J.C. Market Based Approaches to Environmental Preservation: To Environmental Mitigation Fees and Beyond. *Nat. Resour. J.* **2003**, *43*, 837–863.

24. Ahrens, A.; Lyons, S. Changes in Land Cover and Urban Sprawl in Ireland from a Comparative Perspective over 1990–2012. *Land* **2019**, *8*, 16. [CrossRef]

25. Aurambout, J.-P.; Barranco, R.; Lavalle, C. Towards a Simpler Characterization of Urban Sprawl across Urban Areas in Europe. *Land* **2018**, *7*, 33. [CrossRef]

26. DeSalvo, J.S.; Su, Q. The determinants of urban sprawl: Theory and estimation. *Int. J. Urban Sci.* **2019**, *23*, 88–104. [CrossRef]

27. Kim, H.; Kim, S.-N. Shaping suburbia: A comparison of state-led and market-led suburbs in Seoul Metropolitan Area, South Korea. *Urban Des. Int.* **2016**, *21*, 131–150. [CrossRef]

28. Kim, H.; Lee, N.; Kim, S.-N. Suburbia in evolution: Exploring polycentricity and suburban typologies in the Seoul metropolitan area, South Korea. *Land Use Policy* **2018**, *75*, 92–101. [CrossRef]

29. Kim, H.-A.; Park, S.; Kim, H.-J. *A Study on Infrastructure Financing Scheme for Local Public Services: Development Impact Fee System*; Korea Institute of Public Finance: Sejong, Korea, 2004.

30. Ministry of Construction and Transportation. *Management Handbook for Infrastructure Rolling Systems*; Ministry of Construction and Transportation: Seoul, Korea, 2004.

31. Kim, S.-J.; Park, S.-H.; Lee, J.-H. *An Approach on Policy Improvement for Impact Fee Area Considering the Smart Growth*; Korea Research Institute for Human Settlements: Sejong, Korea, 2010.

32. Cheoi, N.-Y. *A Study on Enhancemet and Improvement of Infrastructure Bearing Area System*; Ministry of Land, Infrastructure and Transport: Sejong, Korea, 2013.

33. Cheoi, N.-Y. Determination of the Impact Fee Zone Based on the Grid Analysis of Population Increase. *J. Korean Assoc. Geogr. Inf. Stud.* **2009**, *12*, 74–83.

34. Cheoi, N.-Y. A Grid Analysis to Designate the Zone to Levy the Impact Fee for Infrastructure Provision: The Case of the Industrial Localities. *J. Korean Urban Geogr. Soc.* **2009**, *12*, 65–75.

35. Cheoi, N.-Y. Spatial Designation of Impact Fee Zone based on the Parcel Development Permit Information. *J. Korean Assoc. Geogr. Inf. Stud.* **2009**, *12*, 116–127.

36. Lee, Y.J.; Cheoi, N.-Y. A Method to Use the Land-Use Zoning Information to Extract the DIF Zones. *Korea Soc. Geospat. Inf. Syst.* **2014**, *22*, 89–99.

37. Parish, Y.I.H.; Müller, P. Procedural Modeling of Cities. In Proceedings of the 28th Annual Conference on Computer Graphics and Interactive Techniques; ACM: Los Angeles, CA, USA, 2001; pp. 301–308.

38. Müller, P.; Wonka, P.; Haegler, S.; Ulmer, A.; Van Gool, L. Procedural Modeling of Buildings. In *ACM SIGGRAPH 2006 Papers*; ACM: New York, NY, USA, 2006; pp. 614–623.

39. Stiny, G.; Gips, J. Shape Grammars and the Generative Specification of Painting and Sculpture. In Proceedings of the IFIP Congress, Ljubljana, Yugoslavia, 23–28 August 1971; Volume 2, pp. 1460–1465.

40. Çağdaş, G. A Shape Grammar: The Language of Traditional Turkish Houses. *Environ. Plan. B Plan. Des.* **1996**, *23*, 443–464. [CrossRef]

41. Flemming, U. More Than the Sum of Parts: The Grammar of Queen Anne Houses. *Environ. Plan. B Plan. Des.* **1987**, *14*, 323–350. [CrossRef]

42. Knight, T.W. The Forty-One Steps. *Environ. Plan. B Plan. Des.* **1981**, *8*, 97–114. [CrossRef]

43. Stiny, G.; Mitchell, W.J. The Palladian Grammar. *Environ. Plan. B Plan. Des.* **1978**, *5*, 5–18. [CrossRef]

44. Wang, Y.; Duarte, J.P. Automatic generation and fabrication of designs. *Autom. Constr.* **2002**, *11*, 291–302. [CrossRef]

45. Agarwal, M.; Cagan, J. A Blend of Different Tastes: The Language of Coffeemakers. *Environ. Plan. B Plan. Des.* **1998**, *25*, 205–226. [CrossRef]

46. Pugliese, M.J.; Cagan, J. Capturing a rebel: Modeling the Harley-Davidson brand through a motorcycle shape grammar. *Res. Eng. Des.* **2002**, *13*, 139–156. [CrossRef]

47. Lindenmayer, A. Mathematical models for cellular interactions in development I. Filaments with one-sided inputs. *J. Theor. Biol.* **1968**, *18*, 280–299. [CrossRef]

48. Bruneton, E.; Neyret, F. Real-Time Rendering and Editing of Vector-Based Terrains. *Comput. Graph. Forum* **2008**, *27*, 311–320. [CrossRef]

49. Hnaidi, H.; Guérin, E.; Akkouche, S.; Peytavie, A.; Galin, E. Feature based terrain generation using diffusion equation. *Comput. Graph. Forum* **2010**, *29*, 2179–2186. [CrossRef]

50. Alsweis, M.; Deussen, O. Modeling and Visualization of symmetric and asymmetric plant competition. In Proceedings of the Eurographics, Dublin, Ireland, 29 August–2 September 2005; pp. 83–88.

51. Livny, Y.; Yan, F.; Olson, M.; Chen, B.; Zhang, H.; El-Sana, J. Automatic Reconstruction of Tree Skeletal Structures from Point Clouds. In *ACM SIGGRAPH Asia 2010 Papers*; ACM: New York, NY, USA, 2010; p. 151.

52. Longay, S.; Runions, A.; Boudon, F.; Prusinkiewicz, P. TreeSketch: Interactive Procedural Modeling of Trees on a Tablet. In Proceedings of the International Symposium on Sketch-Based Interfaces and Modeling, Annecy, France, 4–6 June 2012; Eurographics Association: Goslar, Germany, 2012; pp. 107–120.

53. Génevaux, J.-D.; Galin, É.; Guérin, E.; Peytavie, A.; Benes, B. Terrain Generation Using Procedural Models Based on Hydrology. *ACM Trans. Graph.* **2013**, *32*, 143. [CrossRef]

54. Huijser, R.; Dobbe, J.; Bronsvoort, W.F.; Bidarra, R. Procedural Natural Systems for Game Level Design. In Proceedings of the 2010 Brazilian Symposium on Games and Digital Entertainment, Florianopolis, Brazil, 8–10 November 2010; pp. 189–198.

55. Schwarz, M.; Müller, P. Advanced Procedural Modeling of Architecture. *ACM Trans. Graph.* **2015**, *34*, 107. [CrossRef]

56. Smelik, R.M.; Tutenel, T.; Bidarra, R.; Benes, B. A Survey on Procedural Modelling for Virtual Worlds. *Comput. Graph. Forum* **2014**, *33*, 31–50. [CrossRef]

57. Choe, H.; Tai, H.-S.; Jung, Y.; Kim, S.; Yun, S.-J.; Yoo, B.; Kim, S.; Moon, K.; Kang, K.; Kim, D.; et al. *Jeju, the Island of the Commons I*; Jin In Jin: Gwacheon, Korea, 2016.

58. Kwon, Y.; Kim, H.; Yoo, S. Assessment of the conservation value of Munseom area in Jeju Island, South Korea. *Int. J. Sustain. Dev. World Ecol.* **2018**, *25*, 739–746. [CrossRef]

59. Koh, E.-H.; Lee, S.H.; Kaown, D.; Moon, H.S.; Lee, E.; Lee, K.-K.; Kang, B.-R. Impacts of land use change and groundwater management on long-term nitrate-nitrogen and chloride trends in groundwater of Jeju Island, Korea. *Environ. Earth Sci.* **2017**, *76*, 176. [CrossRef]

60. Lee, J.Y.; Yang, W.J.; Jeong, H.J.; Seo, D.J.; Lee, J.C. Distribution and Pollution Assessment of River Sediments Flowing into the Jeju Coast. *J. Korean Soc. Urban Environ.* **2017**, *17*, 409–417.

61. Kam, S.; Paik, B.C.; Kim, K. A Study on the Presence of Perfluorinated Compounds (PFCs) in the Public Sewage Treatment Plants: Case of Jeju Province Sewage Treatment Plants. *J. Korean Soc. Urban Environ.* **2016**, *16*, 35–45.

62. Han, S.; Park, K. The Problem and Solution on the Chinese's Purchase of Jeju -do Real Estate. *Law Policy Rev.* **2015**, *21*, 405–437.

63. Paik, W. Chinese Investment in Foreign Real Estate and its Interactions with the Host State and Society: The Case of Jeju, South Korea. *Pac. Aff.* **2019**, *92*, 49–70. [CrossRef]

64. Yang, S.K.; Jung, W.Y.; Han, W.K.; Chung, I.M. Impact of land-use changes on stream runoff in Jeju Island, Korea. *AJAR* **2012**, *7*, 6097–6109.

65. Song, C. (Jeju Provincial Council, Jeju, South Korea). Interview. 28 March 2019.

66. Southworth, M.; Owens, P.M. The Evolving Metropolis: Studies of Community, Neighborhood, and Street Form at the Urban Edge. *J. Am. Plan. Assoc.* **1993**, *59*, 271–287. [CrossRef]

67. Benedikt, M.L. To Take Hold of Space: Isovists and Isovist Fields. *Environ. Plan. B Plan. Des.* **1979**, *6*, 47–65. [CrossRef]

68. Teklenburg, J.A.F.; Timmermans, H.J.P.; van Wagenberg, A.F. Space Syntax: Standardised Integration Measures and Some Simulations. *Environ. Plan. B Plan. Des.* **1993**, *20*, 347–357. [CrossRef]

69. Freeman, L.C. A Set of Measures of Centrality Based on Betweenness. *Sociometry* **1977**, *40*, 35–41. [CrossRef]

70. Stiglitz, J.E. *The Theory of Local Public Goods Twenty-Five Years after Tiebout: A Perspective*; National Bureau of Economic Research: Cambridge, MA, USA, 1982.

 *land*

Article

Innovating Along the Continuum of Land Rights Recognition: Meridia's "Documentation Packages" for Ghana

Fuseini Waah Salifu [1],*, Zaid Abubakari [2] and Christine Richter [2]

[1]   Ghana Lands Commission, P.O. Box CT 5008, Cantonment, Accra CT5008, Ghana

[2]   Faculty of Geo-information Science and Earth Observation (ITC), University of Twente, P. O. Box 217, 7500AE Enschede, The Netherlands; z.abubakari@utwente.nl (Z.A.); c.richter@utwente.nl (C.R.)

*   Correspondence: wfsalifu@yahoo.com

Received: 7 November 2019; Accepted: 3 December 2019; Published: 9 December 2019

Abstract: Documentation of land rights can ensure tenure security and facilitate smooth land transactions, but in most countries of the global south this has been difficult to achieve. These difficulties are related to the high transaction cost, long transaction times, and procedural rigidity of land registration processes. In response to these problems, innovative approaches of tenure documentation have been conceived at a global level and are being promoted in many countries of the global south. Little is known yet about how such innovative land tenure documentation approaches unfold in various contexts and to what effect. The implementation of innovative approaches is challenging, due to the legal pluralistic nature of land governance and administrative hybridity in many countries of the global south, including the West African region. This qualitative study explores how Meridia, a small for-profit company, develops innovative approaches to register land rights in the form of "documentation packages" within the existing institutional setting of Ghana. In the paper, we describe both the processes of preparing the documentation packages and respective actors involved, as well as the nature of encounters between innovative interventions and existing institutions. Meridia develops specific products in response to both the regional diversity of land tenure, uses, and market demands, as well as in response to the challenges that the institutional context poses to the process of land tenure registration. As such, the case illustrates how innovation evolves in step-by-step fashion through negotiations with existing land institutions. The various documentation packages developed in this manner differ in terms of cost and complexity of preparation, in terms of recognition by customary and statutory institutions, as well as in the usability of the issued certificates and the extent of exchangeability of associated land parcels. Therefore, Meridia's product innovation reflects the continuum of land rights, but it also poses questions for future research regarding the political economy of land tenure certification and regarding the actual uses and benefits of issued certificates.

Keywords: land registration; innovation; customary institutions; statutory institutions; Ghana

---

1. Documenting Land Tenure: Old Needs, New Means

According to Lemmen [1], many developing countries have less than 30% cadastral coverage, which implies that more than 70% of land in most countries is not documented in any formal land register. Similarly, a study by Diop [2] indicated that only 10% of rural lands in Africa is registered and the remaining 90% remains undocumented, which makes it vulnerable to land grabbing and expropriation without adequate compensation. The current methods of data capture are too cumbersome, expensive, inaccessible, complicated, and slow, and they are socio-politically problematic, because they tend to exclude marginalized groups from either getting titles to their farmlands or residential plots [3].

In addition, the conventional methods of tenure documentation do not fit the customary systems of land tenure in many places and may create more trouble than good, including the creation of social inequalities and class differences that did not exist before official registration, which in turn may work to the disadvantage of marginalized groups, such as secondary rights holders, women, and youths [4,5].

Given the problems of conventional methods of land documentation [3], new methods, often described as innovative approaches, are being advocated to enable a shift to the adoption of context-oriented tenure documentation [6]. Innovative approaches for land tenure documentation, summarized under the label of fit-for-purpose (FFP), seek to address the above-mentioned problems of conventional methods of land rights documentation. They seek to replace cumbersome surveying technology and techniques with simpler ones that produce faster results, such as mobile-based digital mapping methods, or automatic feature extraction from remote sensing. They also seek to recognize land rights in a continuum as advocated by UN-HABITAT [6]. In the form of scaled-up variants, fit-for-purpose approaches seek to develop land administration systems that are flexible, participatory, inclusive, affordable, reliable, upgradable, and attainable [7,8]. Innovative approaches are being promoted and implemented in various countries around the globe [9,10], but how and in how far they work out in practice has not yet been studied systematically and with little reference to theory. Barry [11], for instance, calls for the identification of critical success factors in the implementation of FFP approaches to "harmonize the activities of different organizations with different cultures and the way each organization is evaluated in a way that addresses the higher level development planning goals". Many questions invite researchers to contribute both to the evaluation and implementation of tenure documentation and ask the following questions: how do the new approaches for tenure documentation negotiate the existing formal and informal institutions and plural legal frameworks in the course of implementation, how do they bring land tenure documents into official systems of registration, how do they upscale from small-scale projects and initiatives, and how do they affect land governance structures and different groups of people in the longer run?

Specifically, our paper investigates how innovation in land rights documentation actually takes place practically within the plural institutional context of Ghana's land registration. Ghana's land registration primarily aims at ensuring legal certainty for property holders, but the processes of registration cut across both customary and statutory institutions. Currently, various initiatives are underway in Ghana to improve the documentation and certainty of property holdings and tenure rights. One of these is the enactment of the new Land Bill with the aim of consolidating all land laws into one for sustainable land administration. Additionally, in order to facilitate access to land registration services, the Lands Commission under the Land Administration Project (LAP) initiated the establishment of Customary Land Secretariats (CLSs) as an administrative interface for customary authorities and also is in the process of establishing the Client Service Access Units (CSAU) at the regional Lands Commission offices. There are other initiatives being implemented by different organizations in collaboration with statutory actors, customary actors, as well as the communities, whose land rights are being documented. One of these initiatives is promoted by Meridia, a small for-profit organization that has been referred to as an innovative approach to land tenure documentation by Lengoiboni et al [9]. On the basis of fieldwork in Ghana by the first author, our study takes a closer look at the nature of innovation in the case of Meridia and the encounters that ensue between Meridia and the existing customary and statutory institutions during the process of land rights registration as well as the intermediate and longer term outcomes of Meridia's intervention.

After sketching out the land administration scene in Ghana in the following sections, section three describes the methods of data collection and analysis. In section four, we describe the preparation of four documentation packages by Meridia and the kind of challenges that are encountered, due to existing institutional practices and requirements and the contextual dynamics of land tenure and livelihoods. In section five, we discuss the processes as a case of product innovation and reflect on the longer term outcomes of the documentation in terms of recognition and usability of the issued certificates before pointing out future research directions as a conclusion.

2. Ghana's Institutional Framework for Land Administration

Land administration in Ghana is hybrid in nature, cutting across the spectrum of customary and statutory institutions and within the statutory institution, the hybridity cuts across the forms of land registration, title, and deed. The hybridity within the institutions (i.e. the customary and statutory) emerges from the evolved land tenure patterns in Ghana [12,13]. In precolonial Gold Coast (modern day Ghana), customary institutions had control over all land [14,15], but the situation changed during the colonial regime when some land was annexed by the colonial administration for administrative purposes [16]. Following independence and post-independence nation building, the government still acquires land for public purposes. Through these colonial and post-independence developments, two regimes of tenure in Ghana are dominant, customary and state tenure. While customary institutions still hold and control about 80% of land in Ghana [17,18], less than 20% is state land. Article 267(1) of the 1992 Republican Constitution of Ghana mandates that all stool[1] lands in Ghana shall vest in the rightful stool on behalf of, and in trust for the subjects of the stool by customary law and usage. This constitutional provision justifies the role of customary institutions in land administration. Article 258 of the 1992 Republican Constitution establishes the Lands Commission as a statutory body to manage state and vested lands. Besides the Lands Commission, there were other land sector agencies that participated in the land registration process namely, the Land Valuation Board, Survey Department and Land Title Registry. Being statutory agencies, the duties of these agencies are defined. However, the role of customary institutions in land administration was not explicit as per the constitutional provision in Article 267(1), so their involvement in formal land administration was not well streamlined until 2003 when the Ghana Land Administration Project (LAP) made institutional reforms in land administration and established the Customary Land Secretariats (CLSs) to serve as an interface that connects activities of the customary institutions to that of the Lands Commission [19]. The LAP also facilitated the enactment of the Lands Commission Act (Act 767,2008 which turned the four independent land sector agencies into divisions under the new Lands Commission. Currently, land title registration (established by the Land Title Registration Law of 1986) takes place in only the Greater Accra and Ashanti regions of Ghana while the other regions still practice deed registration (established by the Land Registry Act of 1962). The long term plan is to replace the deed registration with title registration.

Current methods of spatial data acquisition in Ghana are field based as required by sections 16 and 17 of the Survey Act, Act 127, 1962 and section 7 (1–2) of the Land Title Regulation 1986 (LI 1341). These regulations require that a permanent beacon (monument) should separate boundaries between two parcels and be maintained in a manner determined by the Director of the Surveys. Connected to the monuments is the rigid observation time for mapping. The Lands Commission has specific required observation times for different land uses. For example after getting a vantage point to pick the coordinates of a boundary, one still has to wait for three minutes for farmlands and 15 minutes for residential properties. Although the land registration laws provide for the registration of diverse land rights, only leaseholds are mostly registered by the Lands Commission due to limitations in administrative capacity, multiple interpretations of land laws and the influences of multiple normative frames [12,13].

---

[1] A stool is a local construct to depict the chieftaincy institution. Section 139 of the Land Title Registration Law, 1986, interprets "Stool" to include "skin" and any person or people having control over skin or community land including family land, as a representative of the particular community. "Stool" is a symbol of authority for chiefs in the southern part of Ghana as such the lands are called "stool lands". "Skin" is a symbol of authority for chiefs in most Northern part of Ghana as such the lands are called "skin lands". Sometime the term "stool" is used to depict both stool and skin.

3. Materials and Methods

3.1. Description of the Study Area

The study was conducted in two regions of Ghana, namely, the Greater Accra Region with Accra as its capital and the Western region of Ghana with Wassa Akropong as the chosen community (see Figure 1). Accra was chosen because most of the national offices are located here, because Accra is the seat of the Government. Meridia's headquarters and the Head Office of the Lands Commission are also located in Accra. The Wassa Akropong farming community was selected as one of the areas where Meridia has implemented their initiatives.

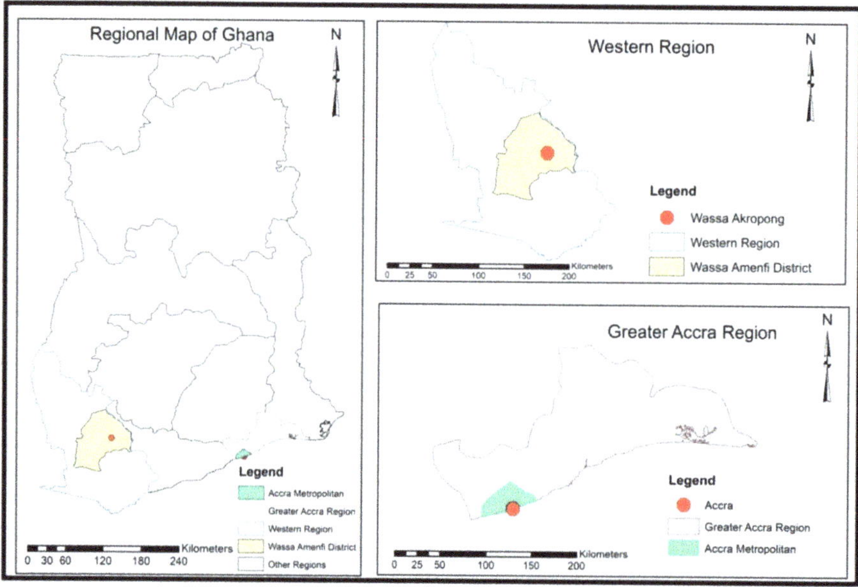

Figure 1. Map of the study area.

Wassa Akropong is a town located in the western region of Ghana, and it is the capital town of Wassa Amenfi East District. According to the 2010 national population census, the town has a current population of 7094. Agriculture is the most dominant economic activity in the area and accounts for about 66 percent of the total population. The Wassa Amenfi East District, where Wassa Akropong is the capital town, is engaged predominantly in cocoa farming and contributes significantly to the overall cocoa production in the country.

3.2. Sampling and Methods of Data Collection

The study employed a non-probability sampling technique dominant in qualitative research, namely purposive sampling and specifically snowballing techniques to obtain primary data for the study using semi-structured interviews and focus group discussions (from October to November, 2017). In purposive sampling units are selected due to their knowledge about and experience with the research topic or case, in direct reference to the research question [20]. To understand the statutory and administrative requirements of land surveying and registration, we interviewed four officials of the Lands Commission. Furthermore, to understand Meridia's operations in land documentation, we interviewed five officials with different ranks and roles within Meridia. To briefly sketch out Meridia's general motivations and approach we draw on an additional interview conducted with Meridia (then called Landmapp) in March 2017 by the third author. Since the processes of land registration in the study areas cut across both statutory and customary institutions, we also interviewed five chiefs and the coordinator of one Customary Land Secretariat to find out the roles they play in land registration and how they deal with Meridia in the registration process. We used purposive sampling to select respondents from the Lands Commission, Meridia, Traditional authorities, and the Customary Land secretariat. The respondents interviewed hold adequate experiences and insight through their long services in their operational institutions. We also organized three focus group discussions with 13 respondents at Wassa Akropong Community, where Meridia implemented their initiative. The participants of the focus group discussions were farmers who had land certificates from Meridia or were in the process of obtaining land certificates. The 13 participants included 5 males who formed one focus group discussion, whilst the 8 females were divided into two separate groups for two focus group discussions consisting of 4 members each. In total, three separate focus group discussions were held at Wassa Akropong to learn about their experiences concerning the implementation of the Meridia initiative. Table 1 below shows the number of people sampled for the purposes of this study and the corresponding data collection methods.

Table 1. Categories of respondents.

Institutions	Total Sampled Size	Data Collection Methods
Lands Commission	4	semi-structured interviews
Meridia	5	semi-structured interviews
Traditional Authorities (Chiefs)	5	semi-structured interviews
Customary Land Secretariats	1	semi-structured interviews
Waasa Akropong Community	13	Focus group Discussion
Total	28	

Aside from the primary data, we also used secondary data which includes the technical instruction of surveyors collected from the Survey and Mapping Division of the Lands Commission. The technical instruction of surveyors contains comprehensive instructions that guide how survey work should be carried out in Ghana.

3.3. Data Analysis and Interpretation

Our analysis consisted of three stages. First, on the basis of the responses from the interviews and focus group discussions, we used process mapping, specifically flowcharting [21], to describe in detail the processes of preparing four so-called documentation packages, which Meridia has designed (see Section 4.1). Then, through thematic analysis [22], the responses obtained from interviews and focus group discussions were sorted into three types of encounters that ensued between Meridia's interventions and the existing institutional arrangements during the processes of registration (see Section 4.2). Finally, we took the analysis to a more interpretive level in Section 5 of the paper. Here, we discuss the nature of Meridia's product innovation and position the certificates issued along three conceptual dimensions of benefit: level of cost/complexity of the registration process, degree of recognition of the certificates, and level of usability of the certificates along with the land parcel's potential exchangeability (see Section 5).

In our study internal validity is established through triangulation. We gathered data from multiple sources, established a sequence of evidence, and analyzed the data across various respondent categories (see Table 1). In the process, data sourced from the different interviewees were cross-checked with others for convergences and divergences. We also used the technical instruction of surveyors to cross-verify the interviews held with the statutory actors as well as the implementers of innovative approaches regarding formal requirements for land surveying. In addition, the first and second authors' experiences of working in Ghana's land administration sector for several years contributed to the internal validity, as this contextual knowledge helped to increase the "level of congruence between concepts and observations" [20] (p. 390). Furthermore, a first draft of this paper was shared with representatives of Meridia for respondent validation [20]; and to ensure the dependability of the study [20], complete records were kept of all phases of the research by means of a data management plan.

4. Results: Meridia's Innovations and Encounters with Land Governance Institutions

This section describes Meridia's innovative approaches to register land. Meridia's initiatives in land tenure documentation take the forms of product and institutional innovations, by which Meridia seeks to meet the needs of landholders and at the same time reasonably negotiate institutional challenges. To achieve the product and institutional innovation, Meridia, among other things, makes use of technological solutions which are embedded in the innovations. Following the first two stages of analysis described above, Section 4.1 below describes the documentation packages designed by Meridia to meet the needs of different categories of landholders and how each requires different processes of registration. Then, Section 4.2. describes the encounters between the existing institutions and the changes sought by Meridia.

4.1. Meridia's General Motivations and Approach

Under its previous name Landmapp, Meridia followed up on land tenure documentation activities by Thomson Reuters in Ghana, a country with an active NGO and consultancy scene and ongoing national and internationally led initiatives to develop land tenure documentation processes. Meridia initially focused on farmers' land tenure security, especially of small holder and cocoa farmers; as the founders of Meridia see farmers as stewards of the environment. The organization later expanded its documentation activities to urban areas. Developing a business model is ongoing, because the process of documentation is influenced by the fees that need to be paid to customary and statutory authorities for signatures and approvals. These fees are not standardized and can vary by region or time. The costs for the training of local mappers and quality checks of the produced data also vary. Besides these variances, the basic premise for the long-term is to be financially self-sustaining and eventually profit-making through the sale of land tenure documentation services to land holders. A customer base is developed step by step following demand pulls. In order to identify this demand and enroll communities into the documentation endeavor, Meridia puts a lot of effort into understanding the socio-economic and political conditions across different regions in Ghana and the local team working on implementation consists of Ghanaian nationals. During initial visits, farmers' willingness to pay for land tenure documents is explored and sometimes down payments are made to indicate commitment to the process. In some cases, several visits are made to communities to gain trust; and before the actual documentation process starts, training and sensitization programs are conducted. In some cases customary authorities, for instances, chiefs have requested for the documentation of specific areas.

4.2. Product Innovation by Meridia: Adjusting the Land Registration Process to Market Demands

Land registration in Ghana typically results in the registration of leaseholds. With this unified form of registration, all categories of landholders, irrespective of location and land use, are required to follow the same process of registration in order to have their holdings recorded. To provide more context-oriented land tenure documentation, Meridia designed a continuum of so-called "documentation packages" through a combination of technological and institutional innovations. These different products provide landholders with the opportunity to engage in different forms of registration ranging from the acknowledgement of customary arrangements to the full registration of a leasehold. Meridia's documentation packages are, in the first instance, designed around different types of land uses. The packages are the FarmSeal, FarmSeal+, HomeSeal, and OrgSeal. The FarmSeal and the FarmSeal+ are more tailored for rural areas where agriculture is the predominant land use with low levels of income among peasant farmers. Given the relatively low costs involved in documenting FarmSeal and FarmSeal+, more peasant farmers are able to have their land rights recorded in some way with possibilities of scaling up to a full leasehold title at a later time. The HomeSeal and the OrgSeal are designed for urban areas which are more cosmopolitan and with relatively higher property values. In these contexts, a higher level of legitimation is often required to validate ownership and transfer, which calls for additional steps at the Lands Commission for official registration. Although in some instances the packages can be adapted according to particular community's or landholders' needs across the urban/rural differentiation, we describe the operations of Meridia in the following first for rural, then for urban areas.

4.2.1. Meridia's Operation in Rural Areas: The Preparation of FarmSeal and FarmSeal+

The steps of Meridia's land tenure documentation process begins with the identification of suitable areas for tenure documentation by Meridia, where Meridia tries to follow demand pulls rather than supply push. Their concept of tenure documentation was first proven in areas with at least some commercial farming, because these farmers can pay for the services. After identifying areas to document, Meridia conducts so called "sensitization activities" with the communities. Meridia informs the community about a date and time to meet for sensitization activities. The sensitization team goes to the community for at least one week to educate the community about the importance of tenure documentation before the actual mapping of land boundaries and rights takes place. After the sensitization is conducted, Meridia interviewers go to farmers' houses to interview them as well as the neighbors to ascertain the oral history of their lands. The initial interview is held with the holder of the land and the grantor and sometimes people from the community. The interviewers obtain some background information concerning how the farmer acquired the land, witnesses that were present during the time of acquisition of the land, and the number of years the farmer had to stay on the land. Later the Meridia mappers also ask the farmers the same questions asked by Meridia interviewers as a means of cross-validation of the information. After the ascertainment of the oral history of the land, the Meridia mappers go to the field with their equipment to carry out the survey. During the survey, the farmers together with a neighbor lead the whole process of defining the boundaries of the farm. They walk around the boundary while the mapper picks the boundary points with GPS. The neighbors are involved in the survey to testify that boundaries, which have been surveyed, are correct and belong to the said owner. The farmers also help in putting the PVC plastic pipes at the place where the coordinates were picked, and concrete is poured into the hole of the PVC pipes to serve as a monument. The farmers also help in the clearing of the boundaries of the farmlands to make it easier for movement during mapping. After the mappers have conducted their mapping, they upload the data into the Meridia integrated end-to-end information system (database linked to mobile application). The Geographic Information Systems (GIS) team in Accra then have access to the data to do computations as well as to eliminate errors and anomalies from the gathered data to produce a farm plan. After the computation is done, the GIS team submits the farm plan and the data to the licensed surveyor for verification and validation. The validated farm plan and deed document are prepared and endorsed by a solicitor and subsequently signed by the traditional authorities (paramount chiefs) and the landholder which are then endorsed by the commissioner of oath at the high court. This marks the end of the documentation for a FarmSeal. For FarmSeal+, the certified farm plan is taken further and submitted to the Regional Lands Surveyor of the Lands Commission for approval. Finally, Meridia sets a date to go to the community to deliver the signed documents. The processes for preparing FarmSeal and FarmSeal+ are shown in Figure 2 below.

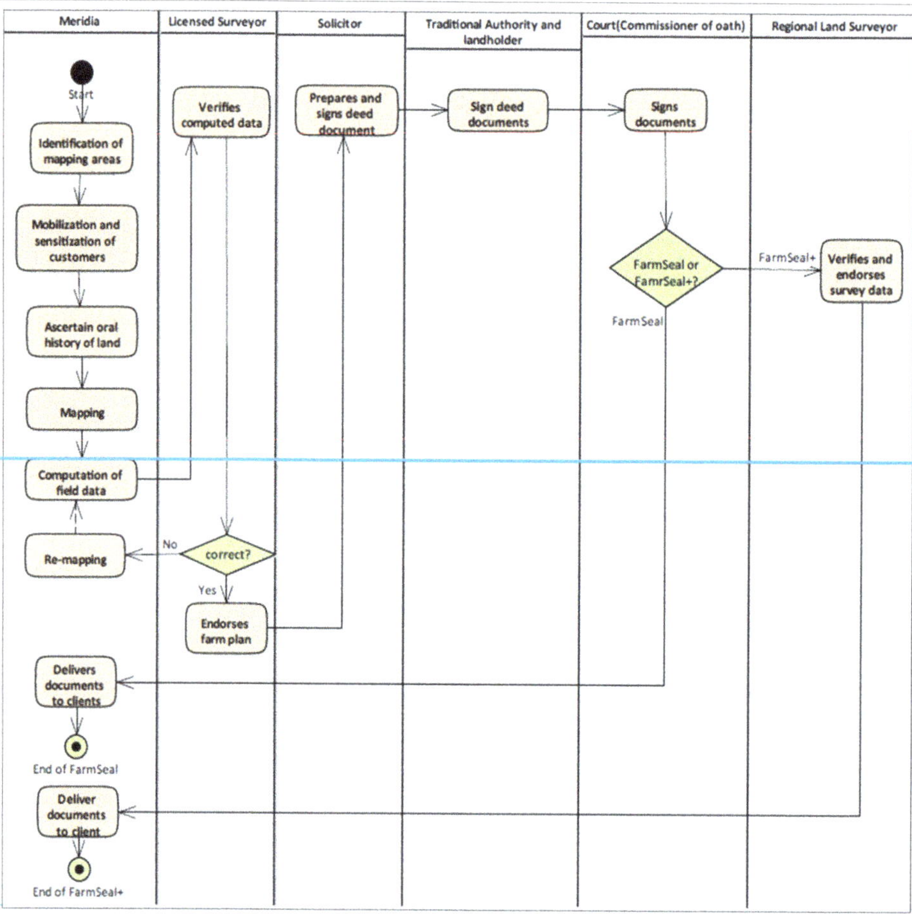

Figure 2. Processes of preparing FarmSeal and FarmSeal+ (constructed on the basis of interviews with Meridia respondents).

The process description above shows that the preparation of the FarmSeal and the FarmSeal+ involves both customary and statutory actors. The chiefs and the commissioner of oath endorse the documents to give it legal backing as well as satisfying land registration requirements set forth by the Lands Commission. The FarmSeal+ constitutes a move into the direction of stronger formal recognition beyond the customary institutional realm in that it also involves approval by the Regional Land Surveyor and certification by the licensed land surveyor as a statutory requirement for further registration with the Lands Commission.

4.2.2. Meridia's Operation in Urban Areas: The Preparation of HomeSeal and OrgSeal

For urban areas, two other documentation packages were designed by Meridia, namely the HomeSeal and the OrgSeal. The HomeSeal itself consists of residential and commercial seals. It is designed to meet the specific conditions of urban areas. It provides home owners and commercial property owners with a certified and approved site plan with a fully signed deed document with an option to continue forward to deed or title registration at the Lands Commission, depending on the client's request. The OrgSeal is similar in this respect, because it may lead to full title registration upon the client's request. While the HomeSeal is for holders of individual land parcels, the OrgSeal is designed for organizations that have large parcels of land and buildings. With the OrgSeal, all parcels and buildings belonging to one organization across the country are mapped and included in an online inventory and processed for title registration.

The documentation process for both the HomeSeal and OrgSeal begins with areas which have already been well planned by the Town and Country Planning Department, the agency responsible for spatial planning. The next step after the preparation of planning schemes is the mobilization and sensitization of potential customers as well as the ascertainment of the oral history of the land in the same manner as described for the FarmSeal and FarmSeal+. After the ascertainment of the oral history of the land, the licensed surveyor applies to the Client Service Access Unit (CSAU) of the Lands Commission for the issuance of regional numbers. These numbers are necessary to identify the survey work to be carried out in that year and area.They are prerequisites needed by surveyors before going to the field. The regional numbers are generated and given to the licensed Surveyor subject to the payment of a fee. After obtaining the regional numbers from the CSAU and related datasets by the licensed surveyor, Meridia's mappers go to the ground to carry out the survey work according to the technical instructions of the surveyors and other relevant laws of the country. The regional numbers are planted on the ground to be able to generate monument numbers for each corner. There are different types of monuments: type A, B, or C, depending on the kind of survey to be conducted. The Meridia mappers usually carry out the mapping using emlid reach GPS devices together with the landowners and neighbors, who help to define the boundaries of the land. After the survey work, the mappers send the data through the Meridia end-to-end integrated information system (database linked to mobile application), which are processed and certified by the licensed surveyor. After the certification of the documents by the licensed surveyor, the documents are submitted to the CSAU of the Lands Commission by the licensed surveyor. The submitted file contains the following: field report or history of the survey, letter of submission, Ex Data (the control points the Licensed Surveyors took from survey and mapping Division), raw field data (rinex format), and point list. The submitted file should also contain the beacon index, computation of bearing and distance, plan data, area computation, a diagram of the survey, total survey record on CD, eight copies of the certified plan and a copy of Land Registration Division's (LRD) request letter. The CSAU verifies the documents against a checklist and either approves or rejects it. After the examination of the content of the file submitted to CSAU, if everything is right, then the file is submitted to the examination section within the survey and Mapping Division of the Lands Commission. The examination section carries out data processing and quality control checks. After the examination of the plan and the documents by the examination section, and given that everything is in good order, the examiner appends his signature and sends the plan for cartographic checks by a department within the Survey and Mapping Division. When they are satisfied with the cartographic aspect of the plan, then the Regional Land Surveyor will append his signature. The plan then comes back to the examination section, where a barcode is placed at the back of the plan indicating that it has been approved. After the plan has been approved, a Solicitor from Meridia prepares a deed document and endorses it. The deed document is then signed by the traditional authorities (paramount chiefs) and the landholder and is also endorsed by the court (commissioner of oath) at the high court. The deed document is submitted to the CSAU, where it is checked for completeness and is sent to the various divisions of the Lands Commission at different stages of processing. Finally, the fully registered deed or title is sent to the CSAU where they are delivered to Meridia who then delivers them to the clients (landholders). These processes are summarized in the flowchart of Figure 3 below.

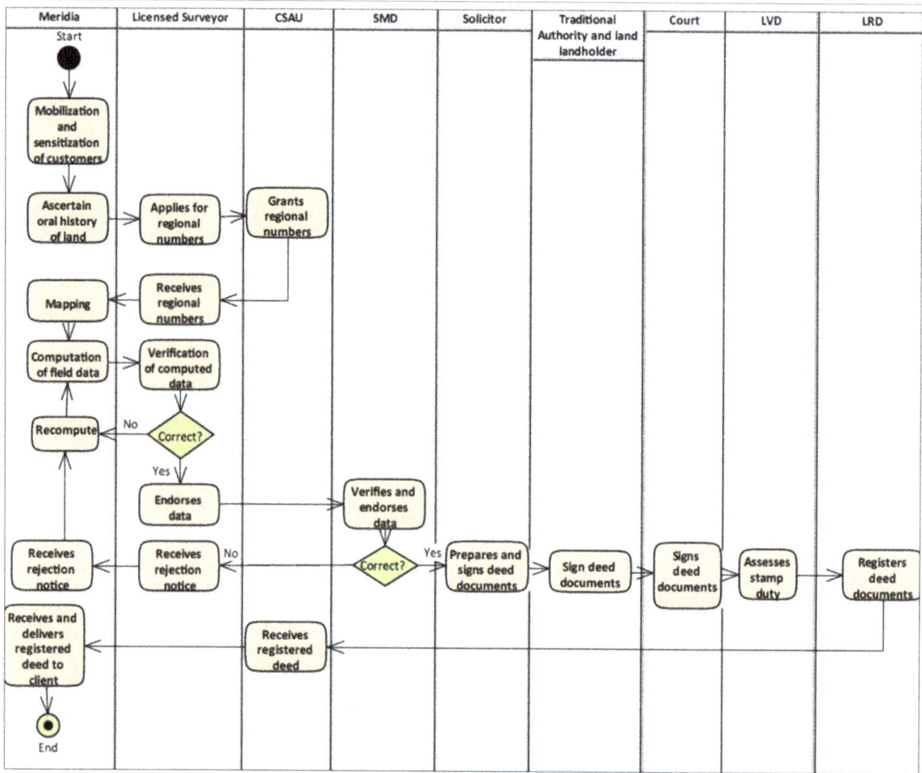

Figure 3. Flowchart of documenting HomeSeal and OrgSeal. This is constructed on the basis of interviews with Meridia. CSAU: Client Service Access Unit. SMD: Survey and Mapping Division. LVD: Land Valuation Division. LRD: Land Registration Division.

Figure 3 and the accompanying narrative above demonstrate that the documentation for HomeSeal and OrgSeal in urban areas involves more statutory actors compared to the preparation of FarmSeal and FarmSeal+ in rural areas. This is partially due to the complexities of land tenure being higher in urban areas, a complexity associated with more land contestations, higher levels of land encroachment, development rates and associated land values, as indicated by Meridia interviewees. The documentation process in the urban areas is therefore relatively more cumbersome, because one needs to contact many stakeholders at higher levels of the administrative hierarchy compared to documentation in rural areas. In addition, statutory actors require stronger adherence to administrative procedures and requirements for surveying and registration work, partially because in urban areas more is at stake with land values and development rates being high and the potential of conflict accordingly higher than in predominantly rural areas with less development and lower land values. Maintaining relationships with these numerous statutory actors in urban areas is time-consuming and financially costly.

In sum, the packages for tenure documentation in rural and urban areas differ in terms of the types of tenure being documented, the labor, costs, type of land holders, and institutional actors involved, but also in terms of aims of documentation and the degree to which a document is recognized by customary only or by both customary and statutory actors. The latter is in turn related to the necessities of adhering to existing surveying standards and requirements that are relatively higher in urban areas for HomeSeal and OrgSeal. The success in the development of these documentation packages as products partly draws from Meridia's innovation (i.e. the use of mobile mappers with an end-to-end integrated information system and use of PVC pipes as monuments). However, in the processes of preparing these various documentation packages, Meridia necessarily encounters existing institutional arrangements (which either aid or impede documentation), and in so doing, Meridia devises institutional innovation to negotiate the challenges of existing institutions. In the next section we discuss these encounters and the institutional innovations devised by Meridia.

4.3. Encounters Between Meridia's Interventions and Existing Institutions

Meridia's innovative intervention in the land registration processes in Ghana diversifies the types of documentation according to different land uses, tenure situations, and landholders' demands and abilities. In so doing, Meridia weaves its initiative through existing statutory and customary institutional arrangements. Meridia develops its documentation processes by exploring the opportunities presented by the existing institutional framework. Such opportunities are afforded specifically by the flexibility of the customary institutions, but also some flexibility on part of statutory institutions, for instance, allowing the mobile mappers for cadastral data collection instead of licensed surveyors and the possibility to use PVC pipes as monuments instead of concrete monuments. At the same time, as Meridia develops its documentation processes, it has to accept and work with some challenges posed by existing institutional arrangements that cannot be changed or easily navigated through an innovative intervention. We identified three forms that the encounters between Meridia and the existing institutions take, namely, encounters with statutory institutional procedures and requirements, encounters with customary institutional practices and requirements, and encounters with the Dynamics of Land and Livelihoods. Each set of encounters is discussed in the following sections.

4.3.1. Encounters with Statutory Institutional Procedures and Requirements

From the point of view of the end product of registration, the Ghana Lands Commission registers only leaseholds in the study areas. By having only leaseholds as a product, landholders have to go through the entire process of a leasehold registration to have their land rights documented. However, Meridia uses a continuum of land rights recognition and recording, whereby different product packages are designed at intermediate stages of land rights recording. This variety of packages which include FarmSeal, FarmSeal+, HomeSeal, and OrgSeal are not only easily accessible to a wide range of landholders but are also scalable to full registration. Meridia's packages are designed to provide flexibility but also to meet the statutory and administrative requirements of registration.

For a land tenure document to achieve full statutory recognition by the Lands Commission, its preparation requires following rather rigid guidelines for the survey work. Especially problematic is the requirement to construct standard monuments for parcel demarcation. According to the technical instructions for the surveyors, these should consist of solid cement concrete 15 cm above ground and with a 30 cm concrete foundation in the ground. The requirement would be impossible to adhere to for Meridia in many rural areas, because some farms have about 70 to 80 boundary points; and monuments are bulky and excruciating to carry. The monuments are also very expensive and add to the cost of surveying and hence making it impossible for farmers to afford documentation. Meridia came up with an improvised solution to overcome this challenge. Meridia, with agreement from the western Regional Lands Commission, uses smaller and lighter PVC pipes for the construction of the monuments for the FarmSeal and FarmSeal+. The PVC pipes are planted into the ground and concrete is poured into the holes of the pipe, and the pillar numbers and other details are inscribed on them. On the other hand, some requirements Meridia cannot circumvent or adjust. For instance, connected to the monuments is the rigid observation time for mapping. The Lands commission has specific required observation times for different land uses. For example, after getting a vantage point to pick the coordinates of a boundary, one still has to wait for three minutes for farmlands and 15 min for residential properties. According to Meridia mappers, "even if you get a good reception, unless the observation time is exhausted, you cannot take a point." Hence, if Meridia has procured a new machine which can pick a point in five seconds, they still cannot use it. Survey instruments are calibrated according to the Lands Commission requirements and for this challenge, Meridia has no solution but has to accept and adapt to the existing situation.

A further challenge encountered by Meridia is the high cost of preparing cadastral plans by surveyors. Meridia improvised ways of cutting down the cost associated with hiring the services of a licensed surveyor by engaging mobile mappers. By law, licensed surveyors and official surveyors are mandated to carry out land surveys in Ghana. Based on their professional training, the fees they charge are very high ($300 per an acre of land) which is beyond the reach of local farmers. To hire a surveyor to carry out survey work, the farmers have to go to the city to look for the surveyor and pay them a daily rate to go to the field to do the work. From the point of hiring a surveyor to the collection of certified survey plan, the farmers incur substantial costs. Meridia is able to cut down the cost of hiring a surveyor drastically by engaging mobile mappers and interviewers. Mobile mappers are part of the Meridia field staff, who are well trained in the use of android tablets with the map of the area for the purposes of mapping the land boundaries. They go to the field to capture the information and then have it cross-verified by the licensed surveyors stationed in the city through Meridia's end-to-end integrated information system (database linked with the mobile application). The cost of Meridia FarmSeal and FarmSeal+ package (site plan and indenture) is about one hundred dollars ($100) for a parcel size of five acres. The Cost of HomeSeal and OrgSeal is about two hundred dollars ($200), which includes a site plan and an indenture.

4.3.2. Encounters with Customary Institutional Practices and Requirements

The land registration processes in Ghana cut across both statutory and customary actors. The processes, although with country-wide and regional variations, involves a series of steps of approvals, certifications, and associated signatures on various documents. When deed documents are prepared, they need to be signed by the grantor chief, for instance. In order to lower the costs of registration, Meridia tries to go for documentation at a scale that is at relatively high volumes of documents. The high volumes of the documents submitted to the chiefs coupled with the numerous customary responsibilities make it extremely difficult for the chiefs to sign the deed documents quickly, which leads to reluctance on part of the chiefs to participate in the process. Here, Meridia came up with a technical solution. Meridia collects the chiefs' signatures and prints them on the documents. The chief's assistant then checks to confirm whether the chief's signature is well embossed. This takes Meridia shorter time to have their documents signed and it saves the chiefs some time.

A challenge emerging from the hybridity of land governance institutions in Ghana (customary and statutory) is the non-recording of some customary land rights such as the customary freehold interest by the Lands Commission in the Western Region due to ambiguity of land laws and related multiple interpretations. The indigenous people in Wassa Akropong are believed to hold customary freehold interest from time immemorial and would not like to sacrifice that interest for a lease, since Meridia land tenure documentation packages are only based on leases. In this case, Meridia has to adjust and adapt to the existing situation by providing customary documentation packages.

4.3.3. Encounters with the Dynamics of Land and Livelihoods

In developing innovative processes to register land rights, Meridia also needs to engage with and adjust to the dynamics outside of statutory and customary land governance institutions per se. These dynamics include the seasonal variations associated with agricultural livelihoods, for instance income fluctuations.

The Wassa Akropong traditional area is a cocoa growing zone where the income of the people varies by season. The main harvesting period for cocoa in Ghana is from October to February and from May to August for the light crop. During the main harvesting period, the farmers have the money to pay for the tenure documentation fees. During the light cropping, the farmers do not have enough money to pay Meridia for their services. Therefore, Meridia collects part payment from those farmers, who cannot afford payment during low income periods. These farmers are then obliged to pay the money during the major cocoa harvesting periods.

But also the distribution of land rights is more dynamic than GPS-based survey logic may anticipate, as farming practices and related use rights vary with natural circumstances and the amount of labor put into clearing and preparing fields. When engaging in mapping activities, it turns out that some of the farmers had reported a different size and boundary during the interviews than what was found later during mapping work. There could be different reasons for these discrepancies. For example, farmers may not know the sizes and boundaries of their lands in acreages. Until recently, the acquisition of land by indigenes was based on one's ability to farm the land. In other words, how much land one farmed and put to productive uses came to constitute one's land size and boundaries. It is the work that makes the boundaries, not the boundaries that allow (give the "right" to) a certain kind of work. Whatever the reasons may be for the mismatch between land sizes as reported during the interviews and the boundaries and sizes measured during the mapping work, the fees to be paid for documentation are charged based on the information provided earlier by the farmer during interview. These discrepancies can therefore lead to substantial problems and delays in the process of documentation. Hence, the way Meridia measures and delineates land using GPS may differ from the logic by which the farmers themselves delineate and measure land. However, there is also a close connection between measurement, reporting of land size, and the negotiations over documentation fees that create a relatively dynamic environment, in which documentation takes place, even in regions that are not characterized by longer term land conflicts or illegal resources extraction activities, which constitute another layer of complex dynamics to land tenure registration.

5. Discussion: Innovating Along a Continuum of Land Rights Recognition

Increasingly, civil society and international organizations in developing countries are making efforts to document land rights, since national governments are slow in capturing and documenting existing land rights. For example, in Zambia, there are about five civil society and private organizations documenting customary land rights in a bottom-up manner [23]. In the case of Ghana, Meridia seeks to fulfil such an augmenting role in land rights documentation. On the basis of our findings in the previous section, Meridia's case allows us to reflect on two elements that are relevant to the current debates to innovate land administration through fit-for-purpose approaches inspired by the continuum of land rights [6]. First, we can reflect on the nature of innovation in the domain of land tenure registration as it presents itself in Meridia's case. Second, we reflect on the documentation packages created and marketed by Meridia according to three dimensions: costs and level of complexity of certification, level of usability of certificate, and the recognition of certificates.

Meridia's product innovation takes place in dialogue with Ghana's existing customary and statutory scene. This process is characterized by incremental learning from the challenges that Meridia and associated actors face and by inventing solutions to these challenges more or less on-the-go. The products developed by Meridia are therefore inventions that become possible only in response to the contingencies of the context. The descriptions of the processes and encounters between innovation and institutions in Section 4 above illustrate that the diffusion and invention of ideas and technologies are always two sides of the same coin [24]. Meridia's is a case of innovation that comes closest to models of "coupling" in Rothwell's models of innovation. According to this model, innovation consists of a sequential process but involves feedback loops and various push/pull combinations with emphasis on the integration of invention, on one hand, and market, on the other [24]. As such, Meridia's interventions appear less disruptive to the existing institutional framework of both the Lands Commission and the traditional authorities. Meridia's activities target only the most problematic aspects of the existing registration processes without setting new operational standards. This approach has somewhat eliminated or reduced the frictions that often ensue between innovation and existing institutions [25]. Research on the adaptation of land administration to the institutional framework of customary tenure in peri-urban Ghana and Sub-Saharan Africa at large highlights the dynamism of customary land tenure institutions [26,27]. According to Arko-Adjei [26] in such a dynamic as well as the complex environment, a non-rigid institutional framework is required to allow innovation in land administration processes to take place. As such Arko Adjei argues for institutional flexibility in the context of both statutory and customary domains to allow innovation to take place. Institutional flexibility denotes the various flexible processes and procedures adopted for land administration [26]. According to Amanor [28], accessibility to land and management of land in a customary land tenure setup is based on negotiations as well as local knowledge, both of which require institutional flexibility. The negotiations also relate to signing fees, or the price for the purchase of land to mention a few. In the case of Ghana, these two authors' position indicates that institutional flexibility is one of the premises for innovation to take place. Meridia's case seems to confirm and illustrate this view. The company's less disruptive strategies are partially dependent on opportunities for flexibility embedded in the existing institutional set-up of statutory and customary tenure in Ghana and also more generally in Sub-Saharan Africa as demonstrated in the study of Sommerville [23] for Zambia. In Section 4, we identify several characteristics that run through the existing statutory and customary institutional scene, which enable innovation to take place. These are mostly related to a degree of flexibility provided by the customary arrangements, for instance the possibility to negotiate signing fees and the presence of an active land governance structure in the study area in form of the Customary Land Secretariats (CLSs) and their mediators. However, also the statutory institutional sphere allows for some flexibility. Monuments are generally used to carry out survey work in Ghana to mark boundary points but are difficult to carry as well as to set up. They are, in this sense, a quite literal and tangible symbol of the rigidity of statutory survey norms. The institutional setting provides some flexibility here regarding the monumentation of farmlands, by allowing Meridia to improvise through the use of PVC pipes as monuments.

The benefits of land rights documentation are many and vary according to different purposes of documentation. Some of the benefits credit access, tenure security, and investment [29,30]. However, our study does not provide empirical insights regarding the uses and benefits of Meridia's documentation packages, because certificate holders only recently obtained the certificates and at the time of fieldwork they had not yet tried to use the certificates to obtain any benefits or services. Some of the perceived uses of Meridia certificates mentioned in focus groups by the Meridia land certificate holders include access to finance, increased legal security, and lastly a source of motivation for farm level investment. However, we may infer potential degree of usability of the certificates from the descriptions of the documentation packages and related process descriptions in Section 4 of this paper. Figure 4 below present a conceptualization of the documentation packages along three axes: amount of cost and complexity involved in preparing the respective certificate, the type of recognition the certificate receives, that is, by customary or both customary and statutory actors, as well as the usability of the certificate and related exchangeability of the associated land parcel. By virtue of the continuum of registration packages, the usability of Meridia's certificates and parcel exchangeability span across a continuum of local to national arenas. For example, when landholders opt for a FarmSeal or FarmSeal+, the potential usability of the certificate in providing security or accessing finance is likely to be higher within the local community than outside of the community, since these certificates are locally verifiable. However, as one moves towards the HomeSeal and OrgSeal, which received also statutory recognition, the usability of the certificates in securing tenure and accessing credit becomes widened, as transacting parties can easily verify the ownership status from the Lands Commission (at the regional level). Similarly, parcel exchangeability is enhanced beyond the community with respect to the HomeSeal and OrgSeal, as compared to the FarmSeal and FarmSeal+. As recognition increases to include recognition of the respective certificate by both customary and statutory institutions, usability of the certificate and exchangeability are likely to increase but so do the costs in terms of finances and labor as well as the complexity of the processes of preparing the package and issuing the respective certificate. The holders of Meridia's certificates show high hopes of benefitting from the documentation in the form of tenure security and credit access from farmer cooperatives and even cocoa buying companies. However, this is yet to be observed in practice.

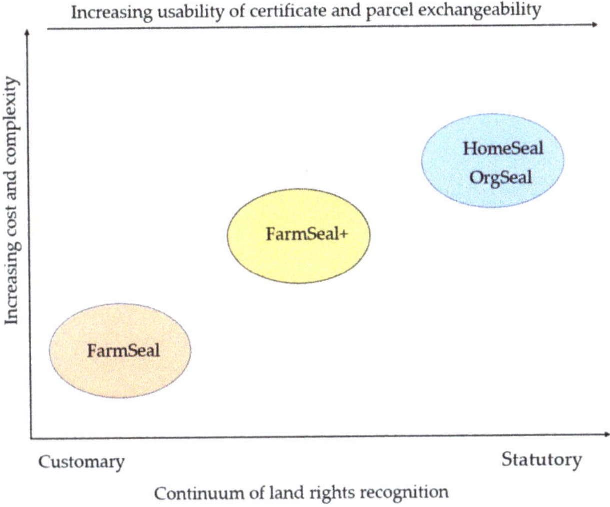

Figure 4. Continuums of cost, usability, transferability, and legitimacy of documentation packages.

At an abstract level then, we can read Meridia's efforts as a kind of product innovation, which diversifies the processes of land tenure registration along the a continuum of land rights, by means of variations in the fee structures and economy of land rights documentation with the two-fold aim of affordability and the company's own financial sustainability.

6. Conclusion and Future Research Directions

Meridia's case shows that at the meeting point of innovation and existing institutions, the latter prompt both in situ and on-the-go solutions through challenges but also enable innovation through institutional flexibilities in both statutory and customary domains. According to Barry's [11] call to identify critical success factors, our study indicates, as one such factor, the identification of "nodes of flexibility" across customary and statutory institutions, which would enable the creation of action nets [31,32] around new ideas and emerging practices. The nature of such nodes of flexibility and how to leverage them in a given case are, of course, context dependent. Meridia found various innovative strategies and solutions of social as well as technical nature [33] to address institutional challenges and constraints, which allow Meridia to distil important elements of the registration processes of the Lands Commission's bureaucratic structure and simplify them. What is also noteworthy in Meridia's case is the indication of a positive correlation between complexity/cost of registration and level of recognition of a certificate across customary and statutory institutions. More research contributing to the implementation of FFP land administration could focus on the kind of trade-offs that are necessary to be made between the affordability and simplicity of registration, on one hand, and the ability to leverage the existing institutional context in such way as to achieve recognition by a broad range of institutional actors and a broad usability of the certificates, on the other.

While innovation generally carries a positive connotation, e.g. increasing performance and effectiveness of organizations [34,35], Taylor [36], for instance, points out that not all innovation is good for everyone and that researchers would do well "to take into account different currents of public resistance to innovation." The challenges of and responses by the existing institutional context therefore also tell us something about the nature of the innovation itself and what kind of innovation is desirable in a given application and development context. New processes and emerging practices of land tenure documentation provide an opportunity for a critical and balanced engagement with the longer term politico-economic effects and implications of such innovation and, as in Meridia's case, provides opportunity for incremental learning and adjustment.

Future empirical research should therefore be conducted on similar processes of implementation at the nexus of innovation and institutional context, also in the form of comparative research. An important criterion for comparison is various financing mechanisms and related to this is the question of sustainability of an initiative. For instance, financing mechanisms of an initiative, especially for-profit, may come to influence land tenure related financial economics through induced changes in fee structures and corresponding beneficiaries. This, in turn, is likely to influence also land holders' political identities, as these are related to the financial mechanisms. For example, depending on the financial set up of land tenure document, a land holder may receive a document either as citizens (in return for taxes paid, for example), as customers of a product, as extended family member, or as the end user of a technology. Each of these roles has different implications on people's relationships to land. Related to this is the need to investigate in how far these initiatives really benefit the intended beneficiaries and/or if the groups who benefit actually change through the process of implementation. Although innovative approaches seem to enhance tenure security, the extent to which land documentation certificates generated through such processes can be used as a legal document is still unclear. The intended beneficiaries of innovation in land documentation need to see and experience benefits of documenting their lands in the long run, also because registration systems work when there are apparent benefits for the intended beneficiaries [37]. Our study does not provide much insight on the specifics of digital data storage and sharing in new initiatives, such as Meridia's, to register land tenure. Given the reliance on digital data technologies in these innovative approaches, questions related to digital data storage need to be addressed as access to storage and uses of data influence power dynamics and actors' roles and agency in land governance. Empirically, and in the longer run, such questions include the following: What happens to the generated data of those initiatives in the long run? Who is currently in possession of and responsible for the generated data of those initiatives and what are the various uses? What are the security issues in relation to data capture and storage?

Addressing these questions in research and evaluation of new initiatives to document land tenure offers the opportunity to chart out ways towards what Winner [38] calls "graceful or benign innovation," which, rather than seeking to be disruptive, is built on the respect for what came before and modifies and renews traditions and practices into fruitful possibilities.

Author Contributions: Conceptualization: F.W.S., Z.A., C.R.; methodology: F.W.S., formal analysis: F.W.S., Z.A., C.R.; investigation: F.W.S.; writing—original draft: F.W.S., review and editing: F.W.S., Z.A., C.R.; supervision: C.R.

Funding: This research received no external funding.

Conflicts of Interest: The authors declare no conflict of interest.

References

1. Lemmen, C. *The Social Tenure Domain Model the Social Tenure Domain Model: A Pro-Poor Land Tool*; FIG Publication 52: Copenhagen, Denmark, 2010.
2. Byamugisha, F.F.K. *Securing Africa's Land for Shared Prosperity: A Program to Scale Up Reforms and Investments*; Africa Development Forum Series 78085: Washington, DC, USA, 2013.
3. Toulmin, C. Securing land and property rights in sub-Saharan Africa: The role of local institutions. *Land Use Policy* **2009**, *26*, 10–19. [CrossRef]
4. Benjamin, S.; Bhuvaneswari, R.; Rajan, P. Bhoomi: 'E-Governance', Or, an anti-Politics Machine Necessary to Globalize Bangalore? CASUM-m Working Paper. 2007.
5. Meinzen-Dick, R.; Mwangi, E. Cutting the web of interests: Pitfalls of formalizing property rights. *Land Use Policy* **2009**, *26*, 36–43. [CrossRef]
6. Secure Land Rights for All-UN-Habitat. Available online: https://mirror.unhabitat.org (accessed on 21 May 2019).

7. Bennett, R.M.; Alemie, B.K. Fit-for-purpose land administration: Lessons from urban and rural Ethiopia. *Surv. Rev.* **2016**, *48*, 11–20. [CrossRef]

8. Enemark, S.; Clifford, B.; Lemmen, C.; McLaren, R. *Fit-For-Purpose Land Administration*; Joint Publication by FIG and World Bank: Copenhagen, Denmark, 2014.

9. Lengoiboni, M.; Richter, C.; Zevenbergen, J. An overview of initiatives to innovate land tenure recordation: 2011 to present. In Proceedings of the World Bank Conference on Land and Poverty, Washington, DC, USA, 19–23 March 2018.

10. McLaren, R.; Fairlie, K.; Kelm, K.; Souza, G.D. New Technology and Emerging Trends: The State of Play for Land Administration. In Proceedings of the World Bank Conference on Land and Poverty, Washington, DC, USA, 19–23 March 2018.

11. Barry, M. Fit-for-purpose land administration—Administration that suits local circumstances or management bumper sticker? Guest Editorial. *Surv. Rev.* **2018**, *50*, 383–385. [CrossRef]

12. Abubakari, Z.; Richter, C.; Zevenbergen, J. Exploring the "implementation gap" in land registration: How it happens that Ghana's official registry contains mainly leaseholds. *Land Use Policy* **2018**, *78*, 539–554. [CrossRef]

13. Abubakari, Z.; Richter, C.; Zevenbergen, J. Plural Inheritance Laws, Practices and Emergent Types of Property—Implications for Updating the Land Register. *Sustainability* **2019**, *11*, 6087. [CrossRef]

14. Cardinal, A.W. *The Natives of the Northern Territories of the Gold Coast: Their Customs, Religion and Folklore*; Routlege: London, UK, 1920.

15. Rattray, R.S. The Tribes of the Ashanti Hinterland: (Some Results of a Two-Years Anthropological Survey of the Northern Territories of the Gold Coast). *J. R. Afr. Soc.* **1931**, *30*, 40–57.

16. Larbi, W.O.; Antwi, A.; Olomolaiye, P. Compulsory land acquisition in Ghana—policy and praxis. *Land Use Policy* **2004**, *21*, 115–127. [CrossRef]

17. Kasanga, K.; Kotey, N. *Land Management in Ghana: Building on Tradition and Modernity*; International Institute for Environment and Development: London, UK, 2001.

18. Abubakari, Z.; van der Molen, P.; Bennett, R.M.; Kuusaana, E.D. Land consolidation, customary lands, and Ghana's Northern Savannah Ecological Zone: An evaluation of the possibilities and pitfalls. *Land Use Policy* **2016**, *54*, 386–398. [CrossRef]

19. Biitir, S.B.; Nara, B.B.; Ameyaw, S. Integrating decentralised land administration systems with traditional land governance institutions in Ghana: Policy and praxis. *Land Use Policy* **2017**, *68*, 402–414. [CrossRef]

20. Bryman, A. *Social Research Methods*, 4th ed.; Oxford University Press Inc.: New York, NY, USA, 2012.

21. Langley, A. Strategies for Theorizing from Process Data. *Acad. Manag. Rev.* **1999**, *24*, 691–710. [CrossRef]

22. Hsieh, H.-F.; Shannon, S.E. Three Approaches to Qualitative Content Analysis. *Qual. Health Res.* **2005**, *15*, 1277–1288. [CrossRef] [PubMed]

23. Sommerville, M.; Bouvier, I.; Chuba, B.; Minango, J. Land Documentation in Zambia: A Comparison of Approaches and Relevance For the National Land Titling Program. In Proceedings of the Responsible Land Governance: Towards an Evidence Based Approached, Washington, DC, USA, 20–24 March 2017; World Bank: Washington, DC, USA, 2017.

24. Godin, B. *Models of Innovation—The History of an Idea*; MIT Press: London, UK, 2017.

25. Govindarajan, V.; Kopalle, P.K. Disruptiveness of innovations: Measurement and an assessment of reliability and validity. *Strateg. Manag. J.* **2006**, *27*, 189–199. [CrossRef]

26. Arko-Adjei, A. *Adapting Land Administration to the Institutional Framework of Customary Tenure: The Case of Peri-Urban Ghana*; University of Twente: Enschede, The Netherlands, 2011.

27. Van Asperen, P. *Evaluation of Innovative Land Tools in Sub-Saharan Africa: Three Cases from a Peri-Urban Context*; IOS Press: Amsterdam, The Netherlands, 2014; ISBN 9781614994442.

28. Amanor, K. The Changing Face of Customary Land Tenure. In *Contesting land and custom in Ghana: State, Chief and the Citizen*; Ubink, J.M., Amanor, K.S., Eds.; Leiden University Press: Leiden, 2008; pp. 55–81.

29. De Soto, H. *The Mystery of Capital: Why Capitalism Triumphs in the West and Fails Everywhere else*; Basic Books: London, UK, 2000; ISBN 0465016154.

30. Feder, G.; Nishio, A. The benefits of land registration and titling: Economic and social perspectives. *Land Use Policy* **1999**, *15*, 25–43. [CrossRef]

31. Czarniawska, B. Organizational Translations. In *Facts and Figures. Economic Representations and Practices*; Kalthoff, H., Rottenburg, R., Wagener, H.-J., Eds.; Metropolis: Marburg, Germany, 2000.

32. Czarniawska, B. On Time, Space, and Action Nets. *Organisation* **2004**, *11*, 773–791. [CrossRef]
33. Bowker, G.C.; Baker, K.; Millerand, F.; Ribes, D. Towards Information Infrastructure Studies: Ways of Knowing in a Networked Environment. In *International Handbook of Internet Research*; Hunsinger, J., Klastrup, L., Allen, M.M., Eds.; Springer: London, UK, 2010.
34. Hult, G.T.; Hurley, R.F.; Knight, G. Innovativeness: Its Antecedents and Impact on Business Performance. *Ind. Mark. Manag.* **2004**, *33*, 429–438. [CrossRef]
35. Verhees, F.J.H.M.; Meulenberg, M.T.G. Market Orientation, Innovativeness, Product Innovation, and Performance in Small Firms. *J. Small Bus. Manag.* **2004**, *42*, 134–154. [CrossRef]
36. Taylor, L. What is data justice? The case for connecting digital rights and freedoms globally. *Big Data Soc.* **2017**, 1–14. [CrossRef]
37. Szreter, S.; Breckenridge, K. Recognition and Registration: The Infrastructure of Personhood in World History. In *Registration and Recognition: Documenting the person in World History*; Oxford University Press: Oxford, 2012; pp. 1–38. ISBN 9780191760402.
38. Winner, L. The Cult of Innovation: Its Myths and Rituals. In *Engineering a Better Future*; Subrahmanian, E., Odumosu, T., Tsao, J.Y., Eds.; Springer: Cham, Switzerland, 2018.

 *land*

Article

Capitalising on the European Research Outcome for Improved Spatial Planning Practices and Territorial Governance

Armands Auziņš

Institute of Civil Engineering and Real Estate Economics, Riga Technical University, Kalnciema street 6—210, LV-1048 Riga, Latvia; armands.auzins@rtu.lv; Tel.: +371-29439004

Received: 24 September 2019; Accepted: 31 October 2019; Published: 1 November 2019

Abstract: If distinguishing between spatial planning systems and practices, the latter reflect on the continuity and perspective of planning cultures and are concerned with the values, attitudes, mindsets and routines shared by those taking part in concrete planning processes. Some recent studies demonstrated comparative assessment of European spatial planning. Thus, the coexistence of continuity and change, as well as convergence and divergence concerning planning practices, was delineated. Moreover, the trends and directions in the evolution of spatial planning and territorial governance were explored when focusing on linkages between diverse national planning perspectives and EU policies. The relevant outcome of European projects met their visionary statements in general and are towards the inspiration of policymaking by territorial evidence. However, it showed a highly differential landscape for territorial governance and spatial planning across Europe in terms of terminology, concepts, tools and practices. Therefore, the paper focuses on how the most relevant outcome of European research may initiate a reasonable in-depth study of concrete planning practices and substantiate an effective planning approach. Mainly based on critical literature review and comparative analysis and synthesis techniques, the overviewed key research results led (1) to agenda-setting for comprehensive evidence gathering (CEG) if exploring spatial planning practices and territorial governance in selected European countries, and (2) to a set of objectives for a values-led planning (VLP) approach to be introduced for improvement of land use management.

Keywords: spatial planning practices; territorial governance; Europe; comprehensive evidence; values-led planning

---

1. Introduction

More than a decade ago, EU aimed to strengthen territorial cohesion, thus gradually encouraged European spatial planning policies (Territorial Agendas (TAs)—TA 2007 and TA 2020), the European Spatial Development Perspective (ESDP) and provided integrated instruments for European spatial development, which have been applied under the European Spatial Planning Observation Network (ESPON) projects [1–3]. Experts in spatial planning of a European Research Group joined for collective work and developed a comparative study on continuity and changes in spatial planning systems and practices of selected European countries [4].

Research towards spatial planning practices in Europe addresses different planning cultures. Some scientific contributions clearly distinguish between planning systems and planning cultures. Reimer and Blotevogel (2012) interpreted *planning systems* as "dynamic institutional technologies, which define corridors of action for planning practice, which may, however, nonetheless display a good deal of variability" [4] (p. 4). While a *planning culture* has sometimes been seen as "equivalent to the values, attitudes, mind-sets and routines shared by those taking part in planning" [5]. The comparative

271

perspective and analysis of planning practices call for studies of spatial planning at the micro level, while national planning systems at the meso level [4] (p. 13). It is concluded that planning practices inherent to the system cannot be drawn from a comparison of legal-administrative framework conditions alone. Therefore, the outcome from the comparative analysis of planning practices (changes in cultures) is essential rather than of planning systems, which are only represented by hierarchies, artefacts and institutional settings.

Based on the studies of Janin Rivolin (2012) and Reimer et al. (2014), the framework of the transformation of spatial planning systems has been proposed [6] (p. 279). At the same time, it was reasoned in the study to apply the designed framework to conceptualize the introduction of a new planning approach and to improve land use and spatial development processes. Spatial planning practices can be identified among five other elements into the framework. Accordingly, it is recognized that spatial planning tools (formal and informal) provide the necessary support to improve planning practices, but positively influenced practices substantiate discourses (e.g., desirable dominating ideas) in spatial planning. To advise adjustments properly, rearranged institutional settings should provide more effective regulations to improve planning practices. However, it is quite obvious that spatial planning practices can be identified and analysed through empirical case studies.

The Commission of the European Communities issued "The EU Compendium of Spatial Planning Systems and Policies" in 1997, which gave some overview about spatial planning systems and traditions of 15 European Member States, as well as enabled the understanding of these systems in operation and identified four "ideal types" [7]. From recent comparative research of Reimer et al. (2014), a "path-dependent evolution" of spatial planning in 12 selected countries can be identified [4]. The continuity and changes in spatial planning systems and practices during the last three decades even show significant shifts from "ideal types" and most of the European countries (i.e., three Baltic countries) were not selected for the mentioned comparative researches.

The objective of the ESPON Cooperation Programme under TA 2020 was to enhance European territorial evidence production through applied research and analyses. Accordingly, the applied research project to contribute to the European territorial and analytical evidence base, through the comparative analysis of territorial governance and spatial planning systems in Europe, has been performed [3]. This comprehensive research covered 39 country cases in total and the results of it have been presented and discussed in the largest world forums of spatial planning (e.g., in AEOSP 2017 and AEOSP 2018). Concerning the outcome of the ESPON applied research project COMPASS, and relevant spatial planning and territorial governance topics that have been critically discussed during international conferences, some high importance research papers have been published as well. Exploring territorial governance and spatial planning systems and trends in European countries in the time period 2000–2016, the study followed an institutional perspective and referred to the concept of socio-institutionalism. Accordingly, *spatial planning* was considered "as the collection of institutions ... "; *spatial planning systems*—defined as "the ensemble of institutions that are used to mediate competition over the use of land and property, to allocate rights of development, to regulate change and to promote preferred spatial and urban form"; and *territorial governance*—comprised "the institutions that assist in active cooperation across government, market and civil society actors to coordinate decision-making and actions that have an impact on the quality of places and their development" [8]. Even if the above-mentioned concepts are evolving and emergent and their definitions have been developed as "working definitions" of the study, those should be perceived and revised critically. However, the study emphasized the relation between spatial planning systems and relevant practices and procedures that might be seen as territorial governance. In this light, *territorial governance* reflects mutual cooperation among key stakeholders and coordination of essential actions in land use management.

If observing the continuity and changes from comparative perspectives, some substantial problems can be emphasized and challenges for the further evolution of spatial planning and territorial governance in Europe discussed. Dimensions and directions of changes are not linear and show multiple trajectories in all observed countries. The principle of sustainable development exists in every planning system, but it appears in planning practice in distinguished contexts. The planning systems are heterogeneous and practices developed differently, which makes it reasonable to choose among spatial planning practices for specific case studies and understand better its perspective from new evidence and place-based knowledge for further improvements into planning.

The research focuses on the main question: What are the preconditions and challenges that we should consider in future to improve spatial planning and the development practice? The purpose of the study is twofold—based on an overview and discussion of significant research outcomes concerning spatial planning and territorial governance in Europe—(1) to set an agenda for comprehensive evidence gathering (CEG), if exploring the spatial planning practice and territorial governance in selected European countries, and (2) to set objectives of a values-led planning (VLP) approach to be introduced for improvement of land use management.

2. Materials and Methods

Most of the relevant European spatial planning research outcomes, policies and instruments since CEC 1997 were reviewed and analysed in the study. Thus, the focus of an unsystematic but critical literature review is on the key research results to identify preconditions and challenges for further methodology development of CEG and discussing the objectives of the VLP approach. This work grounds the knowledge of how the most relevant outcome of European research may initiate a reasonable in-depth study of concrete planning practices and substantiate an effective planning approach.

Comparative analysis and synthesis techniques are employed for collecting information from several studies with different contexts (e.g., country cases, sources, and interpreted results from presented research papers (Section 3.1)). The results of synthesis and the applied logical-constructive method contributed to the development of the framework for case studies (Figure 1 in Section 3.2) as well as to the formulation of particular key questions (Section 3.2) and setting objectives for the VLP approach (Section 3.3). The review of institutional settings promoted (1) the illustration of the implementation of EU policy documents and tools and (2) the explanation of the gap in the study of COMPASS regarding the case of Latvia (Section 3.1). The outcome of previously completed analytical work caused the determination of selection criteria and structuring of case studies as well as the assessment and set of objectives for the VLP approach [6]. It is expected that the results of the study will effectuate empirical research and will sustain decision-making in land use management.

3. Results and Discussion

3.1. European Comparative Perspectives: Lessons Learned from Topical Studies

3.1.1. Presented Scientific Publications and Debated Contributions

The trends and directions in the evolution of spatial planning and territorial governance systems and new typologies in Europe, synergies and/or antagonisms between EU policies and national spatial planning and territorial governance, as well as cross-fertilization between EU cohesion policy and spatial planning practices, have recently been in agendas of European planning communities. These questions were revised and critically discussed at the roundtable *"Trends in European spatial planning systems and linkages between them and EU policies"* [9]. The main outcome of debates on European spatial planning, which is relevant for the promotion of improved planning practice, incorporates conceptual considerations (e.g., the concept of *social innovation*). This concept was debated concerning social change and transformations. A social innovation emerges from a "progressive vision" based on solidarity and reciprocity values. It takes place through the involvement of new constellations of public and private stakeholders, it develops through the creation of new governance arrangements and it spreads within "networks of co-operation between community agents" [9] (pp. 74–75). The research on shaping spaces of interaction for sustainability transitions looks for transformative initiatives that trigger the rise of spaces for the interaction between different stakeholders in an urban environment, which is a key element of participatory city-making. Thus, it connects complexity in planning and transition theory to describe the space of the interaction among different urban stakeholders [9] (pp. 202–206). The authors of the research argue that planning perspective necessitates a more fundamental reflection upon the roles and attitudes of planners, shifting from coordinators or semi-controlled planning processes producing policies and projects towards identifying potential societal challenges and emerging alternatives that can be synthesised, strengthened and empowered to more effectively contribute to desired urban transitions. Besides, it has been emphasised that a stronger role, responsiveness, responsibility and more capable commitment from planners to absorb pressuring issues in planning practice is anticipated [9] (pp. 296–298). When redefining aims and tools in a planner's work and orientation of planning practice, it is important to recognise challenges often conflicting from both: (1) Governance, professional-technical knowledge (efficiency and consensus); and (2) citizenship rights, the ethical dimension of planning knowledge (equity and justice). However, in a conflictive dimension (i.e., obstacles, conflicts, tools, opportunities, capabilities), if opposing it to interests, "the role of values is fundamental". A Brazilian case study [9] (pp. 308–309) follows Deleuze and Guattari philosophical considerations, which take planning as a social process. It provides arguments that the planning process should not be guided by models, ideal visions or prescriptions, but by processes of experimentation, which requires investigating problems, exploring relations between elements "and being open to what might happen if . . . , what differences might emerge". According to the comparative study on spatial planning across European planning systems and social models [9] (pp. 1247–1257) and [10], planning culture framework, key interrelated aspects and approach for methodology development to carry out empirical case analysis can be adapted to introduce an innovation and improve planning practices. Accordingly, when following a *pragmatist approach* in planning, such elements of a planning agenda as (1) social setting, (2) planning process and (3) planning environment should be included and key research questions formulated properly. Observing European planning systems and policies [1,2,7] as well as discussing linkages between European planning systems and policies of EU, collaborative spatial planning did not follow the ESDP or the principles laid out in the TA's.

The main outcome of *AESOP 2018 papers'*, which is relevant for improved spatial planning practices through the promotion of an innovate planning approach, is concerned with the research results of *governance and spatial planning tracks* [11]. Some relevant papers intended for the special session of COMPASS—European and national perspectives. Thus, observing changes and persistence of the German spatial planning system, Reimer and Münter argue that even if the ability to transform institutional patterns of spatial planning in Germany is rather limited and European legislation and policies only marginally influence concrete spatial practices in Germany, some discursive shifts are remarkable and instrumental practices are bound to persistent traditions of acting [11] (p. 132). While observing planning practices in Switzerland and Serbia, Peric provides evidence and advice on how to deal with complex planning problems (e.g., in the brownfield regeneration process). She concludes that the lesson is not in the policy or method but in the practice of comprehending the case accounts and adapting aspects to the demands of new situations [11] (p. 482). De Olde points out the role of urban and rural spaces as a symbolic construct in an urbanization agenda and planning context. His paper casts the urban–rural distinction as a symbolic construct that is part of a planning culture defined as "the result of the accumulated attitudes, values, rules, standards and beliefs shared by the people involved". Through critical discourse analysis, multiple roles of the urban–rural construct in planning are identified [11] (p. 623). Thus, this must be considered when identifying, assessing and discussing the values and attitudes, especially, in the peri-urban context and in designing urban containment. Healey explores a place and its governance, and how citizens are contributing to shaping its future due to initiatives in citizen-based collective actions. He emphasizes the relational dynamics involved through a version of a sociological institutionalist perspective, which focuses on authoritative, allocative and discursive power (rules, resources and ideas), as these play out in specific episodes and come to interact with institutionalised governance practices and broader dynamics of cultural understandings [11] (p. 53). Nadin et al. point out that European spatial planning engages in multiple and sometimes contradictory trajectories at the national, regional and local level. However, their findings identify some common trends in the organisation of spatial planning [11] (p. 652). Here it is important to admit that the first two trends refer to improvements in spatial planning systems, the last two—to sustaining spatial planning practices and all together to the potential for the introduction of the VLP approach.

3.1.2. Applied Research Outcome

The comparative analysis of territorial governance and spatial planning systems in Europe (COMPASS) has been recently completed. This applied research project of ESPON provided an authoritative comparative report on territorial governance and spatial planning systems in Europe [3]. COMPASS covered 39 studies of European countries, including 28 EU member states, four ESPON partner countries and seven candidate and other countries. The project focused on substantial improvement of knowledge based on territorial governance and spatial planning, and in particular, their role in the formulation and implementation of the EU Cohesion Policy [11] (p. 690). The project aimed at answering key research questions: (1) What changes in territorial governance and spatial planning systems and policies can be observed across Europe over the past 15 years? (2) Can these changes be attributed to the influence of macro-level EU directives and policies? (3) What are the best-practices for cross-fertilization of spatial and territorial development policies with the EU Cohesion Policy? (4) How can national/regional spatial and territorial development policy perspectives be better reflected in Cohesion Policy and other policies at the EU scale? Main results and selected recommendations have been presented by G. Cotella at the meeting of the National Cohesion Contact Points in Vienna on 20 September 2018. In general, they meet the visionary statement of COMPASS—they are towards the inspiration of policymaking by territorial evidence. However, the project reports a highly differential landscape for territorial governance and spatial planning in Europe, in terms of terminology, concepts, tools and practices [8]. For instance, after questioning the definition of "spatial planning", this term was interpreted in 24 languages and nearly 100 definitions were recognized as well as, among 32 countries, 255 spatial planning instruments were identified; mostly statutory, with various functions and characters (e.g., multi-purpose tools [11] (p. 627)). The influence of the EU on domestic contexts has been recognised (e.g., affected social learning through laws (structural), policies and funding (instrumental) and experts' knowledge (discursive)). In general, the outcome of the COMPASS study demonstrates relevance to improved planning practice. Thus, some of the recommendations suggest: (1) EU Cohesion Policy as a spatial planning tool, to promote strategic spatial planning approaches, offers also the potential to promote place-based policy development; (2) a higher co-funding rate for place-based actions that fulfil certain criteria (e.g., participatory processes, bottom-up development, explicit reference to spatial planning tools, etc.); (3) make integrated tools to support territorial governance and spatial planning at regional and local levels mandatory; (4) strengthen the capacity building of practitioners (e.g., professional planners, developers, authority representatives and other local stakeholders) by sharing knowledge about the best use of existing, improved and new tools to create added value and potential synergies with other policies, as well as by providing experience and expertise to ensure mutual learning and good practice exchange; (5) develop partnerships under topical objectives, thus partners from different countries and disciplines join to develop and test policy measures and new approaches on how to deal with specific challenges, and find the best possible solutions through cross-sectoral discussions and collaborative work.

The necessity to implement the place-based approach has been emphasized in the applied research at the European scale [2]. *The place-based development policy* was conceptualized in relation to the challenges and expectations of the EU as an agenda for a Reformed Cohesion Policy [12]. This policy was defined as "a long-term development strategy whose objective is to reduce persistent inefficiency and inequality in specific places, through the production of bundles of integrated, place-tailored public goods and services, designed and implemented by eliciting and aggregating *local preferences and knowledge* through participatory political institutions, and by establishing *linkages with other places … "*. The key finding of the applied study pointed out "valuing and reviving territorial identity" as a unique asset and starting point of every place-based initiative. It was also concluded in the study, through the analysed cases, that there is a clear need of changing the mindset of decision-makers, moving from a more administrative and compliance-driven attitude to a more result-oriented and entrepreneurial one in governing territories [2] (pp. 3–4).

Some *developed and provided tools* were suggested to apply for promotion of planning practices and improved territorial governance. The ESPON territorial impact assessment (TIA) tool was designed to assess the territorial impacts of policies and institutions to support efforts to minimize unforeseen negative policy outcomes and maximise territorial potentials. The territorial impact assessment (TIA) is "a method to predict the territorial effects of policies and provide useful insights for both territorial and sector policy-making" [13]. The TIA web tool allows the user to make a "quick and dirty" ex-ante analysis of the potential impact of EU legislation, policies and directives on the development of regions (i.e., NUTS3). The TIA webtool is recognised as a very general model that can help to steer the discussion and cannot replace a thorough assessment of relevant and concrete territorial effects of a policy proposal. Therefore, the question should be addressed to those who care about support to spatial planning and governance at a local territorial level: How could one carry out a territorial impact assessment to improve spatial planning practice when assessing the impact from the implementation of local spatial development strategies, plans and development programs?

Applied research of ESPON on sustainable urbanization and land-use practices in European regions (SUPER) has been carried out since 2019. The SUPER project anticipates an "innovative outcome" through (1) quantitative evidence gathering on land-use developments using latest data sources, (2) qualitative evidence gathering on impacts of interventions to affect urbanisation, (3) empirical evidence-gathering on urbanisation practices carrying out 10 case studies, (4) evidence gathering on possible futures using the land-use modelling technique and (5) engaging stakeholders to ensure project impact by organising workshops and developing a handbook [14]. Accordingly, the conceptual framework of the project shows the acknowledgement of urbanisation and land-use drivers, local practices and outcomes in European regions (i.e., NUTS3), as well as its sustainability assessment. In order to assess the sustainability of development a starting hypothesis was developed to be tested in the course of the project: "compact and denser urban development would lead to less need for transport, less energy use and more open spaces enhancing the quality of the life thus generating benefits and requiring fewer costs—or in other words, enhancing sustainability". Furthermore, a guiding question in this regard was addressed: How and to what extent can territorial governance and spatial planning interventions contribute to more sustainable land use? However, a question rises again: How could one carry out a sustainability assessment to improve spatial planning practice when assessing relevant effects from the implementation of local spatial development strategies, plans and development programs? The approach and aim of the SUPER project seem both quite ambitious and ambiguous when used to address the above question to support a local land management level. The question requires a sustainable land intensification concept to be considered, including such aspects as dynamic changes in land-use patterns (spatial dimension of multi-functionality and synergy aspects), identification and assessment of values (socio-economic, environmental-ecological and institutional dimensions) and stakeholders' preferences of the values. However, it is acknowledged in the case study strategy that only the case studies can provide insight into the local experiences that produce land-use changes in context and the extent to which specific interventions are effective in fostering sustainability (Inception Report of [14], pp. 23–24). From this perspective, both substantive and instrumental questions were developed to have proper responses during the case studies.

3.1.3. Some Critique Outlining the Baltic Perspective and Main Conclusion

Some studies (e.g., COMPASS) include the *planning perspective of Baltic countries*. Adams et al. (2014) characterize it as the one which reflects a "culture of pragmatism", whereby more concrete and specific issues take priority over more abstract and ambiguous ones. At the same time, towards engagement with European spatial planning and the Territorial Cohesion debate, the professional communities of practice have been recognized as institutionally weak and fragmented, local government structures—as fragmented too. The advocacy coalitions identified as quite marginalized and territorial knowledge communities characterized as weak and fragmented [15]. The evolution of the Latvian spatial planning experience was recently explored and discussed in the light of a previously made comparative study [4]. Thus, the key trends and aspects of Latvian policies, spatial planning styles and tools have been presented and, in a more detailed way, discussed when informing about main phases and turning points since 1990 [16]. The author agrees with the arguments provided by N. Adams, that Baltic countries embody a pragmatic view in spatial development planning; However, is critical about an assessment of the domestic spatial planning experience based on study of the spatial planning system (e.g., artefacts, regulations, institutional settings, representation of hierarchies) separately from planning practice, if not considering significant driving forces, which caused qualitative changes in planning practice. *Regional economic planning,* as one of four major traditions of spatial planning in Europe, has been identified already by CEC (1997). In one of the main comparative studies on spatial planning systems in Europe [17], Latvian spatial planning showed to be partly regionally economic. Moreover, in Latvia, a shift towards regionalization in spatial planning competences has been recognized in the COMPASS study, which appeared different from both other Baltic countries [8] (p. 19), [18] (p. 9). After the abolition of administrative regions (26) and thus district planning due to administrative-territorial reform in 2008, planning regions (5) do not provide spatial plans at a regional level. They develop strategies (e.g., spatial development perspective and guidelines, development programs and some plans of thematic character). Already, since the 1990s, there has been strong power at the local governmental (municipal) level in spatial planning; however, inter-municipal cooperation weakened because of changes in regional spatial planning after 2009 in Latvia. Therefore, at least in Latvia, it is difficult to find convincing arguments towards regionalization of the spatial planning agenda. Latvia, like most of the other European countries, pursue regional policy objectives and regional development strategies, but the characteristic of the regional economy is not and has never been evident in spatial planning. The spatial planning system and practice developed gradually with significant changes in institutional settings in 2004 and 2011. The "Spatial Development Planning Law" (2011) determined new institutional settings for the spatial planning agenda and aimed qualitative changes into spatial planning practice. Since then the shift towards a strategic spatial planning approach can be argued, as all three planning levels (national, regional and local) have strategies. Since administrative-territorial reform (2009), physical planning with legally binding parts of the local government plan has been practiced. The planning style can be characterized as decentralized, integrated and comprehensive spatial planning with a tendency of centralization to recognise the priorities at the national and regional scales [16]. However, the spatial planning of five planning regions is of strategical and guiding character that include spatial development perspectives. Thus, the regional level of spatial planning in Latvia comprises a strategical approach, whereas regional development contributes mainly with a statistical approach. Weak cooperation among stakeholders exists and public activity and participation increase slowly. The further shift towards collaborative and consensus-oriented spatial planning may be seen as a big challenge for stakeholders in Latvia.

Concluding about the outcomes of the topical European comparative studies and provided arguments, it seems quite obvious that focus has been directed more to spatial planning systems than practices, as well as to more general scales than local planning experience. Summarizing on the above review and analytical research towards a better understanding of spatial planning and territorial governance in Europe, it is reasonable to continue with methodology building for case studies and structured interviews to explore the spatial planning practices more specifically. It will ground the setting of objectives and the introduction of the VLP approach for improved land use and spatial development processes.

3.2. Empirical Research: Towards Comprehensive Evidence Gathering

Following an institutional perspective, here references to the concept of socio-institutionalism or "sociological institutionalist perspective" [19], that emphasize the complex interplay of governance episodes, processes and cultural (place-specific; cf. [20]) assumptions guiding planning and urban development, are made. The relational dynamics involved through a version of a sociological institutionalist perspective focuses on governing rules and available resources, and dominating ideas like these "play out in specific episodes and come to interact with institutionalized governance practices and broader dynamics of cultural understandings" [11] (Healey: p. 53). Synthesizing from the key study outcome gathered during analytical research, the main characteristics of the planning culture approach were taken as a reference to develop the methodology for empirical research through case studies and to create upon key research questions for qualitative research through semi-structured interviews.

To address the research aims of the VLP approach, it is necessary to perform not only analytical research when examining a range of scientific literature and documentary sources, but also to involve actors engaged in relevant policy processes. Invited competent experts in the spatial planning/land use management field may provide opinions about local governance and relevant processes for CEG. In general, the participants can; therefore, be considered to be versed in the technical language relating to spatial planning/land use management and territorial governance at the local municipal level. Accordingly, it is considered that a CEG based on sufficient participation and targeted challenges and opportunities will contribute to the introduction of a VLP approach into practice when linking scientific achievements with the most feasible practical solutions. CEG is mainly based on the identification of barriers, bottlenecks, good planning implementation practices, values and preferences, governance and collaboration forms, etc. Empirical research is carried out by making case studies at selected and differently-experienced (historical evolution, traditions, institutions and development level) countries. The specific governing administration as a part of a chosen country represents the experience of one of four "ideal types" of spatial planning systems [7], and during the last decades have faced relevant changes in spatial planning practices and territorial governance (discourses/traditions), which is essential to the research context. It is expected that the knowledge of competent experts represents the dominating opinion of local society (stakeholders) to a considerable extent and gives some discursive influence on research, as he/she is well informed about relevant spatial planning/land use management processes.

The *framework* for performing cases studies is proposed and key research questions accordingly are developed (see in Figure 1) if considering the following assumptions:

- A "pragmatic view" reflects on planning culture through the prism of interrelated aspects: (1) Social setting, (2) planning process and (3) planning environment [9] (Peric and Hoch: p. 1250);
- The study of multiple trajectories of European spatial planning points out four common trends in the organisation of spatial planning: (1) Simplification of administrative structures, (2) attempts to integrate planning with other policy sectors, (3) strengthening implementation of plans and (4) engaging more effectively with citizens [11] (Nadin et al: p. 652);
- The logic of the framework of the transformation of spatial planning systems is constructed in the way to provide the guidance when "structures" define "tools" to support "practices" [6] (p. 279).

TOWARDS A VALUES-LED PLANNING APPROACH

ORGANISATION and INVOLVEMENT	**PROCESS and TOOLS**	**ENVIRONMENT and SHARED VALUES**
Administrative structures, policy styles, institutional and social settings, collective actions, social learning	Deliberative plan making, planning modes, formal and informal planning tools, project-oriented techniques	Cultural awareness of stakeholders in planning, shared assumptions, values and preferences of involved parties

Figure 1. The framework for case studies.

If considering the above framework in Figure 1, during the discourse of case studies and further analysis, three *key questions* were formulated. (1) Who and under what circumstances organise planning and are involved in it? (2) What are the peculiarities of the planning process and how is the planning practice supported and improved? (3) What is the environment and how well does the planning absorb the intensions and encourage the actors whose preferences and actions may influence future outcomes? Thus, the first question is concerned more with territorial governance issues and possible arrangements, the second—with procedures, modes, planning tools and techniques, and the third—with a reasonable qualitative assessment of the planning environment and implementation of plans. More specific questions are formulated in the *"Draft schedule for semi-structured interview"* and discussed during the interviews and possible on-site visits, thus the core of the methodological tool for empirical analysis can be recognized.

3.3. A Values-Led Planning Approach: Setting Objectives for Improved Practices

The topicality of the VLP approach to be substantiated and implemented into land use management practice was developed upon an assumption that the creation of positive synergy in managing land-related resources if exploring the territorial capabilities, threats and opportunities (e.g., the effects of urban expansion, multi-functionality of land use, internalization of negative externalities and challenges of a city agglomeration) causes primary necessity for the modern society. Previous studies [6,16] contributed to the conceptual background and feasibility aspects of the VLP approach to be introduced into practice by capitalizing, first of all, on comparative analysis of dynamic spatial planning systems and planning cultures. The recent evolution of planning cultures, its substantial changes during the last twenty years and prospective continuation quite clearly argue towards the VLP approach to be developed and implemented to improve spatial planning as an essential and integrated part of land management. The role of values is fundamental and their assessments and acceptance contribute to reasonability and sustainability considerations when applying the VLP approach. Thus, the domain of the VLP approach is found in ascertaining and acknowledging the values according to their typology and conceptualized participation.

It has been concluded that it is necessary to improve the relevant practice and assess its effects in specific territories based on identified, mainly place-based values and attitudes of primary local stakeholders. Establishing a scientifically-sound framework and providing methodological support will promote not only the internalization of negative externalities, but also enable identification of the synergy that would enhance the balanced socio-economic and environmental impact and improve the governance in the territory. Relevant processes (e.g., formal and informal spatial planning, local development, protection of valuable landscapes and related consequent decision-making) strive for collaborative learning by understanding the values of land-related resources and their most efficient usage. Spatial/land use planners as skilled and capable enough professionals in their positions will face new challenges and need to act as competent advisers to stakeholders. Planning activities without focusing on the planning–implementation relationship should be seen as unprofessionally guided. A discourse towards a consensus-oriented planning style will promote the development and management of sustainable communities when it focuses on win-to-win solutions in planning practices.

A VLP approach contributes, along with the "evaluation and planning–implementation concept" and consequent principles, towards balancing the foremost interests of nature/landscape protection and new development. Theoretically, two main principles provide the grounds for the improved spatial planning practice and value capture: (1) Make the best possible and acceptable use of land and (2) share the profit of land. To reveal these principles, some instruments are effectuated. Usually, the profit may be understood as the ratio of benefits-costs after a particular analysis, but the value should be captured for absorbing value increase and recovering development costs, for instance. The reasonability of the VLP approach is found in dynamics and potential changes in land values and its use. From the spatial planning point of view, if current conditions in a particular territory are found already to be the best possible and they satisfy all stakeholders, then the VLP approach is insignificant, but, if a potential to change anything in the territory is found (e.g., a new development or its restriction, or protection of landscape/land-related resources from external impacts) and the initiative from stakeholders appears, then the VLP approach is essential to apply.

Nowadays *sustainability aspects* should be attributed to challenges and issues towards an intensification of land use—how to manage the growing pressure of human needs (e.g., food, resource exploitation, well-being), while at the same time minimizing the impact on the environment (e.g., ecosystems liveability, resource renewability, biodiversity)? "Sustainable intensification" [21] is a suggested but vague term that needs to be clarified through land-use policies; however, it can be applied to meet the mentioned quite challenging issues. If considering global tendencies, the context of sustainable development is the same as realized since Brundtland's report in 1987; however, the focus and content required nowadays is different. Accordingly, how significantly human needs have to be diminished or changed in order for the impact on the environment and land-related resources to be the smallest possible. Practically, the potential for further spatial development should be assessed and then supported by binding decisions. *Decision-making* needs to be backed with facts, actual data and analysis through empirical evidence (i.e., "measuring and evaluating" as an essential outcome).

The VLP approach requires the organisation of expertise for determining the values and identifying preferences through participatory actions and the consensus-building platform of stakeholders. Therefore, the framework for guiding particular processes should be developed with the main focus on the combination of both values and preferences, if considering their dynamic changes over time. The implementation of the VLP approach needs holistic design and methodology. To understand the general design, the framework of the transformation of spatial planning systems is appropriate to apply [6]. This framework is proposed as it conceptualizes the introduction of the VPL approach and focuses on improved spatial planning practice through (1) organisational formations (governing structures), who provide tools for support and guidance; and (2) properly rearranged and acknowledged institutional settings, which provide more effective regulations. The framework implies also the potential for improving territorial governance.

4. Conclusions

In the light of overviewed European comparative research and analysis of its outcome, it has been found to be reasonable to develop a methodology for CEG to explore the spatial planning practices more specifically. It is concluded that the analytical work and performance of CEG substantiate the setting of objectives for the introduction of the VLP approach.

The research aim for case studies is to discuss and deliberate new knowledge about possibilities to improve the spatial planning practice and territorial governance and thus land use management in general. The framework for case studies to carry out CEG is concerned with three main objectives:

1. To examine the organisation of the planning process and involvement of stakeholders (administrative structures, policy styles, institutional and social settings, collective actions and social learning);

2.   To explore the peculiarities of the planning process and how the planning practice is supported and improved (deliberative plan making, planning modes, formal and informal planning tools, project-oriented techniques);

3.   To examine the planning environment and shared values of the actors whose preferences and actions may influence further outcomes (cultural awareness of stakeholders in planning, shared assumptions, values and preferences of involved parties).

The main objectives for the VLP approach are:

1.   Improved, more supportive and collaborative territorial governance, informal institutions and organisational forms as they significantly support formal spatial planning, social settings driven by common and local, place-based interests;

2.   Ensured spatial planning–implementation relationship, softer, more flexible and complementing planning modes, formal and informal planning tools, project-oriented techniques and integrated assessment instruments;

3.   Balanced planning interests, towards meeting supply and demand in planning, increasing of cultural awareness, shared perception and assumption of values and preferences.

To discuss and promote the introduction of the VLP approach into the planning practice, the objectives should be structured when characterising not only the objectives alone, but also indicate their rationale and tools, which are recommended to apply.

If considering the dynamics in the evolution of spatial planning (e.g., floating discourses, shaping administrative structures, inspiring actors of change) as well as driving forces (e.g., reforms, crises, "Europeanisation"), which influence changes in planning cultures, it is suggested to have some mechanism for systematic assessment of:

1.   Territorial governance (ascertaining the movement between both command/control and consensus-oriented models);

2.   Planning–implementation linkage (ascertaining the movement from just formal institutionalised planning mode towards complementing informal planning arrangements);

3.   Planning environment and shared values (ascertaining the movement between both supply-led planning and demand-led planning styles).

It is hypothesized here that such an assessment will allow for understanding of the impact and integration of national, regional and sectoral policies and priorities into the local spatial planning agenda when meeting the interests of local stakeholders (e.g., housing policies, transport networks and natural protection of coastlines).

Finally, it is considered, for the future, that key stakeholders, including official authorities, landowners, developers, partnerships, advisers and enterprises, have to be sufficiently involved or at least their opinion represented by experts when discussing the guiding methodological solutions and specific tools. Having and analysing timely feedback would demonstrate more demand-driven innovation with sufficient participation and target challenges, and opportunities to introduce a VLP approach into practice when linking scientific achievements with the most feasible practical solutions.

Funding: This work was supported by the European Regional Development Fund within the Activity 1.1.1.2 "Post-doctoral Research Aid" of the Specific Aid Objective 1.1.1 "To increase the research and innovative capacity of scientific institutions of Latvia and the ability to attract external financing, investing in human resources and infrastructure" of the Operational Programme "Growth and Employment" No. 1.1.1.2/VIAA/1/16/161.

Conflicts of Interest: The author declares no conflict of interest. The funder had no role in the design of the study; in the collection, analyses, or interpretation of data; in the writing of the manuscript, or in the decision to publish the results.

References

1. Territorial Agenda of the European Union. *Towards More Competitive and Sustainable Europe of Diverse Regions*; European Commission: Leipzig, Germany, 2007; Available online: http://ec.europa.eu/regional_policy/sources/policy/what/territorial-cohesion/territorial_agenda_leipzig2007.pdf (accessed on 19 December 2017).
2. *Territorial Agenda 2020 Put in Practice. Enhancing the Efficiency and Effectiveness of Cohesion Policy by Place-Based Approach*; Vol.1—Synthesis Report. Regional and Urban Policy; CSIL, European Commission: Brussels, Belgium, 2015; Available online: https://ec.europa.eu/regional_policy/sources/policy/what/territorial-cohesion/territorial_agenda_2020_practice_report.pdf (accessed on 2 August 2019).
3. Comparative Analysis of Territorial Governance and Spatial Planning Systems in Europe. 2020. Available online: https://www.espon.eu/programme/projects/espon-2020/applied-research/comparative-analysis-territorial-governance-and (accessed on 20 June 2019).
4. Reimer, M.; Getimis, P.; Blotevogel, H. *Spatial Planning Systems and Practices in Europe: A Comparative Perspective on Continuity and Changes*; Routledge: New York, NY, USA, 2014; p. 336.
5. Fürst, D. Planning cultures en route to a better comprehension of "planning processes"? In *Planning Cultures in Europe: Decoding Cultural Phenomena in Urban and Regional Planning*; Knieling, J., Othengrafen, F., Eds.; Ashgate: Farnham, UK, 2009; pp. 23–38.
6. Auziņš, A.; Viesturs, J. A Values-led Planning Approach for Sustainable Land Use and Development. *Balt. J. Real Estate Econ. Constr. Manag.* **2017**, *5*, 275–286. [CrossRef]
7. *The EU Compendium of Spatial Planning Systems and Policies*; Office for Official Publications of the European Communities, CEC—Commission of the European Communities: Luxembourg, 1997; p. 192. Available online: https://publications.europa.eu/lv/publication-detail/-/publication/059fcedf-d453-4d0d-af36-6f7126698556 (accessed on 9 November 2017).
8. Comparative Analysis of Territorial Governance and Spatial Planning Systems in Europe. ESPON COMPASS. Final Report. Available online: https://www.espon.eu/planning-systems (accessed on 5 August 2019).
9. *Spaces of Dialog for Places of Dignity: Fostering the European Dimension of Planning*; AESOP, Book of Proceedings (E-Book); University of Lisbon: Lisbon, Portugal, 2017; p. 3327. Available online: https://aesop2017.pt/images/Congresso/proceedings/BookofProceedings20171215.pdf (accessed on 9 November 2017).
10. Getimis, P. Comparing Spatial Planning Systems and Planning Cultures in Europe. *Plan. Pract. Res.* **2012**, *27*, 25–40. [CrossRef]
11. *Making Space for Hope*; AESOP, Abstract Book: Chalmers University of Technology: Gothenburg, Sweden, 2018; p. 762. Available online: http://www.trippus.se/eventus/userfiles/101941.pdf (accessed on 5 August 2019).
12. Barca, F. An Agenda for a Reformed Cohesion Policy: A Place-Based Approach to Meeting European Union Challenges and Expectations, Independent Report. 2009. Available online: http://ec.europa.eu/regional_policy/archive/policy/future/pdf/report_barca_v0306.pdf (accessed on 5 August 2019).
13. ESPON TIA Tool. Available online: https://www.espon.eu/tools-maps/espon-tia-tool (accessed on 13 August 2019).
14. Sustainable Urbanization and Land-Use Practices in European Regions. ESPON SUPER. Available online: https://www.espon.eu/super (accessed on 13 August 2019).
15. Adams, N.; Cotella, G.; Nunes, R. The Engagement of Territorial Knowledge Communities with European Spatial Planning and the Territorial Cohesion Debate: A Baltic Perspective. *Eur. Plan. Stud.* **2014**, *22*, 712–734. [CrossRef]
16. Auziņš, A. Key Trends and Aspects Influencing Changes into Spatial Planning Systems and Practices in Europe. In Proceedings of the 2018 International Conference "Economic Science for Rural Development" No. 48, Jelgava, Latvia, 9–11 May 2018; pp. 26–35.
17. *ESPON Project 2.3.2. Governance of Territorial and Urban Policies from EU to Local Level*; Final Report; ESPON EGTC: Luxembourg, 2006; Available online: https://www.espon.eu/programme/projects/espon-2006/policy-impact-projects/governance-territorial-and-urban-policies (accessed on 13 August 2019).
18. Cotella, G. Editorial: EU Cohesion Policy and domestic territorial governance. What chances for cross-fertilization? *Europa XXI* **2018**, *35*, 5–20. [CrossRef]
19. Gonzales, S.; Healey, P. A sociological institutionalist approach to the study of innovation in governance capacity. *Urban Stud.* **2005**, *42*, 2055–2069. [CrossRef]

20. Reimer, M. Planning cultures in transition. Sustainability management and institutional change in spatial planning. *Sustainability* **2013**, *5*, 4653–4673. [CrossRef]
21. Petersen, B.; Snapp, S. What is sustainable intensification? Views from experts. *Land Use Policy* **2015**, *46*, 1–10. [CrossRef]

 *land*

Review

Towards Responsible Consolidation of Customary Lands: A Research Synthesis

Kwabena Asiama [1,*], **Rohan Bennett** [2]  **and Jaap Zevenbergen** [3]

1   Geodetic Institute, Faculty of Civil Engineering and Geodetic Science, Leibniz University of Hannover, 30167 Hannover, Germany
2   Department of Business Technology and Entrepreneurship, Swinburne Business School, Swinburne University of Technology, Hawthorn 3122, Australia; rohanbennett@swin.edu.au
3   Faculty of Geo-Information Science and Earth Observation (ITC), University of Twente, 7514 AE Enschede, The Netherlands; j.a.zevenbergen@utwente.nl
*   Correspondence: asiama@gih.uni-hannover.de or kwabena.asiama@gmail.com; Tel.: +49-(0)-511-762-2406

Received: 30 September 2019; Accepted: 25 October 2019; Published: 29 October 2019

Abstract: The use of land consolidation on customary lands has been limited, though land fragmentation persists. Land fragmentation on customary lands has two main causes—the nature of the customary land tenure system, and the somewhat linked agricultural system. Since attempts to increase food productivity on customary lands have involved fertilisation and mechanisation on the small and scattered farmlands, these approaches have fallen short of increasing food productivity. A study to develop a responsible approach to land consolidation on customary lands using a design research approach is undertaken and reported here. Based on a comparative study, it is found that three factors inhibit the development of a responsible land consolidation approach on customary lands—the coverage of a land administration system, a land valuation approach, and a land reallocation approach the fits the customary land tenure system. To fill these gaps, firstly, this study developed the participatory land administration that brought together traditional land administration approaches with emerging bottom-up approaches, as well as technological advances that drive these approaches together with the growing societal needs. Secondly, a valuation approach was developed to enable the comparison of the farmlands in rural areas that are without land markets. Finally, a land reallocation approach was developed based on the political, economic and social, as well as technical and legal characteristics of rural customary farmlands. This study concludes that though the land consolidation strategy developed is significantly able to reduce land fragmentation, both physical and land tenure, the local customs are an obstruction to the technical processes to achieve the best form of farmland structures.

Keywords: land consolidation; food productivity; land tenure; land administration; land reallocation; land valuation

---

1. Introduction

Food security as a crucial global challenge has received much attention over the past two decades from international bodies, particularly in relation to sub-Saharan Africa. The importance of food security is highlighted by its elevation from a target of one of the Millennium Development Goals (MDG 1c) to a Sustainable Development Goal (SDG 2). Food security is a multi faceted agenda with several dimensions. Achieving food security means tackling its four dimensions—food availability, food accessibility, food utilisation, and food stability [1]. This study uses agricultural productivity, a component of food availability, as the primary motivator.

The link between food security and agricultural productivity on the one hand and, land and land administration on the other hand has been explored practically and theoretically. Theoretically,

Bennet et al. [2] shows land administration as one of the support systems that lead to increased food security, though it undermines it in some cases. Van der Molen [3] concludes that provision of food security requires the growth of agricultural productivity. Statistically about 821 million people in the world (10.9% of the world population) are undernourished [4]. This is more pronounced in the sub-Saharan African (SSA) region, with 22.7% of the population (224.3 million people) being undernourished [5]. However, even though Africa is estimated to contain 60% of the world's uncultivated land, it is estimated that 65% of Africa's arable land is too damaged to sustain viable food production [6]. This problem points to the need to effectively manage the remainder of the arable land. The institutional and technical approaches to increasing agricultural productivity include, but are not limited to land and water access, access to markets, land tenure security, better roads, mechanisation, and use of fertilizers. However, one factor that is found to impede these institutional and technical approaches to increase agricultural productivity, among others, is the fragmented structure of farms [7–9]. In many cases land consolidation has been touted as an effective solution to land fragmentation [10–12]. This study starts from the endpoint of several studies including Abubakari et al., Blarel et al., Makana, and Takane [13–16], which conclude that conventional approaches to land consolidation are not viable on customary lands. These studies however stopped short of identifying the factors needed to be considered in order to develop a land consolidation approach on customary lands. A deeper analysis of specific cases was deemed necessary. Land consolidation procedures can be generally grouped into three main stages—the administrative preparatory stage, inventory and planning (technical) preparation stage, and the implementation stage [17,18]. This study focuses on the inventory and planning stage, which involve the collection and/or updating of land tenure and spatial information, the valuation of the farms and ancillary lands, and the preparation of the land reallocation and other land consolidation works, as well as the appeals from stakeholders for the plans. This paper summarises and synthesises the results of a study into the development of a responsible approach to land consolidation on customary lands, using Ghana as a case. The following section provides a background to the problem and breaks down the research objectives for the various components of the research.

2. Land Fragmentation and Land Consolidation on Customary Lands—A Background

Land fragmentation is the dispersion of a single farm-holding into several distinct farmland parcels, as well as a discrepancy between land use and ownership [19–21]. Land fragmentation can seriously obstruct agricultural development as it negatively affects mechanisation and reduces productivity. Two forms of land fragmentation are found to exist—physical and land tenure fragmentation. Physical fragmentation is the spatial dispersion of farm parcels over a large area of land (also known as scattering) and the division of farmland parcels into small near-unproductive parcels (sub-division) [7,11]. Land tenure fragmentation is a discrepancy between land use and ownership [21]. Blarel et al. [14] and Netting [22] in studies focused on Ghana, Rwanda, and Switzerland however show that land fragmentation has some positive impacts on farm productivity. McPherson [23] therefore groups the causes of land fragmentation into two causes—supply-side and demand-side. The supply-side causes suggest that land fragmentation is a result of external forces such as population growth and cultural systems which may result in partible inheritance and land scarcity, as seen in most of Western Europe [8]; and a change in government policy that results in a breakdown of common or communal property systems, as in the cases of Central and Eastern Europe and Eastern Nigeria [24,25]. In general, supply-side causes of land fragmentation have largely resulted in negative social, economic, and environmental impacts and outcomes. However, demand-side causes result from farmers' choices, due to the positive impacts and benefits they reap from land fragmentation [26].

Land fragmentation has always been prevalent in the agricultural system of customary lands, however its articulation as a problem is a recent occurrence [27–29]. Despite this, recent studies examining food productivity in customary lands rather focus on the mechanisation of farms and fertilizer use than dealing with land fragmentation [30–34]. Land fragmentation on customary lands has two key causes—the customary land tenure (a supply-side cause), and the agricultural system (a demand-side cause) [35].

Customary land is defined in several ways depending on the origins. However, there are three fundamental elements. The first is that land is held on the basis of locally evolved native land tenure; secondly, the basis of the land holding includes group and individual rights, with the former superseding the latter; and thirdly, the mechanisms for obtaining, using, distributing and disseminating these rights arise from accepted practices based on the group's customs and traditions [36–38]. Customary lands may also be referred to as community lands, communal lands, indigenous lands, traditional lands, among others [36,39–41]. Customary land tenure reflects the socio-cultural and spiritual bonds among generations—the many who have passed on, the living few, and the countless generation yet unborn. The basic tenet of customary land administration is that the current generation is a mere caretaker of the land meant to protect it as the legacy of their ancestors and safeguard it for the future generation [42].

The nature of customary land tenure systems, together with the changing agricultural system of customary lands, also presents another key cause of land fragmentation [29,40]. Shifting cultivation, the predominant agricultural system of customary lands, allows for the tilling of the farms one after the other, gradually causing land fragmentation. The fragmentation of parcels is not a problem when population and demand for food is low: The farmer is able to take advantage of the fragmented parcels to deal with seasonal labour bottlenecks [43,44]. However, the increase in demand for food in urban areas, in tandem with the supply of fertilizer, causes the adoption of more intensive agricultural systems such as the annual cultivation and the multiple cropping farming systems which require simultaneous cultivation of the farm parcels, intensive weeding and ploughing [45–47]. Higher returns to labour offered by the industrial and service sectors, as against the farming sector, also substantially reduce the available pool of labour that can be hired, resulting in the farm labour being determined by the household size [44]. The labour shortage necessitates the adoption of large farm machinery, to keep up with the increasing urban food demand, which is difficult with small, scattered farms. The simultaneous farming of the fragmented parcels, use of rudimentary farming equipment, and application of fertilizer, still results in a less optimum productivity than experienced with the shifting cultivation [46,48]. This makes it necessary to deal with the land fragmentation situation.

Land consolidation has been successfully used to curb land fragmentation and increase food productivity, and further develop rural areas in Europe and to some extent in Asia [11,49]. However, the majority of land consolidation attempts in customary lands in sub-Saharan Africa have either failed or broken down the customary land tenure in the areas [16,33,50,51]. Despite the un-supportive land tenure and agricultural systems, attempts were made at land consolidation, predicated on the assumption that land consolidation was needed as an approach to developing the agricultural sector [15,34,52]. Makana [15] notes that land consolidation yielded rather positive results on some customary lands results in terms of increase in food production, though the customary land tenure system in those areas broke as a result. The results advanced for the successes and the failures of these land consolidation schemes include the nature of the participation of the parties involved, and the failure to adapt the land consolidation scheme to the conditions of the customary lands [51,53]. Malawi and Kenya provide examples. In Malawi, in the 1940s, although the colonial government successfully consolidated 81,000 hectares of farmlands, complete with infrastructural improvements, the programme still failed because it was solely run by the colonial government, without local participation, after being prematurely rolled out without consideration for local factors and conditions [33]. Kenya's land consolidation, also started by the colonial government, led to the complete overhaul of the land tenure system, to do away with the customary land tenure and replace it with individual titles as a major objective. The colonial government saw the customary system as a militating factor against the benefits of land consolidation and a well-functioning land market [50]. This notwithstanding, the land consolidation planning was participatory, with the plans being drawn by the government officials together with the clan elders. However, the last step of the plan was to grant individual titles, thus effectively ending the coverage of customary land in these areas. The most recent of the land consolidation activities in sub-Saharan Africa is from Rwanda, which undertook a new form of land use consolidation [54–56]. With the prime objective of increasing agricultural production, the reasoning behind this is to be able to undertake a land consolidation programme that does not alter the land tenure relations [57]. The success of the Rwandan Land Use Consolidation, and the failure of the land consolidation approaches in Malawi and Kenya, coupled with the general sentiment towards some requirements for consolidating lands across sub-Saharan Africa, shows the need to investigate the knowledge gap between the development of land consolidation and customary lands with consideration for the local societal context through using a responsible approach.

Responsible approaches and policies apply broadly to a paradigm shift from traditional, and general approaches and policies to solving problems, to more societally and contextually based approaches and policies. The term "responsible" was mostly used in government and public administration circles to describe the system of accountability. Land consolidation as a land development tool dwells within a broader context. The adoption of responsible approaches to land consolidation is needed to be able to align the land consolidation approaches to the conditions that exist on customary lands [26]. There is therefore the need to comparatively study the areas that have already undertaken land consolidation on customary lands, to be able to identify their commonalities and peculiarities before a responsible land consolidation approach for customary lands can be developed. The technological advances in land administration that have paved the way for land administration to be aligned to customary lands and used as an aid to combat the problem of inadequate land information and the absence of land value. It is acknowledged that certain characteristics of customary lands cause land fragmentation and that land fragmentation can be reduced by land consolidation. However, attempts to undertake land consolidation on customary lands have largely failed in the face of inadequate land administration processes on customary lands. There is therefore the need to adapt responsible approaches to land consolidation. The concepts relating to the problems and associated in knowledge gaps in the development of a responsible land consolidation process are summarised in Figure 1.

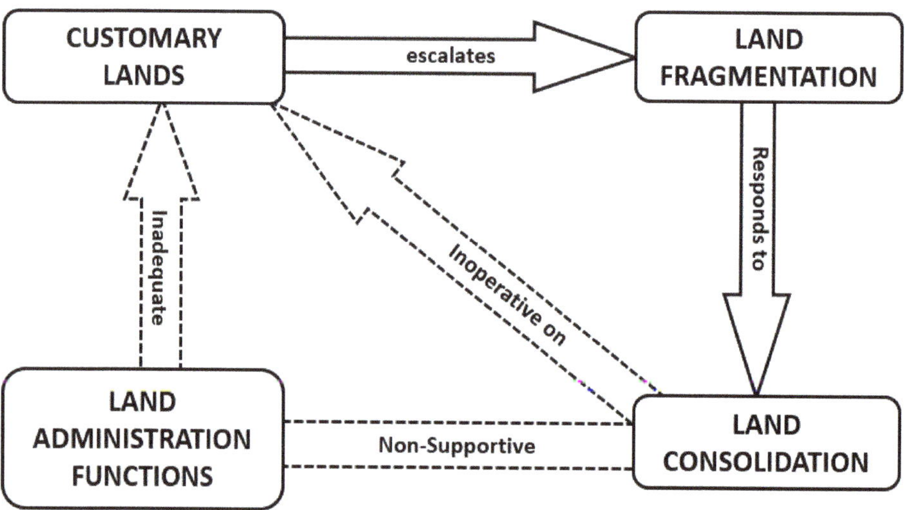

Figure 1. Conceptual framework showing the gaps to be researched.

It is known that certain conditions on customary lands (land tenure and the farming system), escalate the occurrence of land fragmentation. It is also known that land fragmentation responds to land consolidation. However, it is seen that land consolidation has not been operative on customary lands. Therefore, to connect land consolidation with customary lands, it is necessary to improve the inadequate land administration functions on customary lands that cannot support land consolidation.

The development of a responsible approach on customary lands took four steps; first the factors that need to be addressed to develop a responsible land consolidation approach for customary lands were explored. An approach for collecting land information to support responsible land consolidation on customary lands was then developed and assessed. Furthermore, a land valuation approach to support responsible land consolidation on customary lands was also developed, and the above were applied to a process model for a land reallocation approach to support land consolidation on customary lands.

3. Methodology and Research Approach

The choice of the research methodology is largely driven by the nature of the research problem, the objectives, as well as the questions asked to reach the objectives. When the research seeks a method that emphasizes the solving of problems through the combination of methods from different paradigms that allow for the generalizable and quantifiable results by answering questions related to how much? (qualitative methods), and those that allow for the rich and deep understanding of the situation, answering questions related to the who, what, and how (qualitative method) related to information systems, the design research is found to be most appropriate [58]. Design research is preferable because it allows for the use of diverse research strategies needed when the research seeks to deal with real world complexities. The design research is operationalized in five steps. First step is the problem identification; second is understanding and agreement about the problem is generated; thirdly, the options for the development of the system is explored and the system is developed; fourth, the designed system is then implemented, and finally the implemented system is evaluated. The design research methodology is summarized in Figure 2.

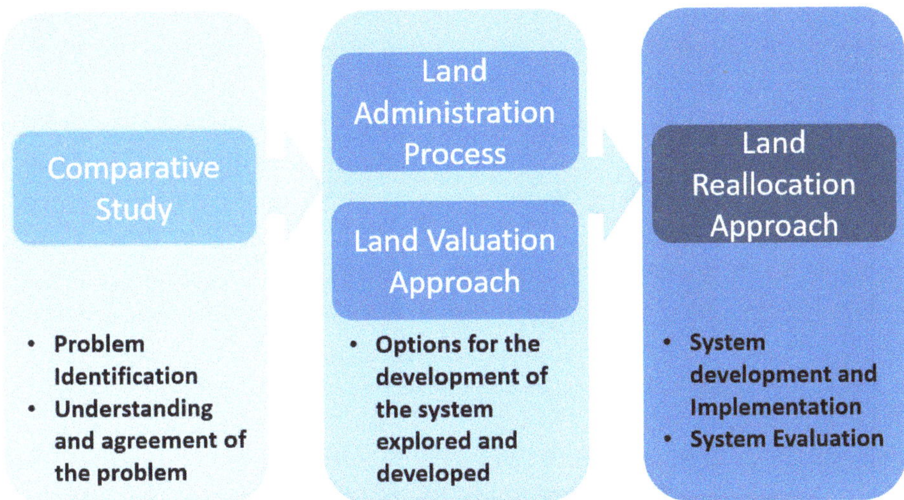

Figure 2. The design research approach.

The first step requires the exploration of how land consolidation's factors need to be addressed on customary lands, encompassing the first two steps of the design research. A comparative case study approach is adopted. In this vein, an analytical framework for understanding the reasons different land consolidation strategies are developed and adopted or adapted in different contexts, from existing literature, to form a scientific basis for the comparison. Using Van Dijk's [59] model of comparative analysis for cross-country exporting of knowledge, three countries with existing land consolidation strategies are selected, observed, and compared to Ghana's rural customary lands. This model is grounded in the reasoning that in transferring development and planning approaches across international borders, it is necessary to understand how and why the approach was developed in the original context. The goal of using this model is to first understand the local contexts, and then to examine land consolidation factors and how they influenced the selection of the land consolidation strategy. The selected countries included the Netherlands, Lithuania, and Rwanda. Data from the Netherlands, Lithuania, Rwanda, and Ghana was collected through a document review—scientific literature, government policies, laws, and technical reports; and supplemented with interviews with land and agricultural sector officials, and farmers. In Ghana, further interviews were conducted with the traditional authorities and local government leaders.

The second step used the experimental case study with a Living Lab approach. A Living Lab is based on two main concepts—first is the involvement of users early in the innovative process, and the second is experimentation in real-world settings, aimed at integrating the social structure and governance, as well as user participation in the innovative process [60–62]. The stakeholders of the experiment were identified, and the process of mapping and recording the land rights was developed with the assistance of the Traditional Authority (the Nanton-Na and the family heads in the area), the leaders of the Farmers' Association of the area and the Lands Commission. Two technologies were adopted for the experiment—a smartphone app and satellite imagery. The smartphone app used was Esri's Collector for ArcGIS. The satellite image was a February 2016 GeoEye-1 satellite image of the area of interest was freely acquired from DigitalGlobe Foundation, and printed at a scale of 1:4000, which is within the range of scales recommended by Byamugisha et al. [63] for mapping rural agricultural land parcels with medium density.

The third step developed a valuation approach for land consolidation. Here, the multiple attribute decision-making (MADM) method is used based on the general land valuation approach. MADM methods are flexible and can be adapted with ease to the development of indices being represented by a set of parameters, where the aim is to evaluate an object compared to a standard for which the application is concerned. In the case of this study, the standard is the most appropriate land parcel for farming. This approach is used because it is about to achieve quid pro quo values that can be used for land consolidation.

The fourth step is achieved using the process modelling method that details the steps of the approach taking into consideration the social, economic, cultural, technical, and political considerations on customary lands. The process model developed in this paper is a meso-micro-level procedural model. The meso-micro-level procedural model conveys best practices intended to guide real-world situations by providing prescriptive guidelines for a design and/or problem-solving activity with a focus on individual steps as well as end to end flows of the activity, where each step establishes objectives, and constraints for the next, with feedback loops between the steps for the possibility to re-work undesirable outcomes.

4. Overview of Study Area and a Background on Ghana's Customary Lands

The study focuses on Ghana, an agriculturally dominant country. The choice of Ghana is made because it is one of the two countries that undertook efforts to adopt customary land tenure laws that were derived from an African angle, expend state influence out into the customary domain and strengthening the governance structures already in place right after independence [64,65]. The other country is Botswana. However, compared to Ghana, Botswana has a low land productivity that can still be improved and is one of Africa's smallest agricultural economies [66]. About 49% of the population of Ghana lives in rural areas, with 45% of the country's labour population (15 years and above) being engaged in agriculture [67]. Agriculture contributes to 54% of the Ghana's gross domestic product, and accounts for over 40% of its export earnings, whilst at the same time providing over 90% of the food needs of the country. Out of the 258,539 km sq. area that Ghana covers, 57% is classified as agricultural land area.

Customary lands are recognized by the 1992 Constitution of Ghana (Article 38) and cover 80% of the lands in Ghana with the remaining 20% being public lands vested in the President in trust for the people of Ghana [68]. The main interests in customary land tenure that relate to farming are the Allodial Title, the Customary Law Freehold or Usufructuary interest, and Tenancy (Figure 3) [37,69,70]. The allodial title is held by the community and managed by its leaders under customary law, free from any restrictions and obligations, except such imposed by the laws of Ghana. The allodial interest cannot be transferred as this is restricted by the 1992 Constitution of Ghana and the customs, and it is exclusive to the community or tribe that holds the rights. The Usufructuary interest is exercised by individual members of a community to take possession of vacant land of which the community is the allodial owner subject to certain restrictions and obligations, upon payment of nominal consideration or free of charge [42]. The Usufructuary interest is transferable within the allodial land owning group under certain strict circumstances. The Tenancy can be acquired by any person, indigene or otherwise, based on specific prior agreed terms, usually share cropping or an annual payment, usually for a term of one farming season. The tenant holds the land for the term exclusively, but subject to rules of the allodial title holder and/or the usufruct and cannot transfer his rights without the consent and concurrence of the landlord. Although the modes of acquiring the Usufructuary interest include the clearing of an unencumbered land followed by uninterrupted settlement, or as a gift or purchase; inheritance is currently the most common means of land acquisition [37]. The Usufructuary interest is held in perpetuity except for situations of abandonment, forfeiture, or want of successor; in which case, the land reverts to the allodial title holder [42,71]. The nature of the Usufructuary interest restricts farmers from expanding, as contiguous parcels' holders are unwilling to sell their parcels in order to hold the land for the future generations. This causes land fragmentation because to expand their operations, farmers move to parcels further away from their primary parcels.

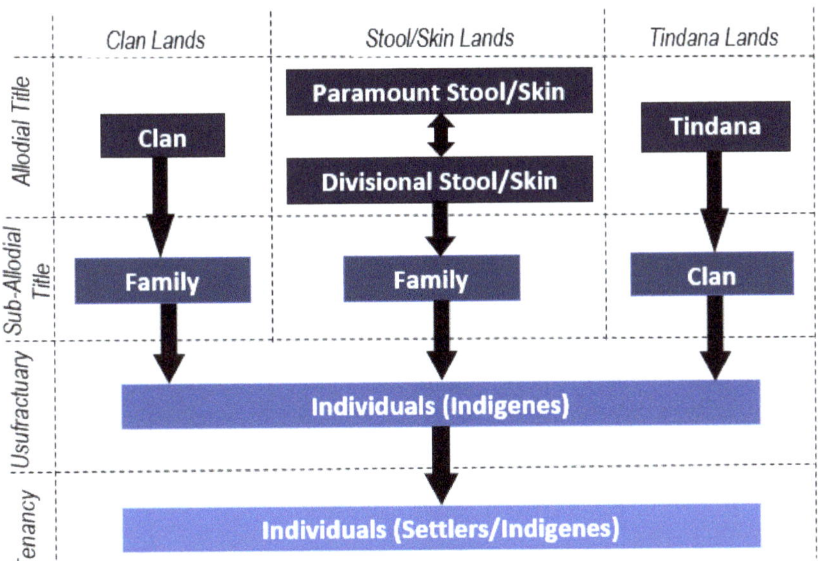

Figure 3. Hierarchy of land tenure types in Ghana [35].

Farming in Ghana varies according to the seven agro-ecological zones—the rainforest, deciduous forest, semi-deciduous, transition, and the savannah zones (Guinea, Sudan, and Coastal) (Figure 4). In the forest zones, plantation and tree crops such as cocoa, oil palm, coffee and rubber are pre-dominant. The savannah and transition zones are characterized mostly by annual crops such as maize, roots, sorghum, and cowpea. In terms of rain, the forest and coastal savannah areas have bimodal rainy season, giving rise to two farming seasons per year—a major and a minor farming season. In the Guinea and Sudan savannah, and transition zones, there is one rainy season.

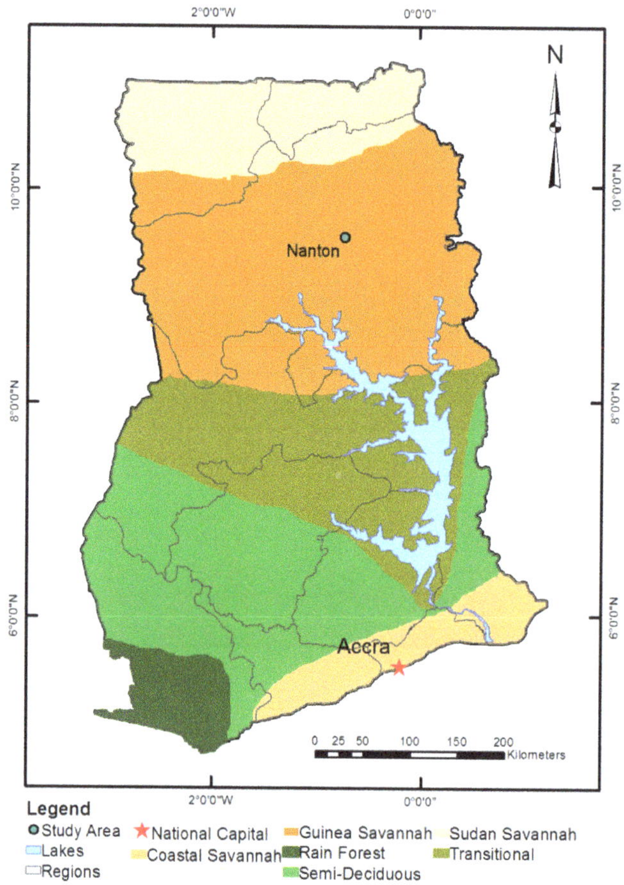

Figure 4. Location of the study area in Ghana.

This study is conducted in a farming community in one of the savannah zones of Ghana. These areas were chosen for two reasons. The first relates to the agro-ecological characteristics of the area. The savannah areas are characterized by tall grasses and a few trees (mostly shea, acacia, baobab, and mango) dotting the landscape. These conditions are favourable for the use of GNSS in this study, as the absence of tree cover will reduce the likelihood of multipath errors when using GNSS. The second reason relates to the tree crops grown. The growing of annual crops allows a certain amount of flexibility when dealing with the manipulation of farmland parcel arrangements. The area used for this study, Nanton is in the Guinea savannah agro-ecological zone, with the land tenure being held on the basis of the skin lands.

5. Outcomes of the Aspects of Responsible Land Consolidation

This section outlines the main results of the project, according to the research objectives. The results have been summarised in Figure 5 with respect to the gaps in relationships between the concepts shown in Section 2.

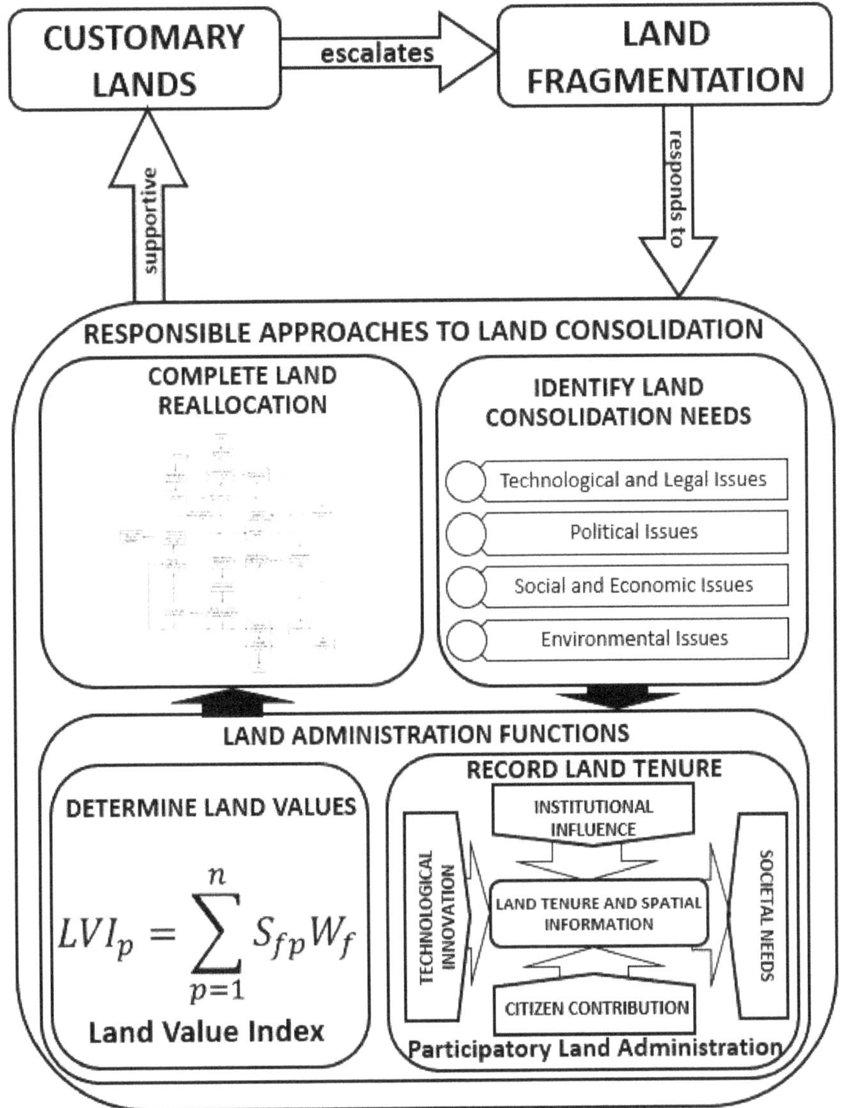

$$LVI_p = \sum_{p=1}^{n} S_{fp} W_f$$

Figure 5. Summary of the results and components of responsible land consolidation.

5.1. Factors that Influence the Selection and Development of a Responsible Land Consolidation Approach

To identify how local factors, affect the selection of a land consolidation approach, three countries with contemporary land consolidation approaches were identified—the Netherlands, Lithuania, and Rwanda. The Netherlands was found to have land consolidation approaches that have evolved over five centuries, from Voluntary Land Exchange, to Land consolidation by agreement and Land Development. Lithuania has developed the Voluntary and Simple Land Consolidation over the past fifteen years, and Rwanda developed its own form of land consolidation, the land use consolidation in 2008. A harmonisation of the land consolidation approaches in these three countries shows that generalising the development of land consolidation approaches in a continuum from simple and voluntary approaches to comprehensive and compulsory approaches, as is done in certain studies, based on the development level of the locality, does not result in development of a responsible land consolidation strategy, but the local needs and societal makeup is key in the selection of the land consolidation approach. The results showed five key areas around which the development of the land consolidation approaches centre—the government support for and role in land management activities; the land market and land mobility; land tenure, land fragmentation and farming technology; the coverage of a land information system, as well as environmental and ecological considerations.

Comparing these influences, it was found that the state of the economy, the type of land fragmentation, ecological considerations, and the level of farming technology in Ghana was similar to at least one of the countries with an existing land consolidation approach. The conditions that did not bear any similarities with existing land consolidation strategies were the low influence of the government in land management activities, the absence of a land market, the inadequate coverage of a supportive land information system, and the customary land tenure. However, it was found that the conditions that did not adequately match the countries with existing land consolidation approaches require a substantial change to the social, economic, and cultural structure of the communities, in order to align them with the existing approaches. These conditions therefore require innovative and responsible interventions to enable response to the requirements of land consolidation. The detailed results of this objective may be found in Asiama et al. [35].

5.2. Participatory Land Administration: An Approach to Collecting Land Information

Land administration processes in Ghana have been found to be slow and expensive in relation to the urgency of the results, and out of reach of most of the citizens. Furthermore, they have failed to integrate all forms of land tenure arrangements especially secondary and customary land rights. It was found that the innovative approaches to land administration on customary lands in Ghana which include the systematic titling by the Millennium Development Authority, the Paralegal Titling Project and the Community-based Land Survey Tool, all had the same problem of being slow, expensive, and concentrated in urban areas and on large-scale farms. Here, participatory land administration (PLA) that sits at the nexus of the drivers of technological innovation and approaches to development studies; where traditional land administration approaches, deeply rooted in western views, together with bottom-up emerging approaches that challenge traditional approaches, as well as technological advances that drive these approaches together with the growing societal needs.

The experiment into PLA in Nanton, Ghana was assessed in terms of reliability, affordability, local participation, and attainability. In terms of reliability, it was found that both technologies, the smartphone app and satellite image were easy for the farmers to use, as the majority of them were users of smartphones. The accuracy of the mobile app ranged between 1 and 3 m, which even though it is not adequate for the land title registration in Ghana, is enough for the recording of land rights in rural areas. No boundary disputes were encountered. The mobile app was further able to capture all 230 farmland parcels in the area, though the identification on the satellite yielded 143 parcels (Figure 6). The former was further able to identify and collect information on all the customary land rights that are related to farming. In terms of affordability, the two technologies used together are found to be cheaper to use than the current approaches on customary lands. Whereas the current conventional and innovative approaches in Ghana cost at least GH¢ 500 (EUR 125) and GH¢ 200 (EUR 20) per parcel, this approach is estimated to cost GH¢ 36.83 (EUR 9.24) per parcel. This cost will reduce with scaling up. In terms of local participation, it was found that the local people were involved in every step of the approach. This according to them gave them a sense of ownership of the data and the process. The involvement of the Trusted Intermediaries further created a layer of check for the information collected. In terms of attainability, the experiment took 10 working days, roughly 20 minutes per parcel. This would however reduce when the interviews and focus group discussions for the assessment of the process is excluded. The process is therefore fast. The use of locally acquired and accessible materials further boosted the ability of the local people to replicate the process.

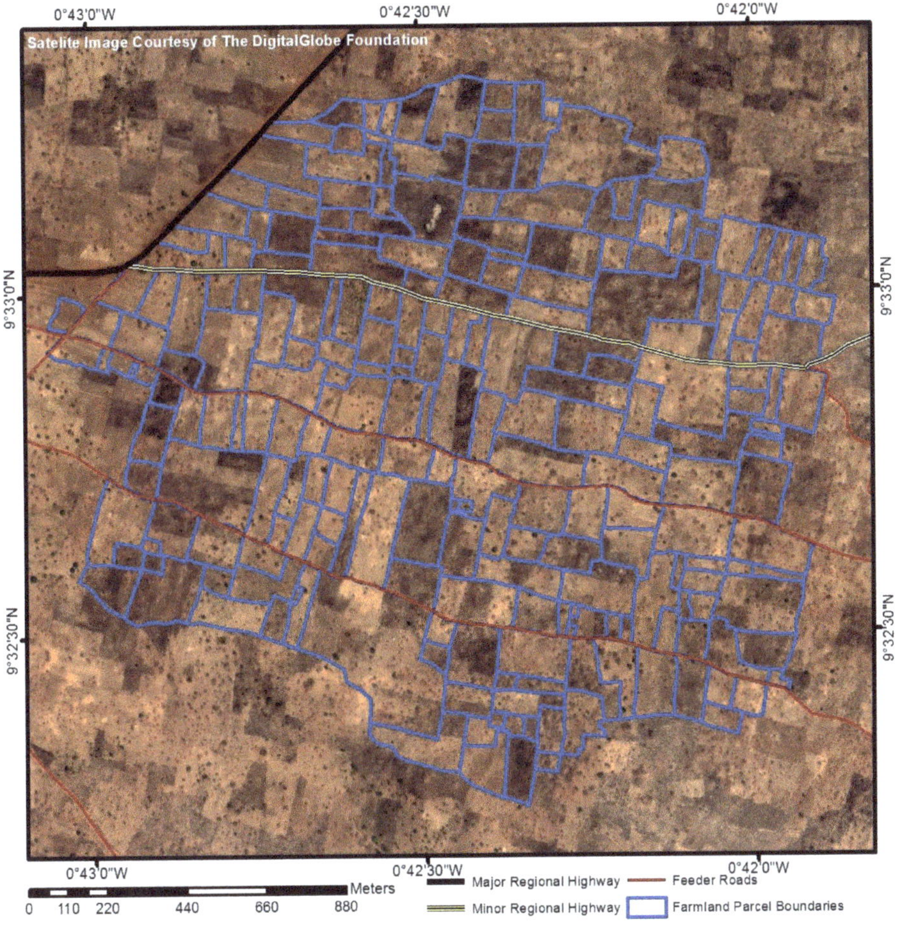

Figure 6. Parcel arrangement in the area of interest—Nanton collected by PLA.

Even though the experiment did not set out to undertake a full land consolidation, it is found capable of capturing all customary land rights as well as other information relevant to land consolidation. The detailed results of this objective may be found in Asiama et al. [72].

5.3. Valuation of Farmlands for Land Consolidation

Land value is not explicit on rural customary lands, mostly because the social, cultural, and spiritual bonds with land inhibit the free operation of a land market. Land reallocation in land consolidation relies on land valuation to describe and assign a value to the farmlands that will be reflective of the farmers' perception of their farmland values. The traditional valuation approaches, including the cost, investment, and comparative methods, are used to value customary lands. However, in rural areas, it is found that even though the sales of land are very uncommon and unlikely, where land is rented out, the money that exchanges hands is a flat rate that is charged regardless of the nature of the farmland parcel.

There are two approaches to land valuation in land consolidation—the agronomic value, with its basis being the soil productivity and quality, and the market value. Alternatively, market value has been touted as the better approach with studies pointing out the deficiencies in the agronomic value approach. However, the market value approach cannot be used in sub-Saharan Africa's customary lands due to the limited land market. Here, a framework was developed for an approach for assigning values to customary rural farmland parcels that reflects the local people's view of land value. Land value indices are used here instead of scores to allow for continuous values in comparison instead of discrete values through a flexible and content-specific approach that allows replication in other contexts. The approach is knowledge-based, using local and expert judgement through value functions. LVIs measures how far land value factors LVFs), that influence land value, deviate from the most suitable farming conditions, here denoted as one, or the worst, here being zero. To identify and understand the factors that affect farm-land value on rural customary lands, the factors found in previous studies relating to the valuation of other types of land were identified. The LVFs are first assigned scores through a quantitative method for the continuous variables, or a categorical rating method for discrete variables using the appropriate ordinal scale. These scores are derived from expert and local judgement. The scores of the factors were standardized using the direct value rating to allow for comparison on the scale.

The land value index (LVI) for each parcel is calculated by multiplying the standardised score of each factor (S_{fp}) by the corresponding weight of the factor (W_f), and summing for each farmland parcel, as depicted in the equation function below;

$$LVI_p = \sum_{p=1}^{n} S_{fp} W_f \tag{1}$$

It was found in the case study of Nanton that key land value factors that determine land values relate to the physical attributes, legal conditions, agricultural productivity, locational factors, and the planning scheme of the farmland parcels (Figure 7). These factors were weighted by the local community according to their perception of what affected their choice of farmland parcels. The weights were integrated into the framework that produced the land value index (LVI) for each land parcel in the area of study. The results showed that in a scenario analysis, a change in weights affected the land value indices at a scale that could change the comparative basis of the land parcels. The sensitivity analysis however showed that the LVIs were not significantly sensitive to the changes in the weight of the factors. However, a prime weakness of this framework is that it is more expensive to use than automatic valuation models. The results demonstrate that it is possible to place relative quid pro quo values on rural agricultural farmlands that are not part of a land market. These quid pro quo values will serve as a basis for the reallocation of the farmland parcels. The detailed results of this objective may be found in Asiama et al. [73].

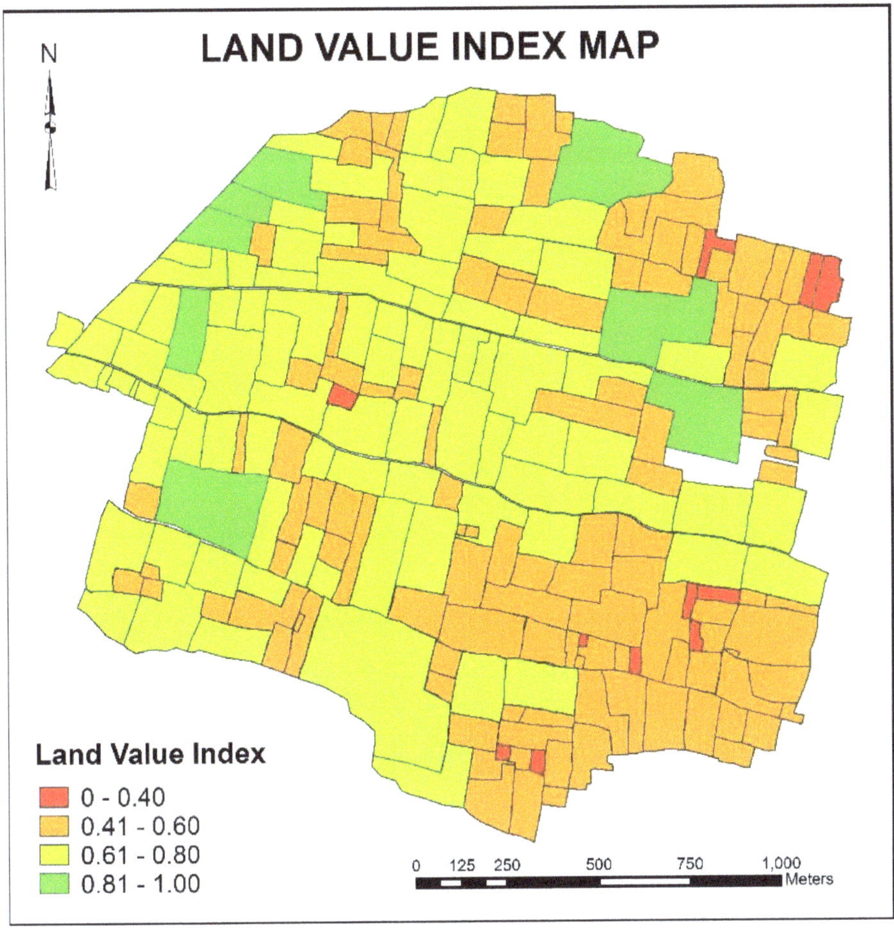

Figure 7. Land value index map of the area of interest.

5.4. Land Reallocation for Customary Lands

The results of the previous two sections were synthesised to develop the framework of a land reallocation model. Land reallocation is seen as the most important stage in the land consolidation process, where property rights are exchanged, and farmland parcels are redistributed and reorganised. A model of land reallocation should therefore consider all related land information (spatial, rights, and value) and the wishes of the involved land holders. The framework for the model is developed using a process model taking into consideration the social, economic, cultural, technical, and political considerations on customary lands, through the steps of analysis, synthesis, and evaluation.

The undertaking of land reallocation generally has three key requirements and considerations. Politically, land reallocation requires a mediating authority to act as an arbitrating force during the planning and implementation, because of the disputes that land reallocation may spark. Similar to land consolidation in general, land reallocation also requires a level of land mobility that will allow for the exchange of farmland parcels, in this case related to social land mobility, i.e., land mobility based on the social and cultural norms in the community. The development of a land reallocation model further requires a consideration for the land tenure system and the land fragmentation situation. Customary lands characteristics that are relevant to land reallocation relate to the rules that relate to the transfer of land between two parties. Here, even though it is generally accepted that customary lands cannot be transferred, it is found that customary land tenure rules do indeed allow for the transfer of land, but with strict restrictions. The framework of the land reallocation model is built around the legal and technical aspects of land reallocation, taking into consideration the levels of landholding (individual, family/clan, village, etc.). The framework for the land reallocation is focused on the family level as Section 5.1 shows that transfer of lands within families involves only the individuals concerned. However, where land is transferred outside the family, it requires the consent of the two families. Figure 8 below shows the flowchart of the land reallocation framework.

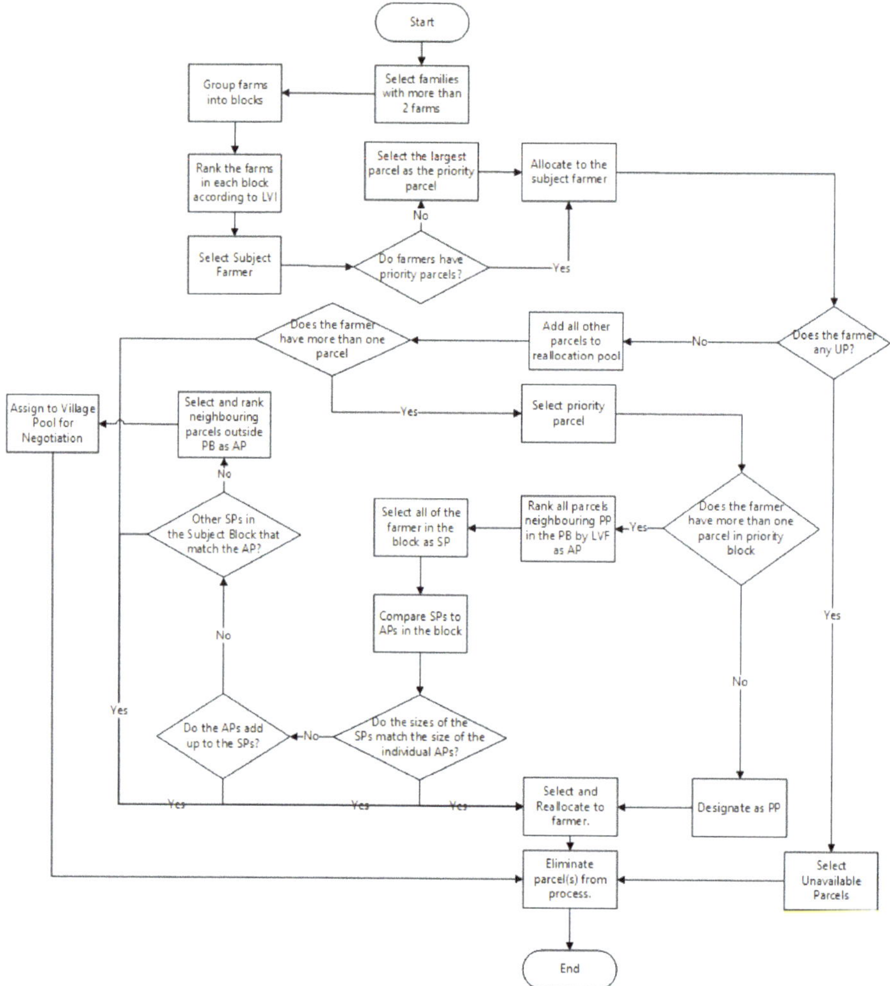

Figure 8. Flowchart of the land reallocation framework for customary lands.

When the model framework was applied to the area of interest, it was found that physical land fragmentation was significantly reduced, with a reduction in the number of farmland parcels, an increase in the parcel sizes, a reduction in the land tenure fragmentation, an increased accessibility to key lines of transportation, and slight improvement in the parcel shapes in the area, even though this was not a goal of the approach. The most appropriate central mediating authority in the area was found to be the traditional authority in the area, much different from the other areas in the world where land re-allocation has been done. Figure 9 shows a change detection map of the area of interest before and after the land reallocation. Table 1 also summarises the effects of the land reallocation.

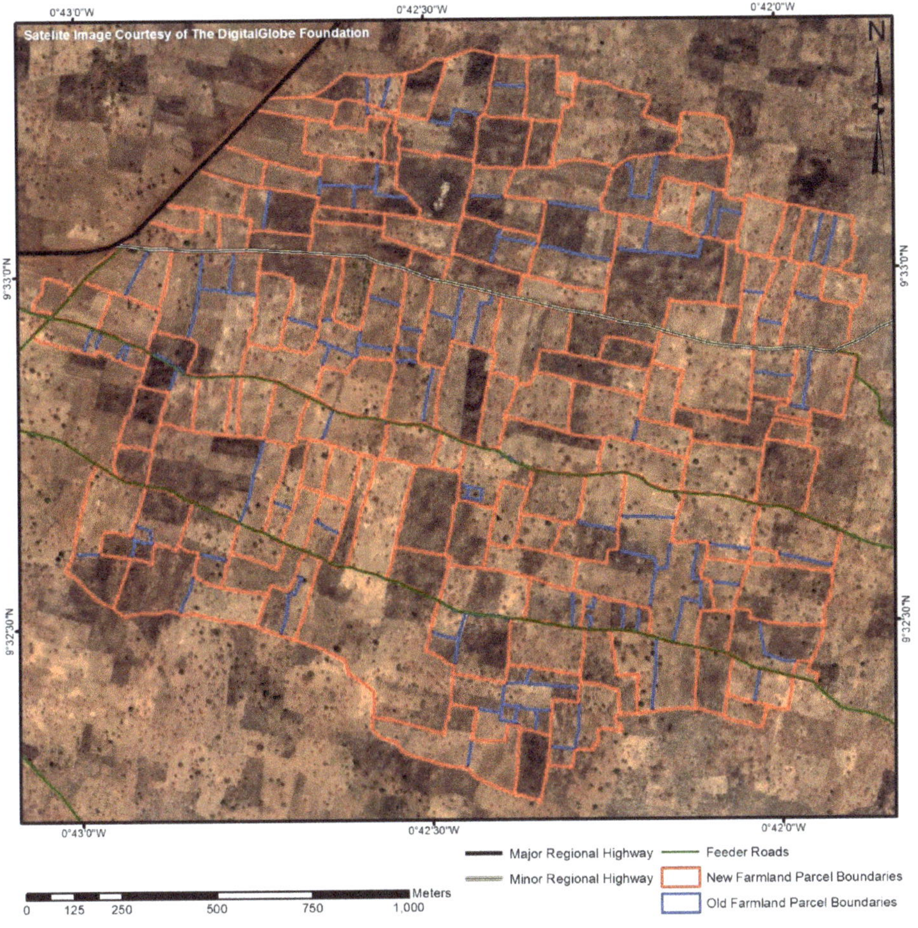

Figure 9. Parcel arrangements before and after land reallocation.

In terms of the land tenure system, local customs, and land mobility, the study found the relationship between the tenant and the usufruct to be a key cause of land tenure fragmentation. However, the study further showed that land tenure fragmentation would be reduced with the application of the approach. With regards to the land reallocation between families, it was found that the developed approach could not handle this as the local people were vehemently against families parting with their sub-allodial interests in land, even when it is swapped for a similar parcel of land. The only seeming solution was to rent out the family land to serve the purposes of re-allocation. However, although this would reduce the physical land fragmentation, the land tenure fragmentation would worsen. The detailed results of this objective may be found in Asiama et al. [74].

Table 1. Summary of the effect of the approach of the land reallocation.

Category	Initial	Final
Number of individual farmland parcels in area of interest	230 Parcels	130 Parcels
Number of Farmers in the area of interest	95 Farmers	95 Farmers
Average number of farmland parcels per Farmer	2.4 Parcels	1.37 Parcels
Number of Family Lands	42 Farm holdings	42 Farm holdings
Average number of farmland parcels per Family	5.4 Parcels	3.1 Parcels
Area of largest farmland parcel	10.79 Hectares	10.79 Hectares
Area of smallest farmland parcel	0.07 Hectares	0.22 Hectares
Average farmland parcel size	1.25 Hectares	2.20 Hectares
Standard deviation of farmland parcel areas	1.08	1.68
Maximum Parcel Shape Index	1.0	1.0
Minimum Parcel Shape Index	0.00	0.02
Average Parcel Shape Index	0.66	0.76

6. Synthesis and Discussion of Results

This section synthesises the results from the four standalone research results summarised in the previous section and discusses the implication of the results to current land consolidation strategies, to land and food policy formulation and implementation, meeting other societal challenges and needs, and to the study area of Ghana.

6.1. To Current Land Consolidation Strategies

As shown in the different sections, land management activities are very much dependent on the local context in which they are being applied. However, there is very little literature on the considerations for the transfer of land management activities into other areas. This work, using land consolidation as a land management activity, explores the factors that must be considered and how to achieve those factors when transferring the processes to Ghana's rural customary lands. This is important because even though the problems in two areas may be similar, the response to those problems may differ, as shown in Section 5.1, and therefore need comparative analysis. The results in this work further contribute to the literature and scientific knowledge on how land management activities could be transferred from one part of the world to another, in this case from western countries to sub-Saharan Africa. Thus, building upon the works of Masser and Williams [75], and Van Dijk [59], where the latter explored the transfer of land consolidation knowledge from Western Europe to Eastern and Central Europe.

This work further builds upon other works in land consolidation such as Hartvigsen [25] and Van Dijk [8] that looked at the land consolidation approaches and policies in Central and Eastern Europe viz-a-viz Western Europe at a macro-level, and Demetriou [17] who looks at the development of systems to support the undertaking of the individual stages of land consolidation at a micro level. These three studies however dwell on areas with already existing and functioning land administration systems. Hence a lot of core and key steps of land administration functions in land consolidation are not considered. This study is developed in three steps—a land administration process, a land valuation approach, and a land reallocation model, contributing to the scientific knowledge of land consolidation at a micro-level (Figure 2). All these three steps exist in literature, the difficulty in using them elsewhere stems from the inadequacy of a general framework for all of them that can be adapted to a local context. This is because each of these three processes are developed to fit a particular local and legal context. In this work, the general requirements and processes for each approach were explored and defined, before the case specific adjustments were developed. This means that some processes had to be deconstructed as they assume certain minimum characteristics in the conventional approaches such as land mobility and a land market which did not exist. In this regard, the processes developed in this work can be applied to any area, with consideration for the local context. This work therefore contributed the knowledge of developing general processes for land administration processes that are needed to support land consolidation.

6.2. To Land and Food Policy Formulation and Implementation

Policies form one of the bases for land management (the other two being Land Information Infrastructures and country context) [76]. The results in Section 5.2 demonstrated the inability of the collected land information to support land management activities for sustainable development. The need for land policies to consider the gap between land information collection or the building of a cadastre on the one hand and sustainable development on the other, is shown in that section. The results show that merely collecting land information is not enough, but the land information should be meant for a particular purpose. Such an approach as collecting land information is not immediately obvious when looking at western countries, however, it is more obvious with SSA countries. Therefore, this work can help with the formulation and improvement of land policies to re-orient them towards gearing land information to land management activities to support sustainable development.

The formulation of food and agricultural policy can also be influenced by the results of this work. As shown in Section 5.1, policies on increasing food productivity focused mostly on mechanisation and fertilisation, rather than looking at land availability, the size of farmland parcels, and the land tenure security of farmers. The result in Section 5.4, describes the framework for the land reallocation model and its application in a case area, shows how the application of the approach can be applied to increase the size of the farmland parcels and reduce the fragmentation of land tenure. This research therefore enriches the need for a stronger link between food policy and land, especially in terms of food productivity.

6.3. To Other Societal Challenges and Needs

Societal challenges such as climate change adaptation, poverty alleviation, food security, post conflict nation building, and tenure security have a land footprint. Land has been viewed as a key driver for sustainable development. Therefore, its effective management will contribute to meet the aforementioned challenges and needs towards sustainable development. Though the focus of this research is on food security, and more specifically food productivity, the findings can further impact on the other societal challenges and needs.

The results show the development of innovative land administration processes that may assist in land management activities that are geared towards meeting the identified societal needs and challenges. In Section 5.2, the participatory land administration (PLA) approach developed can be used to collect land information to support other activities, such as large-scale land acquisitions, disaster risk management, and post conflict nation building, with the goal of land tenure security. PLA may also aid with economic and infrastructure development and increasing investments in property by providing land documents to aid in the procurement of loans for property investments within the legal framework. This further contributes to food security, as farmers are more likely to invest in their farms when they are more tenure secure. In Section 5.3, the land valuation approach developed does not only apply to rural customary lands and land consolidation. This land valuation approach is applicable for large-scale land acquisitions, by the government or by private entities, especially in areas without land markets. This will ensure that the values arrived at bear close resemblance to the market value. Furthermore, the land valuation approach may be used by the government for the fair assessment of taxes and the payment of fair and adequate compensation for compulsory land acquisition.

6.4. To the Study Area of Ghana

The three areas of interest selected in Ghana were based on the agro-ecological characteristics, the types of crops grown, and the land tenure system. The first two bases of selection were chosen because of their commonalities in the three areas of interest; however, the land tenure system was chosen because of its variety in the three areas. The findings in Section 5.1 demonstrate that all three tenure types have common underlying basic principles, therefore, one area of interest was adequate for the remainder of the work. Nanton was the area of interest chosen for the remainder of the work, because, with its skin lands, it has the most complex land tenure structure among the three (Figure 3). This implies that the results of this work can be directly applied in Nanton. In the remaining two areas of interest the results need to be adjusted according to the complexity of the land tenure system. The results of this work can be further extrapolated to other areas of Ghana with skin, family, or Tindana lands. However, this work did not cover stool lands as they have similar land tenure characteristics as skin lands. Therefore, minimum adjustments would be expected to be made to apply the results on stool lands.

Furthermore, cases from the Netherlands, Lithuania, and Rwanda are useful for the areas of interest and Ghana as a whole, especially regarding the evolution of land consolidation approaches overtime and land management activities in general.

7. Conclusions

This study aimed at developing a responsible approach to land consolidation on customary lands, using Ghana as a case. The study found that in a comparison between countries with a responsible land consolidation (Rwanda, Lithuania, and the Netherlands) on one hand and a country with customary lands but without a land consolidation (Ghana) on the other hand, there were three areas that needed attention to develop a responsible land consolidation the land administration processes, the land valuation approach, and the land reallocation approach. The participatory land administration (PLA) was developed to bring together traditional land administration approaches, deeply rooted in western views, together with bottom-up emerging approaches that challenge traditional approaches, as well as technological advances that drive these approaches together with the growing societal needs. A valuation approach was then developed to enable the comparison of the farmlands in rural areas that are without land markets. Finally, a land reallocation approach was developed based on the political, economic and social, as well as technical and legal characteristics of rural customary farmlands. This study finds that though the land consolidation strategy developed is significantly able to reduce land fragmentation, both physical and land tenure, the local customs are an obstruction to the technical processes to achieve the best form of farm structures. However, the consideration of all aspects of the society and technology being a basic tenet of responsible approaches, the changes to the local customs is beyond the scope of this study.

A further comparative study can be undertaken on other SSA countries' rural customary lands to further understand the differences, in terms of the requirements of land consolidation. In addition, future work should focus on further developing land valuation and land reallocation approaches by automating them using Computer Assisted Mass Appraisal (CAMA) systems with GIS, and Spatial Decision Support Systems (SDSS), respectively. This is because the processes developed in this study were generalised processes that exist in other regions of the world. The valuation approach was developed in the rural farming context, it can therefore also be developed further looking at urban land to put it in a broader context. This will further deepen and enrich the use of market information in the valuation of urban lands, especially for slum areas for non-market values. Furthermore, as shown in Section 5.1, customary lands are independently managed in each community, saved for the national legal framework that tries to harmonise their management. This means that a single land consolidation approach will not fit the whole country. Further research should therefore be conducted into the legal framework of Ghana, vis á vis land consolidation in order to develop an integrated, flexible, and inclusive framework for customary lands towards land consolidation. Further research also needs to be done in the implementation, through active research in the conduct of a pilot land consolidation process in the customary areas, to further ascertain the limitations that the approaches have in other areas. This, in tandem with the scaling up approach by further establishing workflows, will enable the testing of the approach with a wider coverage.

Author Contributions: K.A. collected and prepared the data for this research article; R.B. and J.Z. contributed to and supervised the analysis and interpretation of the data. The manuscript was written by K.A., with contributions from R.B. and J.Z.

Funding: This research received no external funding.

Acknowledgments: The authors thank Seth Opuni Asiama and Andre Da Silva Mano for the technical support offered in parts of this work. The authors also thank the DigitalGlobe Foundation for the satellite photos used in this study. This paper is part of the PhD Thesis, Asiama, K. O. (2019). Responsible Consolidation of Customary Lands (PhD Thesis). Retrieved from https://library.itc.utwente.nl/papers_2019/phd/Asiama.pdf.

Conflicts of Interest: The authors declare no conflict of interest.

References

1. Pinstrup-Andersen, P. Food security: Definition and measurement. *Food Secur.* **2009**, *1*, 5–7. [CrossRef]
2. Bennett, R.; Rockson, G.; Haile, S.A.; Nasr, J.; Groenendijk, L. Land Administration for Food Security. In *Advances in Responsible Land Administration*; Zevenbergen, J.A., de Vries, W.T., Bennett, R., Eds.; CRC: Boca Raton, FL, USA, 2015; pp. 37–52.
3. Van der Molen, P. Food security, land use and land surveyors. *Surv. Rev.* **2016**, *49*, 147–152. [CrossRef]
4. FAO. The State of Food Security and Nutrition around the World. 2018. Available online: http://www.fao.org/state-of-food-security-nutrition/en/ (accessed on 11 October 2018).
5. FAO. *Regional Overview of Food Security and Nutrition in Africa 2017*; FAO: Rome, Italy, 2017.
6. Panel, M. *No Ordinary Matter: Conserving, Restoring, and Enhancing Africa's Soils*; Agriculture for Impact: Dakar, Senegal, 2014.
7. King, R.; Burton, S. Land Fragmentation: Notes on a Fundamental Rural Spatial Problem. *Prog. Geogr.* **1982**, *6*, 475–494. [CrossRef]
8. Van Dijk, T. Dealing with Central European Land Fragmentation: A Critical Assessment on the Use of Western European Instruments. Ph.D. Thesis, Delft University of Technology, Delft, The Netherlands, 2003.
9. Bentley, J.W. Wouldn't you like to have all of your land in one place? Land fragmentation in Northwest Portugal. *Hum. Ecol.* **1990**, *18*, 51–79. [CrossRef]
10. Jürgenson, E. Land reform, land fragmentation and perspectives for future land consolidation in Estonia. *Land Use Policy* **2016**, *57*, 34–43. [CrossRef]
11. Bullard, R. *Land Consolidation and Rural Development*; Ashgate: Chelmsford, UK, 2007.
12. Niroula, G.S.; Thapa, G.B. Impacts and causes of land fragmentation, and lessons learned from land consolidation in South Asia. *Land Use Policy* **2005**, *22*, 358–372. [CrossRef]
13. Abubakari, Z.; van der Molen, P.; Bennett, R.; Kuusaana, E.D. Land consolidation, customary lands, and Ghana's Northern Savannah Ecological Zone: An evaluation of the possibilities and pitfalls. *Land Use Policy* **2016**, *54*, 386–398. [CrossRef]
14. Blarel, B.; Hazell, P.; Place, F.; Quiggin, J. The Economics of Farm Fragmentation: Evidence from Ghana and Rwanda. *World Bank Econ. Rev.* **1992**, *6*, 233–254. [CrossRef]
15. Makana, N.E. Peasant Response to Agricultural Innovations: Land Consolidation, Agrarian Diversification and Technical Change. The Case of Bungoma District in Western Kenya, 1954–1960. *Ufahamu J. Afr. Stud.* **2009**, *35*, 1–18.
16. Takane, T. Customary Land Tenure, Inheritance Rules, and Smallholder Farmers in Malawi. *J. South. Afr. Stud.* **2008**, *34*, 269–291. [CrossRef]
17. Demetriou, D. The Development of an Integrated Planning and Decision Support System (IPDSS) for Land Consolidation. Ph.D. Thesis, University of Leeds, Leeds, UK, 2014.
18. Vitikainen, A. An Overview of Land Consolidation in Europe. *Nord. J. Surv. Real Estate Res.* **2004**, *1*, 25–44.
19. Binns, B.O. *The Consolidation of Fragmented Agricultural Holdings*; Food and Agriculture Organization: Rome, Italy, 1950; p. 11.
20. Burton, S.P.; King, R. Land fragmentation and consolidation in Cyprus: A descriptive evaluation. *Agric. Adm.* **1982**, *11*, 183–200. [CrossRef]
21. Van Dijk, T. Scenarios of Central European land fragmentation. *Land Use Policy* **2003**, *20*, 149–158. [CrossRef]
22. Netting, R.M. Of men and meadows: Strategies of Alpine land use. *Hum. Ecol.* **1972**, *4*, 135–146. [CrossRef]

23. McPherson, M.F. *Land Fragmentation: A Selected Literature Review*; Harvard University, Harvard Institute for International Development: Cambridge, MA, USA, 1982; pp. 4–8.

24. Udo, R.K. Disintegration of Nucleated Settlement in Eastern Nigeria. *Geogr. Rev.* **1965**, *55*, 53–67. [CrossRef]

25. Hartvigsen, M. Land Reform and Land Consolidation in Central and Eastern Europe after 1989—Experiences and Perspectives. Ph.D. Thesis, Aalborg University, Aarlborg, Denmark, 2015.

26. Asiama, K.O.; Bennett, R.; Zevenbergen, J.A. Land Consolidation for Sub-Saharan Africa's Customary Lands—The Need for Responsible Approaches. *Am. J. Rural Dev.* **2017**, *5*, 39–45.

27. Eastwood, R.; Lipton, M.; Newell, A. Farm Size. In *Agricultural Economics*; Pingali, P., Evenson, R.E., Eds.; Elsevier: Burlington, NJ, USA, 2010; pp. 3324–3394.

28. Headey, D.; Jayne, T.S. Adaptation to land constraints: Is Africa different? *Food Policy* **2014**, *48*, 18–33. [CrossRef]

29. Pingali, P.; Bigot, Y.; Binswanger, H.P. *Agricultural Mechnization and the Evolution of Farming Systems in Sub-Saharn Africa*; The Johns Hopkins University: Baltimore, MD, USA; London, UK, 1987.

30. Baudron, F.; Sims, B.; Justice, S.; Kahan, D.G.; Rose, R.; Mkomwa, S.; Kaumbutho, P.; Sariah, J.; Nazare, R.; Moges, G.; et al. Re-examining appropriate mechanization in Eastern and Southern Africa: Two-wheel tractors, conservation agriculture, and private sector involvement. *Food Secur.* **2015**, *7*, 889–904. [CrossRef]

31. Binswanger, H.P.; Pingali, P. Technological priorities for farming in Sub-Saharan Africa. *J. Int. Dev.* **1989**, *1*, 46–65. [CrossRef]

32. Houmy, K.; Clarke, L.J.; Ashburner, J.E.; Kienzle, J. *Agricultural Mechanization in Sub-Saharan Africa*; FAO: Rome, Italy, 2013.

33. Nothale, D.W. Land Tenure Systems and Agricultural Production in Malawi. In *Land Policy and Agriculture in Eastern and Southern Africa*; Arntzen, J.W., Ngcongco, L.D., Turner, S.D., Eds.; United Nations University: Tokyo, Japan, 1986.

34. Thurston, A.F. *Smallholder Agriculture in Colonial Kenya: The Official Mind and the Swynnerton Plan*; African Studies Centre: Cambridge, UK, 1987.

35. Asiama, K.O.; Bennett, R.M.; Zevenbergen, J.A. Land consolidation on Ghana's rural customary lands: Drawing from The Dutch, Lithuanian and Rwandan experiences. *J. Rural Stud.* **2017**, *56*, 87–99. [CrossRef]

36. Van Gils, H.; Siegl, G.; Bennett, R. The living commons of West Tyrol, Austria: Lessons for land policy and land administration. *Land Use Policy* **2014**, *38*, 16–25. [CrossRef]

37. Arko-Adjei, A. Adapting Land Administration to the Institutional Framework of Customary Tenure. Ph.D. Thesis, Delft University of Technology, Delft, The Netherlands, 2011.

38. Kalabamu, F.T. Divergent paths: Customary land tenure changes in Greater Gaborone, Botswana. *Habitat Int.* **2014**, *44*, 474–481. [CrossRef]

39. Asiama, S.O. Current Changes in Customary/Traditional Land Delivery Systems in Sub-Saharan African Cities—Ghana. In *Housing the Poor through African Neo-Customary Land Delivery Systems*; Mattingly, M., Durand-Lasserve, A., Eds.; DFID: London, UK, 2004; pp. 41–57.

40. Migot-Adholla, S.E.; Hazell, P.; Blarel, B.; Place, F. Indigenous land rights systems in sub-Saharan Africa: A constraint on productivity? *World Bank Econ. Rev.* **1991**, *5*, 155–175. [CrossRef]

41. Quiggin, J. Common property in agricultural production. *J. Econ. Behav. Organ.* **1995**, *26*, 179–200. [CrossRef]

42. Ollennu, N.A. *Principles of Customary Land Law in Ghana*; Sweet and Maxwell: London, UK, 1962.

43. Fenoaltea, S. Risk, transaction costs, and the organization of medieval agriculture. *Explor. Econ. Hist.* **1976**, *13*, 129–151. [CrossRef]

44. Ohene-Yankyera, K. Determinants of Farm Size in Land-Abundant Agrarian Communities of Northern Ghana. *J. Sci. Technol.* **2004**, *24*, 45–53. [CrossRef]

45. Pingali, P. Agricultural Mechanization: Adoption Patterns and Economic Impact. In *Handbook of Agricultural Economics*, 1st ed.; Evenson, R., Pingali, P., Eds.; Elsevier: Amsterdam, The Netherlands, 2007; Volume 3, pp. 2779–2805.

46. Abunyewa, A.; Osei, C.; Asiedu, E.; Safo, E. Integrated Manure and Fertilizer Use, Maize Production and Sustainable Soil Fertility in Sub Humid Zone of West Africa. *J. Agron.* **2007**, *6*, 302–309.

47. Kuusaana, E.D.; Bukari, K.N. Land conflicts between smallholders and Fulani pastoralists in Ghana: Evidence from the Asante Akim North District (AAND). *J. Rural Stud.* **2015**, *42*, 52–62. [CrossRef]

48. Heisey, P.W.; Mwangi, W. *Fertilizer Use and Maize Production in Sub-Saharan Africa*; Mexico, D.F., Ed.; CIMMYT: Mexico City, Mexico, 1996; pp. 96–101.

49. Shuai, Y.; Chao-Fu, W.; Xin-Yue, Y.; You-Jin, L. The Ecological Compensation of Land Consolidation and Its Evaluation in Hilly Area of Southwest China. *Energy Procedia* **2011**, *5*, 1192–1199. [CrossRef]
50. Coldham, S. The Effect of Registration of Title Upon Customary Land Rights in Kenya. *J. Afr. Law* **1978**, *22*, 91–111. [CrossRef]
51. Taylor, D.R.F. Changing Land Tenure and Settlement Patterns in the Fort Hall District of Kenya. *Land Econ.* **1964**, *40*, 234–237. [CrossRef]
52. Swynnerton, R.J.M. *The Swynnerton Report: A Plan to Intensify the Development of African Agriculture in Kenya*; Government Press: Nairobi, Kenya, 1955.
53. Abubakari, Z. Investigating the Feasibility of Land Consolidation in the Customary Areas of Ghana. Master's Thesis, University of Twente, Enschede, The Netherlands, 2015.
54. GoR. *Organic Law Determining the Use and Management of Land in Rwanda*; UrbanLex: Zaragoza, Spain, 2005.
55. Muhinda, J.J.M.; Dusengemungu, L. *Farm Land Use Consolidation—A Home Grown Solution for Food Security in Rwanda*; Rwanda Agricultural Board, Ministry of Agriculture and Animal Husbandry: Kigali, Rwanda, 2013.
56. USAID. *Literature Review for Land Use Consolidation and Crop Intensification in Rwanda*; USAID: Kigali, Rwanda, 2013.
57. Musahara, H.; Nyamulinda, B.; Bizimana, C.; Niyonzima, T. Land Use Consolidation and Poverty Reduction in Rwanda. In Proceedings of the 2014 World Bank Conference on Land and Poverty, Washington, DC, USA, 24–27 March 2014.
58. Mingers, J. Combining IS Research Methods: Towards a Pluralist Methodology. *Inf. Syst. Res.* **2001**, *12*, 240–259. [CrossRef]
59. Van Dijk, T. Export of Planning Knowledge Needs Comparative Analysis: The Case of Applying Western Land Consolidation Experience in Central Europe. *Eur. Plan. Stud.* **2002**, *10*, 911–922. [CrossRef]
60. Almirall, E.; Wareham, J. Living Labs: Arbiters of mid- and ground-level innovation. *Technol. Anal. Strateg. Manag.* **2011**, *23*, 87–102. [CrossRef]
61. Liedtke, C.; Welfens, M.J.; Rohn, H.; Nordmann, J. LIVING LAB: User-driven innovation for sustainability. *Int. J. Sustain. High. Educ.* **2012**, *13*, 106–118. [CrossRef]
62. Pallot, M.; Trousse, B.; Senach, B.; Schaffers, H. Future internet and living lab research domain landscapes: Filling the gap between technology push and application pull in the context of smart cities. In Proceedings of the eChallenges e-2011 Conference Proceedings, IIMC International Information Management Corporation, Florence, Italy, 26–28 October 2011.
63. Byamugisha, F.F.K.; Burns, T.; Evtimov, V.; Santana, S.; Zulsdorf, G. *Appraising Investments and Technologies for Surveying and Mapping for Land Administration in Sub-Saharan Africa*; World Bank: Washington, DC, USA, 2012.
64. Cotula, L.; Chauveau, J.-P.; Cissé, S.; Colin, J.-P. *Changes in Customary Land Tenure Systems in Africa*; FAO and IIED: Rome, Italy, 2007.
65. Knight, R.S. *Statutory Recognition of Customary Land Rights in Africa*; FAO: Rome, Italy, 2010.
66. Benin, S.; Nin-Pratt, A. Intertemporal Trends in Agricultural Productivity. In *Agricultural Productivity in Africa: Trends, Patterns, and Determinants*; Benin, S., Ed.; International Food Policy Research Institute: Washington, DC, USA, 2016.
67. MoFA-SRID. *Agriculture in Ghana—Facts and Figures 2015*; Ministry of Food and Agriculture: Accra, Ghana, 2016.
68. Kasanga, K.R.; Kotey, N.A. *Land Management in Ghana: Building on Tradition and Modernity*; International Institute for Environment and Development: London, UK, 2001.
69. Chimhowu, A.; Woodhouse, P. Customary vs. Private Property Rights? Dynamics and Trajectories of Vernacular Land Markets in Sub-Saharan Africa. *J. Agrar. Chang.* **2006**, *6*, 346–371. [CrossRef]
70. Woodman, G.R. Land title registration without prejudice: The Ghana land title registration law, 1986. *J. Afr. law* **1987**, *31*, 119–135. [CrossRef]
71. Kalabamu, F.T. Land tenure and management reforms in East and Southern Africa—The case of Botswana. *Land Use Policy* **2000**, *17*, 305–319. [CrossRef]
72. Asiama, K.O.; Bennett, R.M.; Zevenbergen, J.A. Participatory Land Administration on Customary Lands: A Practical VGI Experiment in Nanton, Ghana. *ISPRS Int. J. Geo-Inf.* **2017**, *6*, 186. [CrossRef]
73. Asiama, K.O.; Bennett, R.M.; Zevenbergen, J.A.; Asiama, S.O. Land valuation in support of responsible land consolidation on Ghana's rural customary lands. *Surv. Rev.* **2018**, *50*, 288–300. [CrossRef]
74. Asiama, K.O.; Bennett, R.; Zevenbergen, J.; Da Silva Mano, A. Responsible consolidation of customary lands: A framework for land reallocation. *Land Use Policy* **2019**, *83*, 412–423. [CrossRef]

75.   Masser, I.; Williams, R. *Learning from Other Countries*; Geo Books: Norwich, UK, 1986.
76.   Enemark, S. Understanding the Land Management Paradigm. In *Innovative Technologies for Land Administration*; FIG: Madison, WI, USA, 2005.

MDPI

St. Alban-Anlage 66

4052 Basel

Switzerland

Tel. +41 61 683 77 34

Fax +41 61 302 89 18

www.mdpi.com

Land Editorial Office

E-mail: land@mdpi.com

www.mdpi.com/journal/land